ANTHROPOLOGY AND CLIMATE CHANGE

Second Edition

ANTHROPOLOGY AND CLIMATE CHANGE

From Actions to Transformations

Second Edition

Edited by

Susan A. Crate and Mark Nuttall

NEW YORK AND LONDON

Second edition first published 2016

by Routledge
711 Third Avenue, New York, NY 10017

and by Routledge
2 Park Square, Milton Park, Abingdon, Oxon, OX14 4RN

Routledge is an imprint of the Taylor & Francis Group, an informa business

First edition published by Left Coast Press, Inc. 2009

Library of Congress Cataloging in Publication Data

Anthropology and climate change: from actions to transformations / edited by Susan A. Crate and Mark Nuttall. – Second edition.
pages cm
Includes bibliographical references and index.
ISBN 978-1-62958-000-5 (hardback : alk. paper) – ISBN 978-1-62958-001-2 (pbk. : alk. paper) – ISBN 978-1-315-53033-8 (consumer ebook)
1. Climatic changes. 2. Climatic changes–Social aspects. 3. Ethnology. 4. Anthropology. I. Crate, Susan Alexandra, editor. II. Nuttall, Mark, editor.
QC981.8.C5A63 2016
304.2'5–dc23

2015036535

ISBN: 978-1-62958-000-5 (hbk)
ISBN: 978-1-62958-001-2 (pbk)
ISBN: 978-1-315-53033-8 (ebk)

Typeset in Sabon
by Zaza Eunice, Hosur, India

Contents

Illustrations

Figures

Tables

Introduction: Anthropology and Climate Change

Susan A. Crate and Mark Nuttall

Our 2009 edited volume, *Anthropology and Climate Change: From Encounters to Actions,* emerged from our increasing engagement with climate change—in the field, in our conversations with colleagues, in our participation in regional and global assessments of climate change, and also from our experiences of interfacing our research with other social scientists and with natural and physical scientists. The main objective was to advance the nascent work in the field of anthropology and climate change so that social and cultural anthropologists could more fully contribute to perhaps our most pressing global plight. To this end, the first volume assessed the field by bringing together a number of scholars to reflect on how anthropologists encountered climate change, to consider the action they took in terms of theory, method and practice, and to suggest some possible trajectories for further work. It was reviewed favorably in a number of scholarly journals, and we have been pleased to learn from colleagues and students that it has been a useful and widely used text and one that has found a readership beyond anthropology. Since then, much has changed. Many areas of that nascent work have come into full bloom. Today anthropologists are engaging research that has a concern with resilience, vulnerability, adaptation, mitigation, anticipation, risk and uncertainty, consumption, gender, migration, and displacement. Anthropologists have developed significant work on the politics of climate change, inequality, health, carbon markets and carbon sequestration, and water and energy. They have turned their attention to studying the assumptions and practices of meteorologists and climate scientists and to forecasting, scenarios, global circulation models and cultures of prediction, and the atmosphere. There are valuable studies of perceptions and observations of climate change, and of local impact, based on long-term ethnographic research in specific geographical places and localities, but our understanding of "the field" and the scope of anthropological enquiry has never been confined or constrained by lines of latitude and longitude and restricted to ethnographic regions or to localities. Anthropologists work on the human rights aspects of climate change;

---

they assess and evaluate the vulnerability and resilience of communities (for example, Hastrup 2009)—as well as critiquing the ways that concepts such as vulnerability, resilience, and adaptive capacity are deployed and used—they examine rapid landscape change and transformation (for instance, Orlove, Wiegandt, and Luckman 2008), and they investigate the nature of risk and the political ecology of disasters, hazards, and population displacement (for example, Oliver-Smith 2009). They also consider the social construction of climate change and climate change knowledge in terms of decision making, politics, and power (for example, Pettenger 2007). Some anthropologists are investigating the scientific construction of climate change scenarios and model-making and also the social relations and social and cultural contexts within which climate models are imagined, conceived, assembled, and used to frame scientific practice (for instance, Lahsen 2005; Hastrup and Skyrdstrup 2013). They work in contexts and situations where anthropological knowledge and practice help to inform, shape, or critique policy (for example, Fiske 2009), and they do ethnographic research on international congresses, processes, policy dialogues, and initiatives that deal with the politics and science of climate change (for example, Skrydstrup 2009). Anthropologists are increasingly finding greater roles to play in regional and global assessments of climate change (the work of some of our contributors to both editions has been referenced in recent reports of the Intergovernmental Panel on Climate Change and other regional and global change assessments) and in our understanding of the planetary biosphere, and they work on theorizing and conceptualizing how to bridge temporal and spatial scales and illuminating ways of understanding how to disentangle the effects of natural variability and change from those of human action. A much-needed emerging area of study concerns how people know what they know about climate change (for example, Rudiak-Gould 2011). In short, since the 2009 publication of the first edition there has been a flurry of activity and development and refinement of theoretical and methodical approaches, all of which is reflected in this new volume.

Today, after some 30 years of testimony by scientists and 25 years of high-level scientific reports, it is largely accepted that contemporary climate change is real and primarily human-driven. This belief is borne out not only by research but also by the discursive way *climate change* is being abandoned in the scientific literature in favor of a much more emphatic term, *anthropogenic climate forcing*. However, humans are not only changing the global climate in an unprecedented way but also transforming the world's ecosystems, including their animal, plant, and cultural diversity and their overall ecological balance, which in combination is having a significant influence on the planetary biosphere. Recent work by theorists who have put forth and work with elaborating and refining the "planetary boundaries framework" suggests that climate change and biosphere integrity—two core boundaries—have the potential to drive the Earth system into a new state if they are persistently, substantially, and significantly transgressed (Steffen et al. 2015). Because it is posited that humans are the main driving force in global change, many scientists and commentators argue that we have entered a new geological epoch, the Anthropocene (for example, see Steffen, Crutzen, and McNeill 2007; Steffen et al. 2011). The term *Anthropocene*, credited to chemist Paul Crutzen and biologist Eugene Stoermer (Crutzen and Stoermer 2000), suggests that "the Earth has now left its natural geological epoch, the present interglacial state called the Holocene. Human activities have become so pervasive and profound that they rival the great forces of Nature and are pushing the Earth into planetary *terra incognita*" (Steffen, Crutzen, and McNeill 2007: 614). Steffen and colleagues (2011) are careful to talk about an "Anthropocene concept," arguing that the Anthropocene should be understood as not merely a geological classification but also as a conceptual framing of the place of humans on Earth and their actions, roles, relations, influences, and transformative practices. Although some in academia and the general public disagree about the term's applicability and, even if the term is widely accepted, where the lines should be drawn between the Holocene and the Anthropocene—and while some geoscientists question how far the concept "can be validated with a global stratigraphic marker" (Autin and Holbrook 2012: 60)—the debate nonetheless shifts the focus of climate change directly to humans as the main agents in shaping geologies and influencing geophysical change (for example, Price et al. 2011). It draws attention to anthropology as an area of enquiry concerned with, among many things, understanding how the world has been transformed by human action in a constant process of engagement between people and the environments they inhabit. Anthropologists, as practitioners of a discipline that has long questioned the idea of the "natural world," play unique and critical roles in understanding these human-environment interactions, transformations, and their influences; the effects and consequences; and how humans and the diversity of the planet's ecosystems are affected, respond, and implicate the future.

At the turn of the 21st century, anthropological work on climate change did not figure prominently in the scholarly literature or in the regional and global assessments conducted by international bodies such as the Intergovernmental Panel on Climate Change (IPCC). Although anthropologists were working on climate change and climate variability and their social and cultural dimensions in a variety of historical and contemporary settings (for example, Rayner and Malone 1998; Orlove, Wiegandt, and Luckman 2008; Strauss and Orlove 2003), and some were participating in climate change assessments, including the first reports from the IPCC, their studies and publications did not markedly influence or impress a field of research dominated by the natural and physical sciences. Although there were contributions by anthropologists and other social scientists, the absence of a concerted effort to set out an agenda for anthropological action was noticed. For example, anthropologist Steve Rayner (1989) and geographer Simon Batterbury (2008) were critical of, and surprised by, the lack of professional anthropological engagement with climate change. However, the broader field of climate science was dismissive, skeptical, and even ignorant of the precise roles anthropologists could play, thereby frustrating the social scientists actively involved in climate change research at the time.

Anthropological interest in climate is not new, of course (for instance, see Dove 2014, for a survey), and anthropologists have always seen it as fundamental (but not in a deterministic way) to understanding human-environment relations. Our point is that extensive and intensive anthropological engagement with climate change in a diversity of ethnographic and policy settings is relatively recent; now we can talk of an anthropology of climate change or a climate change anthropology that has emerged with its own theoretical, methodological, and policy applications and marked by a sophisticated and steadily growing literature (for example, Crate and Nuttall 2009; Hastrup 2009; Baer and Singer 2014). The last decade especially has seen significant change, and for that we have the earlier, seemingly unrecognized work to credit and value as the foundation for current anthropological involvement with anthropogenic climate change research. In the 1990s, anthropologists, most prominently Rayner and colleagues, designed the conceptual architecture for social science contributions to climate science and to understanding climate change (see Rayner and Malone 1998). In tandem, archaeology and historical ecology enhanced awareness of the human modification of regional and global environments (for example, Crumley 1994). Climate and weather, which have preoccupied anthropologists since the early foundations of the discipline, were being reframed in the anthropogenic climate change context, and new ground was being broken by its pioneering scholars (Strauss and Orlove 2003). However, it is only in the last 10 years or so that anthropologists have been responding in greater numbers to a call to position themselves at the forefront of research on climate change (Batterbury 2008; Milton 2008).

Today, anthropology meetings and conferences are flush with panels on climate change exploring anthropological engagement, practice, and policy applications. The last decade has also seen a surge in publications, including not just peer-reviewed journal articles, edited volumes, and monographs but also teaching texts, popular writings, blogs, and the like. Researchers in cognate disciplines such as sociology, human geography, and archaeology have been similarly busy (for example, Hulme 2009, 2014; Urry 2011; Redclift and Grasso 2013; Van de Noort 2013), making it difficult to keep up with the amount of work being produced. The activities of working groups and focused conferences to enliven and enact the field are also increasing; by way of example, the American Anthropological Association (AAA) set up a task force on global climate change, and the authors, several of whom are contributors to this volume (including one of its editors, Susan Crate), produced a comprehensive report—*Changing the Atmosphere*—intended as "a guiding document on anthropology and climate change in its broadest sense, including anthropology's contributions to, and concerns about, climate change and climate change policy and discourse; to provide commentary on interdisciplinary research and relationships; and to identify research frontiers for anthropology with respect to climate change" (Fiske et al. 2014: 5). Outside the discipline, there has been not only an increasing interest in the place of anthropology in climate change understanding and policy actions but also a broader acceptance that anthropologists have a key role to play in finding solutions to climate change and its effects (Barnes et al. 2013). The recent reports of the IPCC have demonstrated the importance (and acceptance) of interdisciplinary processes in which the social sciences and the humanities are critical collaborators with the natural and physical sciences in the production of detailed assessments of regional and global climate change. Interdisciplinarity in global change research is rapidly maturing and emerging as an arena to integrate not only the natural and social sciences effectively but also the humanities (Castree et al. 2014).

The emergence of a distinctive anthropological contribution has also been in part the result of anthropologists working hard to promote their research and argue for their discipline's relevance. Yet challenges remain, particularly with regard to how anthropologists (together with archaeologists) can move beyond descriptions of social and cultural effects of climate change, in both the past and the present, and make contributions to refining climate change models concerning human dynamics—models that have been criticized as being economically reductionist and authoritative forms of climate prediction (Hastrup and Skrydstrup 2013; Roscoe 2014).

In archaeology, a discipline long interested in how climate change has affected societies and cultures, studies and projects on contemporary climate change are today extensive and multifaceted (see, for example, McGovern 2013, 2015). But it is not this volume's objective to detail archaeological engagement with climate change, since our focus is on approaches in social and cultural anthropology, and such detail would necessitate a volume in its own right. However, a brief overview of important aspects and current status is crucial.

Recent archaeological work on contemporary climate change has done much to develop a climate change archaeology (Van de Noort 2008, 2013) that "views the past as a repository of ideas and concepts that can help build the resilience of communities in a time of rapid climate change" (Van de Noort 2013: vii). Van de Noort argues that, while archaeology has researched climate change for the past 150 years, studies of how climate change has affected human societies and the environment and how those societies have adapted "have made no contribution to the debates on the ways in which humanity will need to adapt to climate change in the future" (ibid.: 2). Archaeologists are more than willing to contribute, but contemporary debates of climate change science and climate policy tend to treat the past as unable to provide few lessons for the future, because it is argued that anthropogenic climate change and contemporary global environmental mismanagement have no precedent and therefore cannot inform the present. In his wide-ranging book, Van de Noort scrutinizes this attitude and argues the case for understanding the past via the contributions archaeology makes to understanding and responding to climate change. In addition to providing "insights into long-term geological, geographical, and environmental processes, and occasionally technical solutions that have been forgotten," the study of the past is especially important, he argues, because "the knowledge gained connects in a very direct way to people's sense of place and their self-reflected place in a world that needs to adapt to climate change. In other words, the past and long-term perspectives on adaption to change can act as a source of inspiration for modern communities" (ibid.: 2–3). Sandweiss and Kelley (2012) discuss how the archaeological record often incorporates important, sometimes unique, proxy records of climate and environment and of the changes in both. Significantly, these data are often important for the techniques and approaches used in modeling future climatic and environmental change. In fact, Sandweiss and Kelley note a growing use of archaeological proxy data by climate scientists. The challenge for archaeology, which it meets very well, is also to show how past societies were not just affected by climate change, or how they responded or adapted, but how they demonstrated agency (Hassan 2009).

How we approach climate change, its temporal and spatial aspects in all its local, regional, and global dimensions—that is, how we identify it, define it, study it, contextualize it, think about its past, present, and future effects, and contribute to policy processes—challenges how anthropology and its practitioners study and evaluate the impacts of climate systems on society, culture, and everyday human life and how it contributes to how anthropologists' advocacy and policy positions are defined. We can define *climate change* as a variation in climatic parameters attributed directly or indirectly to human activities, the growing use of technology, industrialization and the burning of fossil fuels, deforestation, resource depletion, environmental degradation, and consumer lifestyles, all of which is entangled with natural variations in climate. Yet we seem to think about it, reflect on it, worry about it, and understand it through a complexity of different narratives, explanations, discourses, and myths that all say something about our lifestyles, social arrangements, cultural practices, and economic systems and about our relationships to the planet, to one another, to the future and our place in it. While anthropology solidly contributes to understanding past and present human adaptive strategies and the effects of climate change, how humans observe and perceive these changes, and how they think about and relate to the weather, increasingly anthropologists are called on to predict future occurrences and adaptations. In other words we are being asked to move toward anticipation, to peer into the future and to suggest what it may look like and what our adaptive responses to future climate change should be. An anthropological focus on anticipation may help to shed light on how people think about the world around them, how they engage with it and move around in it, and how they create and enact change within a world that is also undergoing a constant process of becoming, remaking, and reshaping (Nuttall 2010).

In our own respective work, which has focused primarily on a number of regions in the circumpolar North, such as Greenland, Siberia, Canada and Finland, we have also sought to make a contribution to this emerging literature and to refinement of anthropological approaches. For both of us, our initial anthropological encounters with northern places coincided with the emergence of the Arctic and Subarctic as international political regions at the end of the Cold War. In the late 1980s and early 1990s, the circumpolar North was beginning to be seen less as a militarized and strategic space and more as a space for scientific and political concern about long-range pollutants and contaminants, research on thinning sea ice and thawing permafrost, and the increasing articulation of indigenous peoples' experiences about a threatened and changing environment, contributing to the discursive construction of the Arctic as an environment at risk (Nuttall 1998).

Today, scientists and the media regularly report about the consequences of anthropogenic climate change, most immediately observable in the Arctic. Owing to the phenomenon of polar amplification, wherein warming is not uniform across the globe but rather increases toward the poles, since the 1950s average annual temperatures in some parts of the Arctic have increased by 2 to 3°C (IPCC 2013), or twice as much as the rest of the globe since 1980 (ACIA 2005). Greenland's inland ice is melting rapidly (Chen, Wilson, and Tapley 2006; Hanna et al. 2008), and summer sea ice is decreasing at a rate faster than that foreseen by any climate model—meaning that by 2050, or even sooner, the Arctic Ocean may be completely ice free in summer (Overland and Wang 2013). These changes are already affecting biodiversity and other environmental conditions in the circumpolar North (Chapin et al. 2005; CAFF 2013), having considerable social and economic implications (Nuttall et al. 2005; Cornell et al. 2013) and affecting the cryosphere (AMAP 2011). "Ice-dependence" and the implications of shorter and shorter ice seasons can be seen across the Arctic (Crate 2012), and the local and regional consequences of ecosystem transformation have far-reaching implications for human security, people's rights, and human-environment relations (Nuttall 2013). For the peoples of the Arctic, climate change is a lived experience, and anthropological research is seeking to understand not only how people live with such change but also how they reflect on past changes to negotiate present circumstances and anticipate future conditions. Migration, an age-old adaptation in Arctic communities, is coming into new forms and timings as climate change effects plant and animal ranges and seasonal variability (Crate 2013; Marino 2015).

Our respective work has also pointed out that we must attend to how we disentangle the various effects of climate change and social change and how we understand climate change in context. For example, in the Arctic, indigenous and local communities have experienced, and are experiencing, stress from a number of different forces that threaten to restrict hunting, fishing, and herding activities and sever human-environment relations. The Arctic regions are tightly connected politically, economically, and socially to the national mainstream of the world's northern circumpolar states and are inextricably linked to the global economy, just as all other areas on the planet are. Rapid social, economic, and demographic change (particularly over the past 50 to 60 years), resource development, trade barriers (particularly affecting products from the hunting of whales, seals, and polar bears), and animal-rights campaigns have all had effects on hunting, herding, fishing, and gathering activities. Climate change perhaps magnifies these issues, brings them into sharper focus and exacerbates them. At the same time, scientific models and scientific research on climate change tend to ignore these other drivers of change. Climate change is often framed as the sole force shaping and influencing human life in the world's high northern latitudes. Anthropological work focused on climate change encourages understanding of how climate change is one of several drivers of change affecting local ecosystems and cultures (Crate 2014; Nuttall et al. 2005).

Whereas the Arctic has been perhaps viewed as the frontline of global climate change, other places around the world are also being unprecedentedly affected. Although it could be argued that all world regions are climate-sensitive, there are places, not unlike the Arctic, that are more sensitive to climate change effects because their conditions are in some way extreme—for example, high-altitude areas where, much like high-latitude regions, temperature and seasonal changes are amplified (Orlove, Wiegand, and Luckman 2008). Near-sea-level ecosystems also show amplified effects, since one of the cascading effects of climate change is sea-level rise. Anthropologists have written about climate change effects on sea-level rise, especially in places such as the South Pacific (for example, Rudiak-Gould 2013), but it is now gaining more pronounced attention in such regions as Europe and North America (Montgomery 2014). Two other ecosystems showing amplified effects are arid and humid ones, which are tending to be more extreme in those qualities as climate change alters the global water cycle. The recent California drought has brought climate change home to North Americans by providing a visceral understanding of how climate change is not happening only at what are often viewed as being the "edges" or "margins" of the world (Nesbit 2015).

From Actions to Transformations

In response to the flurry of activity and a quickly developing and expanding field, and at a time when climate science appears saturated with references to tipping points and thresholds, planetary boundaries and the Anthropocene, we present a second edition of *Anthropology and Climate Change*. This was not an easy task. Because of the rapidity of change in the field and the ongoing relevance and application of the first edition's content, we developed this book not as a simple update of the first. Although it may be marketed as a "second edition," this book is atypical to the extent that it does not replace or make obsolete the first edition but rather builds on it as a stand-alone volume that can also be used in tandem with the first edition.

This approach is warranted not only because anthropology has come into a new understanding of climate change but also because the anthropologists who are moving work on this subject forward are also evolving in their own research and teaching trajectories. And although much in the field has indeed changed, our first volume is not obsolete—much of it remains highly relevant. Therefore, our second edition builds on the first edition both by describing the development of the field and by providing insight into the evolving paths of anthropologists. In this new book, leading scholars demonstrate the extensive nature of anthropological engagement with climate change and the advances in theory and method. A number of the first edition's contributors present updated or entirely new work, and they are joined by more than a dozen new authors.

To these ends, Chapters 1–4 of the first edition (in the section Climate and Culture") remain foundational to the discipline—and so they are not republished or updated in this volume but rather can be read in the first edition. In this second edition we present five new foundational chapters (Chapters 1–5) to complement the original four; these new chapters forefront the main emergent precedents in the anthropology of climate change. In addition to these new theoretical/disciplinary background chapters, authors have contributed chapters that pull out core thematic areas that are common threads throughout the book and represent key concepts/memes/trajectories in climate change research.

Part 1: Building Foundations of Anthropology and Climate Change

Part 1 of this book lays the groundwork for how anthropological concern with climate change has come into being in the last 10 years, moving away from the anthropological documentation of local observations in climate sensitive places to seeing the interconnections, interdependencies, and complexities of ways of knowing about the diversity of change that climate change ushers in. It emphasizes how adaptation, vulnerability, and resilience are used extensively in climate science and policy to address this diversity and complexity, how one of anthropology's theoretical frames, Cultural Theory, works to unpack climate alarmism, provides a nuanced example of how sociology has a similar explanatory power in climate change discourse as anthropology does, and expounds on the place of anthropological investigations and collaborations in the world of interdisciplinary and international policy efforts.

Kirsten Hastrup sets the stage in Part 1 with an extensive chapter on how climate knowledge is assembled, how climate knowledge has been contextualized historically and socially, and how climate knowledge is configured in multiple ways, with scientific climate models being but some of them. Because of worldwide concern about climate change, people living with manifest environmental changes and seeing old landmarks disappear or livelihoods disintegrate increasingly interpret this situation in terms of climate change and construct their own scenarios. Yet climate scenarios are deeply social, she argues, incorporating past experiences and responding to future concerns. Hastrup explores this knowledge production process with reference to three key themes: assemblage, anticipation, and action. She discusses the assembling of knowledge and how it is made and authorized, shows how such knowledge sustains platforms for anticipating nature, and clarifies how action is never simply a reaction to an antecedent but also a response to anticipation and how the future is therefore performed into being. In reflecting on all this, she draws on extensive fieldwork in northwest Greenland and illustrates how anthropology contributes to a shared and interdisciplinary field of climate concerns.

The concepts of adaptation, resilience, and vulnerability play central roles in academic and policy discourses on climate change effects on human communities and were explored in depth in several chapters of the first edition. However, as Anthony Oliver-Smith argues in Chapter 2, their use and application, particularly as policy instruments, have proven to be imprecise and flawed, potentially distorting or obscuring important questions. His chapter interrogates these concepts and their deployment in research, analysis, and policy making regarding the significant potential for large-scale population displacement and resettlement associated with climate change. Oliver-Smith pays particular attention to recent policy initiatives from the international community relating to the limits of adaptation and programs for compensation of losses and damages resulting from climate change for which neither mitigation nor adaptation will be sufficient.

In Chapter 3 Clare Heyward and Steve Rayner identify some of the political factors behind the rise of tipping point rhetoric, with warnings about climate tipping points that are commonplace today. Much of the language and terminology used to construct the tipping points narrative is apocalyptic, emphasizing abruptness, irreversibility, and catastrophe. They argue that this rhetoric's success can be explained by its consonance with two enduring tropes, millenarianism and appeals to a delicate "balance of nature." Tipping point rhetoric was initially used to advocate greater action on mitigation, but it—and the millenarian narrative in particular—has been used to argue for an entirely different solution to climate change: that of geoengineering. Thus Heyward and Rayner draw attention to how the rise of tipping points rhetoric has political roots and how the invocation of tipping points is a political strategy in the discourse on climate change, particularly in terms of climate emergency rhetoric.

Following from this, in Chapter 4 sociologist John Urry examines multiple interdependent processes in environment, economy, climate, food, water, and energy. In particular he considers how in the final years of the last century, neoliberalism ratcheted up the global scale of movement within the global North enabled by oil. This travel of people and goods became essential to most social practices that depend on and reinforce a high carbon society. This extravaganza, Urry argues, came to a shuddering halt when oil prices increased in the early years of this century. Suburban houses, especially those found in far-flung oil-dependent locations, could not be sold. Financial products and institutions were rendered worthless. This story of oil in the United States provides a bleak vision of the future as "easy oil" runs out by the middle of this century. Urry develops a broad analysis of how "complex" energy systems and energy crises can bite back with interest and may portend some bleak futures.

Completing the foundational chapters, in Chapter 5 Eduardo Brondizio discusses how climate change research has become inherently collaborative in nature and scope. At the center of global environmental change (GEC) more broadly, since the 1980s the field has mobilized an unprecedented global network of scholars through the establishment of international GEC programs. Brondizio's chapter provides an overview of the history of GEC programs as they have evolved from predominantly disciplinary physical sciences to interdisciplinary communities increasingly influenced by social scientists. Brondizio goes on to describe the evolution of the new Future Earth Program around questions and concepts such as codesign and transdisciplinarity, stakeholders' engagement, and transformation toward sustainability. He discusses the need for new conceptual frameworks to support understanding complex multiscale problems that transcend scales and disciplines and reflects on anthropology's challenges, opportunities, and prospects for such engagement.

Part 2: Assessing Encounters Old and New

With these foundational chapters as a base, we next launch into encounters old and new. Herein some of the first edition's authors revisit their work not only to show how these encounters have changed over time but also to reflect on how their approaches to the field have evolved. In addition, new authors bring to light how other world areas and their inhabitants are interacting with the changes due to climate and other drivers. We begin with seven chapters of previous authors followed by five chapters written by new authors.

In Chapter 6 Susan Crate builds on her continued research with Viliui Sakha communities in Siberia to reveal how change, in its many forms, has brought a new reality to rural areas—the relatively rapid end of cows and kin, the main adaptive strategy Viliui Sakha households adopted after the sudden dismantling of their Soviet-period agroindustrial farms. As Crate describes, the environmental changes due to the local effects of climate change are making cow-keeping increasingly challenging—flooded haylands, disrupted seasonal timing, an unpredictable precipitation regime, and the like. However, a deeper probing shows that other forces of change are at work in tandem, most strikingly the relative abundance of Western products available in village stores due to economic globalization and the out-migration of youth to regional and urban centers. Crate explores these issues and contemplates the possible future implications, documenting both the trend to being "cow-free" and the foresight of several households who refuse to abandon household-level production in light of the unpredictability of the market. Her chapter also entertains the extent to which these three forces are at work in other rural areas of the globe.

In Chapter 7 Sarah Strauss revisits the ethnographic account she presented in the first edition. Reporting on her research in the Swiss Alpine village of Leukerbad, she describes how there have been many different kinds of reactions to the questions opened by the possibilities and constraints of a warming world since she began working there more than 15 years ago. In the original version of her essay, Strauss described the movements of glaciers back and forth as climate cycles shifted and the risks inherent in living near them. Her new chapter considers the responses to climate change in Leukerbad and examines the ways that traditional narratives of engagement with environmental threats posed by glaciers or avalanches are connected with modern reactions to the risks of a changing climate. However, there is some pessimism running through her work. In studying how climate change models are constructed, and what the limits of the modeling process are, Strauss came to learn that we do not possess what most local mountain communities need in the face of dramatic change, and she concludes that it is unlikely that climate science will be able to provide such information in the near future.

Chapter 8 also revisits an earlier case study. Timothy Finan and Ashiqur Rahman describe how, in the world of climate science, Bangladesh is treated as a poster child of vulnerability. With a large part of its 170 million people living close to sea level, with its masses of impoverished citizens struggling on the edge of survival, and with its history and traditions of social exclusion and differentiation, the country, they argue, faces the "perfect storm." Climate change and Bangladesh seem ready to be paired in word association games. With Bangladesh arriving at nearly a half-century of optimistic modernization strategies, large-scale earthworks, increasing inequality, and shaky governance, nature has laid down the ultimate gauntlet. At the local level, climate change has already provided glimpses of the future in the form of rising sea level in the Bay of Bengal, extreme floods along the three giant rivers that meet south of its capital, powerful cyclones fueled by warm waters, changing ecologies of the water bodies, and unexpected winter droughts in prime agricultural zones. Finan and Rahman focus on the role of anthropological research in understanding the nature of vulnerability and adaptation to climate change pressures, the village level experience with climate change, and the influence of unequal power relations and governance on the vulnerability of local populations. They also demonstrate how an anthropological perspective of local livelihood systems and their adaptive capacities, of power and governance, can contribute to a greater understanding of what a climate change future will bring and what adaptation pathways are viable.

In the highlands of Papua New Guinea, extreme El Niño-Southern Oscillation (ENSO) events create droughts and frosts that detrimentally affect the local tropical subsistence food base. In Chapter 9 Jerry Jacka shows how people, keen observers of climatic norms and anomalies, respond with a suite of ecological and social practices to mitigate extreme climatic events. Jacka documents these practices and compares oral testimonials of climate change with local climatic data. In doing so, he updates his original chapter with more recent findings from climate change assessments and with new information on ENSO in the western Pacific. He draws on new material documenting people's responses to climatic extremes and compares local perceptions of climate with temperature and rainfall records obtained from the Porgera Gold Mine. In his chapter in the first edition he explored the cosmological implications of climate change; this current chapter places emphasis on actual practices and knowledge of mitigation. Overall, local perceptions of climate change closely mirror temperature and rainfall records from official sources. However, while strong anomalies in both wet and dry climate events affect subsistence, only extreme drought events enter into the local discourse and knowledge that prompt mitigation measures. Jacka argues that understanding the kinds of varying local responses to climatic anomalies are essential for future global adaptation strategies designed to mitigate the impact of a changing climate.

In Chapter 10 Elizabeth Marino and Peter Schweitzer update their contribution to the 2009 volume on discourse and knowledge in the Arctic around climate change. They previously argued that using the term *climate change* was detrimental in field sites where long-term, situated local knowledge had alternative epistemologies for understanding the way weather worlds and environmental systems interact with one another and with social systems. This new chapter expands on those arguments and uses more robust field data to demonstrate that media discourses and local discourses about climate change also differ significantly. Ultimately they argue that, while climate change discourses are important tools for furthering global action, a nuanced understanding of what local knowledge can offer this debate is still needed.

Chapter 11 is Donna Green's update of her 2009 essay. She outlines the sociocultural inequalities that can lead to the reduction of resilience to environmental change, such as climate change, in Australia's Aboriginal and Torres Strait Island communities. Green touches on the indigenous response to climate change, adaptation and mitigation strategies, and how traditional knowledge and environmental justice have been used in the Australian case. Using the case study of her work in the Torres Strait Islands, Green explores how Ailan Kastom (Islander culture) is being threatened by climate change. She reviews some of the recent national and state government responses to this situation, and discusses how Islanders have taken to social media to engage mainland Australian support to advance their adaptation planning and strategy.

In 2009 Heather Lazrus addressed how people in the low-lying island communities of Tuvalu view and manage their vulnerability to the effects of climate change. Chapter 12 now extends this topic to discuss discourses of the future, including migration and relocation. As is the case in many places facing extreme climate change and creeping environmental problems, including sea-level rise, migration and relocation are possible outcomes for the affected populations. Lazrus shows how an anthropological perspective on migration in response to climate change situates population mobility culturally and historically, disentangles local views on migration from externally imposed assumptions and discourses, and examines community-based needs and priorities in response to climate change.

Kicking off chapters by new authors on encounters, in Chapter 13 Michael Sheridan draws on recent work on the intersections of culture and power to examine the history of contestation over the social organization of rainmaking, sacred forests, and ecocosmologies in North Pare, Tanzania. Using ethnographic and historical research Sheridan shows how the local narrative of declining rainfall over the 20th century is both a metaphor for changing terms of resource entitlement and the ambiguities of power, morality, and social relationships in the postcolonial state and a description of a geophysical process. He argues that to fully grasp this process requires a closer look at the historical course of social change, the cultural roots of environmental narratives, and the political relationships between powerful institutions (such as governments and development agencies) and the rural populations on the periphery of the global economic system. By showing how these shifting ideologies of power, legitimacy, and value shape social relations and land management, he proposes a more nuanced vision of both "power" and "politics" within the existing political ecology paradigm that can account for how culture affects responses to climate change.

In Chapter 14 Tori Jennings considers the unexamined social relevance of a phenomenology of light in ecological research. She draws on historical and ethnographic research in Cornwall to show how light has played an important role in both the perception and economics of climate in this part of England. Beyond her fieldwork, however, Jennings analyzes the practical relevance of a phenomenology of light in anthropology. Following Tim Ingold, she examines the epistemological bias of "visualism" and the implications it has for environmental research. How researchers regard vision is the outcome of a specific historical trajectory, one that prejudices us against a variety of perceptual experiences that exist in the world. Through a critique of the Western ontological tradition of separating light and vision, Jennings argues that experiences within light are an important nexus of study for understanding climate and weather from a cross-cultural perspective.

One thing that links many of the chapters in this book is that while global climate change affects just about everybody in the modern world, the way people understand its causes and account for the role humans play in it varies considerably. In Chapter 15 Karsten Paerregaard reviews ethnographic field data gathered in an Andean community during the past 30 years. He explores how the villagers account for climate change and examines how global imaginaries of climate change challenge their ideas about nature and culture. He argues that, although the villagers are increasingly concerned about the impact of climate change and have adopted the global vocabulary on climate change, they do not believe people in other parts of the world are causing the environmental problems they currently face. The chapter concludes that, even though the villagers' reluctance to subscribe to the idea of an anthropogenic world makes them look like the companions of conventional climate skeptics, their skepticism is of a very different kind.

Peter Rudiak-Gould argues that although academic and governmental attempts have been made to rebrand climate change as economic, political, military, cultural, moral, religious, and so forth, in the Western world the dominant framing of the issue remains "environmental." Understanding climate change as a problem in the environment is valid but also limiting, both as a platform for change and as a theoretical framework for the anthropological study of climate change. He illustrates these points in Chapter 16 with the case of the Marshall Islands, a low-lying "small island state" that has been struck not only by rising sea levels, droughts, increased temperatures, and other physical phenomena possibly attributable to global climate change but also by a powerful set of scientific, activist, and humanitarian discourses surrounding climate change, usually centered on visions of imminent catastrophe, inundation, and relocation. Marshall Islanders thus encounter climate change as much through discourse as through the lived experience of local "environmental" change. Further challenging the understanding of climate change as "environmental," Marshall Islanders usually speak of climate change as a social and moral crisis of disappearing tradition, or as a theological issue of divine intent, rather than a perturbation in "nature." With the continued habitability of their homeland at stake, Marshall Islanders understand climate change as something far beyond the merely environmental, an understanding that climate change anthropologists, researchers, and activists ought to take seriously in their own investigations and endeavors.

In Chapter 17 David Rojas describes how a prominent Brazilian environmentalist once explained to him that, for Amazonian populations, climate change "is not science fiction." By this he meant that people in the region have already experienced profound socioenvironmental disruptions. Moreover, he invited Rojas to consider that, unlike conventional science fiction stories, Amazonian narratives of climate change do not regard technological deployments as capable of bringing climate crises under control. Rojas follows this environmentalist's advice and explores Amazonian accounts of climate change offered by rogue entrepreneurs. He shows how land speculators and landholders in the area link ongoing environmental transformations with previous economic projects in Amazonia that led to social and ecological disruptions. Rojas argues that these Amazonian stories of climate change are analogous to extreme science fiction narratives that portray futures in which people and groups create—and learn to navigate—inhospitable worlds.

Part 3: Refining Anthropological Actions

Next, as we did in Part 3 of the first edition, we position anthropological work focused on actions that affect policy, facilitate community and social change, affect activism in institutions, and the like. Here we also include first-edition authors who reflect on how their actions and, in some cases, the orientation of their actioned approaches, have changed. As a reflection of how the field has evolved, we also include new authors who write on the pedagogy, evolving institutional arrangements, and technical tools of climate change.

We begin with an author whose essay had been in Part 2 but whose work has clearly turned from encounters to action. In Chapter 18 Ann Stevens Henshaw, who since the first edition has moved from academia to the world of nonprofits, explores the ways in which anthropologists can take a critical and active role in advocacy, here by serving as a program officer for a private foundation. Henshaw's transition into philanthropic work from her witnessing and research within academia gives her powerful insight into the on-the-ground realities for which funding is being sought. Such monetary resource allocation plays an important role in advancing community resilience and self-determination to communities who are struggling with having a greater voice in decisions about the natural resources that make their lives possible, which are additionally challenged by diminishing sea ice, among other climate change effects. Henshaw draws links between resilience theory and philanthropy, arguing that they feed off and complement each other to the extent that they both tend toward fostering innovation and ideas. With anthropological and other "bottom-up" insights, such philanthropy can be directed more successfully toward adaptive management approaches.

In Chapter 19 Richard Wilk takes a different approach from the more mainstream consumer culture literature by questioning the category of *consumption* itself and suggesting that the term is profoundly cultural and also structured by metaphor. He emphasizes that there is a great deal of discussion in the global change literature on "drivers" and "forcing," which are a form of technospeak that allows some misdirection on the actual causes of the increased emissions of greenhouse gases. Wilk's long-term anthropological engagement with consumer culture has given him unique insights that suggest that we ask more fundamental questions about the cause of climate change rather than simply writing it off as a consequence of human culture. We need to talk about it using new metaphors and models in order to reframe environmental issues, including climate change, in ways that give us some realistic expectations of changing behavior.

In her chapter for our 2009 edition Shirley Fiske described the political trajectory of "cap and trade" bills in the U.S. Congress with optimism; her chapter predicted that the United States would finally have a national policy on mitigation. She also described the growing market and use of carbon sinks for environmental governance (carbon offset projects) and the anthropological insights into their problematic adoptions worldwide. Here, in Chapter 20, Fiske takes up the policy and anthropological actions story where it left off—explaining the apparent failure and the changing politics surrounding cap and trade in subsequent years, along with the Obama administration's pivot toward regulatory actions and special funds as the only arena for climate change activity. Fiske describes the deep politicization of the climate change concept (again, as developed in the United States) with an emphasis on epistemic challenges to climate change science.

In Chapter 21 Kristina Peterson returns with new coauthor, Julie Maldonado, to describe how long-standing communities and their places of inhabitation are disappearing at an alarming rate in the southeast Louisiana delta region of the United States owing to sea-level rise. For years communities have been adapting in creative ways to the changes occurring in their locales and their livelihoods. Yet there are limits to the adaptation communities can make to address the rising waters. The desire to keep cultural continuity has led to cultural rejuvenation in many of the communities, including coastal tribes. Some communities have decided that they no longer have adaptive measures that will maintain the cohesiveness of their community and so have opted to explore relocation. The complex decision to relocate necessitates many years of planning for the actual relocation. Peterson and Maldonado explore the critical, complicated issues faced by communities when adaptation isn't enough, and they examine how communities are living and working "in the between of the now and then."

In the first edition Mark Nuttall drew attention to the importance of grappling with the interplay of climatic, social, economic, and political factors when anthropologists work in indigenous communities confronting climate change—how the "regional texture to climate change means [that] changing environments are perceived and experienced differently" (Nuttall 2009: 296). Based on his long-term research in Greenland, Nuttall illustrated how climate change means different things for different people. Now, in Chapter 22, Nuttall develops some of these themes and argues that the extent of societal vulnerability and resilience to climate change, and the nature of anticipatory knowledge, depends not only on social and cultural aspects and ecosystem diversity but also on political, legal, and institutional rules and institutions that govern resource use. With reference to a research project on narwhals and narwhal hunting in northwest Greenland, a place where climate change is apparent in the thinning sea ice and receding coastal glaciers, Nuttall examines the potential effects of oil exploration in the form of seismic surveys, the management of resources, and the effectiveness of governance institutions. He ponders how governance mechanisms being developed in Greenland help or hinder people in negotiating and managing the effects of climate change, allowing, or not, local and regional authorities to act on policy recommendations and improving the chances for local communities to meet the challenges and opportunities of a changing environment. His chapter points to the need for understanding climate change within the context of other changes and societal and economic transformations in Greenland, including resource development and extractive industries. Rapid social, economic, and demographic change, wildlife management and resource development, trade barriers, and conservation policies all have significant implications for human security and sustainable livelihoods in the Arctic. This underscores the point we made earlier in this Introduction that, in many cases, climate change magnifies existing societal, political, economic, legal, institutional, and environmental challenges that people experience and negotiate in their everyday lives.

Nicole Peterson, originally coauthor of a foundational chapter of the first edition, returns to contribute with Daniel Osgood, in Chapter 23, to report on their work in Ethiopia. Technological and financial tools are often expected to help agricultural communities adapt to climate change. Although most studies of these tools focus on technical details and implementation, studies of microinsurance in Ethiopia suggest that social and political aspects can be even more important. Peterson and Osgood discuss how interview and focus group data from an agricultural community in northern Ethiopia show how trade-offs between insurance and other coping strategies are factored into farmers' decisions and the importance of understanding the needs of female heads of household, landless men, and differences among farmers.

In Chapter 24 Chris Hebdon, Myles Lennon, Francis Ludlow, Amy Zhang, Lily Zeng, and Michael Dove argue that anthropologists teaching about climate change today face the challenge, in the spirit of the sociological imagination, of connecting biography with history and grappling with problems of ethnocentrism. When it comes to climate change, almost everyone has something to teach and to learn. Pedagogies that can connect personal experiences to wider politics, ignite senses of purpose, and encourage effective action must be collaborative, requiring not only an engagement with a range of forms of knowledge but also a repertoire of communication skills. Hebdon and colleagues highlight multiple forms of pedagogy, drawing on examples across scales and space to address climate change and education writ large, including questions about misinformation. Examples range from pedagogy in schools, including the need for transparadigmatic thinking, to the international level, including issues of epistemic cultures. They note the importance of practical knowledge, social movements, and tactical uses of humor for addressing serious climate problems.

Noor Johnson's concern is with how institutions play a central role in organizing, documenting, and disseminating knowledge about climate change and in organizing societal responses. In Chapter 25 she considers the role of translation in climate change governance and politics, focusing on the work of actors embedded in a wide variety of institutions in mobilizing particular framings and understandings of climate change or in working to creatively facilitate responses that may lie outside dominant climate change framings. Drawing on her experience conducting fieldwork and collaborating with interdisciplinary networks of researchers, global science bodies, and Arctic indigenous peoples' organizations, Johnson discusses some of the various roles for anthropologists in bridging social, ecological, and institutional worlds and the various perspectives on climate change that they encounter.

Part 3 concludes with Chapter 26, an essay by Werner Krauss in which he argues how anthropological fieldwork can and must significantly contribute to a broader and more realistic understanding of global climate change. Instead of taking climate change as a monolithic fact threatening humanity, Krauss traces climate change as it appears in science, in public discourse, and in initiating societal responses concerning mitigation and adaption. Using examples from his fieldwork in a research institute, in the climate blogosphere, and in an emerging coastal "climate-scape," Krauss shows that an actor-network oriented approach helps to identify specific controversies and activities related to climate change. Tracing climate change via multisited ethnography, Krauss argues, helps to keep the challenges imposed by climate change more real in order to "compose" carefully, in Latour's sense, a common world and thus to enable a global response.

To some extent this second edition has been a much larger project than was the first, because of the rapid expansion and development not only of the field of anthropology and climate change but also of the sobering reality of the phenomenon and its effects on our planetary system. We also pause for reflection on how much has changed since our first writing. Clearly it is time to move forward whole-heartedly in anticipation of transformative change.

The parts of the second edition, not unlike those of the first edition, constitute a whole that certainly is an evolving assessment of the state of anthropology and climate change in the second decade of the 21st century. From the first edition's preliminary foundation, framing anthropology's nascent entry into contemporary climate change and inviting increased collaboration and dialogue, we now move into a survey of the crystallizing field of anthropology and climate change not only to gauge its rapid rise and continued expansion but also, and perhaps more important, to spark the continued dialogue within and outside the discipline. A decade after the initial ideas for the first book were identified at the 2006 Society for Applied Anthropology meetings, where just one panel mentioned climate change, the topic has reached into every subdiscipline of our field, and we now can move forward to bring our work into effective inter- and transdisicplinary contexts and applications to do our part in the work of transformation. In spite of enormous change, our sentiment in the first edition still holds here, with an additional qualifier: "It is also clear that we have much to learn and explore across the scope of this volume, from encounters to actions *and now to transformations*" (Crate and Nuttall 2009: 34).

Acknowledgments

Susan Crate owes great thanks, first and foremost, to her research partners in the Viliui Regions of in western Sakha-Yakutia, Russia, who first quickened her understanding of how contemporary climate change was interacting with their lives and their unique ways of adapting to their environment. She also acknowledges subsequent research collaborators of communities in other world areas, including Labrador, Canada, Maryland's Eastern Shore, rural Wales, Mongolia, Kiribati, and Peru. She is indebted to the National Science Foundation, Office of Polar Programs, for their support in research and to George Mason University for providing her the time, graduate students, resources, and overall institutional support to follow this research path. Additionally she thanks her mentors, colleagues, and academic collaborators for their support and for sharing in the ongoing adventure of trying to make a difference in the world. Finally, she expresses sincere gratitude to her daughter Katie, aspiring anthropologist herself, for her support and continuing presence in life.

Mark Nuttall thanks the Department of Anthropology at the University of Alberta and the Greenland Climate Research Centre, Greenland Institute of Natural Resources, and Ilisimatusarfik/University of Greenland in Nuuk for institutional support and for continuing to give the time, space, and resources to carry out research on a number of projects on climate change in Greenland and other parts of the circumpolar North—as well as colleagues, students, and community research partners who are too numerous to mention—for discussion, conversation, and reflection. He is grateful to research grant support from the Henry Marshall Tory Chair research program at the University of Alberta, from the Academy of Finland, the European Union, and from the Greenland Climate Research Centre. In particular, he thanks Anita and Rohan for support, encouragement, and for being there.

In addition, our thanks go to our contributors for their outstanding work, to Jennifer Collier, Jack Meinhardt, and Ryan Harris at Left Coast Press, Inc., our previous publisher, and to Staccy Sawyer for her excellent copyediting, as well as to the reviewers who gave generously of their time and who offered critique and insight.

References

ACIA. 2005. *Arctic Climate Impact Assessment*. Cambridge: Cambridge University Press.

AMAP. 2011. *Changes in Arctic Snow, Water, Ice, and Permafrost*. Oslo: Arctic Monitoring and Assessment Programme.

Autin, W. J., and Holbrook, J. M. 2012. Is the Anthropocene an issue of stratigraphy or pop culture? *GSA Today* 22(7): 60–61.

Baer, H., and Singer, M. 2014. *The Anthropology of Climate Change: An Integrated Critical Perspective*. London: Routledge.

Barnes, J., et al. 2013. Contribution of anthropology to the study of climate change. *Nature Climate Change* 3: 541–44.

Batterbury, S. 2008. Anthropology and global warming: The need for environmental engagement. *The Australian Journal of Anthropology* 19(1): 62–65.

CAFF. 2013. *Arctic Biodiversity Assessment: Status and Trends in Arctic Biodiversity*. Akureyri: Conservation of Arctic Flora and Fauna.

Castree, N., Adams, W., Barry, J., Brockington, D., Buscher, B., Corbera, E., Demeritt, D., Duffy, R., Fely, U., Neves, K., Newell, P., Pellizzoni, L., Rigby, K., Robbins, P., Robin, L., Rose, D., Ross, A., Schlosberg, D., Sorlin, S., Whitehead, M., and Wynne, B. 2014. Changing the intellectual climate. *Nature Climate Change* 4: 763–68. Doi:10.1038/nclimate2339.

Chapin, S. F., Berman, M., Callaghan, T. V., Convey, P., Crépin, A.-S., Danell, K., Ducklow, H., Forbes, B., Kofinas, G., McGuire, A. D., Nuttall, M., Virginia, R., Young, O., and Zimov, S. V. 2005. 'Polar Systems' in millennium ecosystem assessment. *Ecosystems and Human Well-Being: Current State and Trends*. Washington, D.C.: Island Press, pp. 717–43.

Chen, J. L., Wilson, C. R., and Tapley, B. D. 2006. Satellite gravity measurements confirm accelerated melting of Greenland Ice Sheet. *Science* 313(5795): 1958–60.

Cornell, S., Forbes, B., McLennan, D., Molau, U., Nuttall, M., Overduin, P., and Wassman, P. 2013. Thresholds in the Arctic, in Arctic Council, *Arctic Resilience Interim Report 2013*. Stockholm: Stockholm Environment Institute/Arctic Council.

Crate, S. A. 2012. Climate change and ice dependent communities: Perspectives from Siberia and Labrador. *Polar Journal* 2: 61–75.

———. 2013. Climate change and human mobility in indigenous communities of the Russian north. *Brookings Institute*, www.brookings.edu/research/papers/2013/01/30-arctic-russia-crate

———. 2014. An ethnography of change in northeastern Siberia. *Sibirica* 13(1): 30–74.

Crate, S. A., and Nuttall, M. (Eds.). 2009. *Anthropology and Climate Change: From Encounters to Actions*. Walnut Creek, CA: Left Coast Press.

Crumley, C. (Ed.). 1994. Historical Ecology: Cultural Knowledge and Changing Landscapes. Santa Fe, NM: School of American Research Press.

Crutzen, P. J., and Stoermer, E. F. 2000. The Anthropocene. *Global Change Newsletter* 41: 17–18.

Dove, M. 2014. *The Anthropology of Climate Change: An Historical Reader*. Colchester: Wiley.

Fiske, S. 2009. Global change policymaking from inside the Beltway: Engaging anthropology, in S. A. Crate and M. Nuttall (Eds.), *Anthropology and Climate Change: From Encounters to Actions*. Walnut Creek, CA: Left Coast Press: 277–91.

Fiske, S. J., Crate, S. A., Crumley, C. L., Galvin, K., Lazrus, H., Lucero, L., Oliver-Smith, A., Orlove, B., Strauss, S., and Wilk, R. 2014. *Changing the Atmosphere: Anthropology and Climate Change*. Final Report of the AAA Global Climate Change Task Force. Arlington, VA: American Anthropological Association.

Hanna, E., Huybrechts, P., Steffen, K., Cappelen, J., Huff, R., Shuman, C., Irvine-Fynn, T., Wise, S., and Griffiths, M. 2008. Increased runoff from melt from the Greenland Ice Sheet: A response to global warming. *Journal of Climate* 21: 331–41.

Hassan, F. 2009. Human agency, climate change, and culture: An archaeological perspective, in S. A. Crate and M. Nuttall (Eds.), *Anthropology and Climate Change: From Encounters to Actions*. Walnut Creek, CA: Left Coast Press: 39–69.

Hastrup, K. 2009. *The Question of Resilience: Social Responses to Climate Change*. Copenhagen: Det Kongelige Danske Videnskabernes Selskab.

———. 2013. Anthropological contributions to the study of climate: Past, present, future *WIREs*. *Climate Change* 4(4): 269–81.

Hastrup, K., and Olwig, K. F. 2012. *Climate Change and Human Mobility: Global Challenges to the Social Sciences*. Cambridge: Cambridge University Press.

Hastrup, K., and Skyrdstrup, M. (Eds.). 2013. *The Social Life of Climate Change Models: Anticipating Nature*. London: Routledge.

Hulme, M. 2009. *Why We Disagree about Climate Change: Understanding Controversy, Inaction, and Opportunity*. Cambridge: Cambridge University Press.

———. 2014. *Can Science Fix Climate Change? A Case against Climate Engineering*. Cambridge: Polity Press

IPCC (Intergovernmental Panel on Climate Change). 2013. *Climate Change 2013: The Physical Science Basis*. Bern: IPCC Working Group.

Kerner, S., Dann, R., and Bangsgaard, P. 2015. *Climate and Ancient Societies*. Copenhagen: Museum Tuscalanum Press.

Lahsen, M. 2005. Seductive simulations: Uncertainty distribution around climate models. *Social Studies of Science* 35: 895–22.

McGovern, T. 2013. Circumpolar Networks: Understanding Cultural and Socio-Environmental Connections in the North Atlantic on a Millennial Scale, *http://ihopenet.org/circumpolarnetworks/*

McGovern, T. 2015. Global Environmental Change Threats to Heritage and Long Term Observing Networks of the Past, *http://ihopenet.org/global-environmental-change-threats-to-heritage-and-long-term-observing-networks-of-the-pas/*.

Marino, E. 2015. *Fierce Climate, Sacred Ground*. Fairbanks: University of Alaska Press.

Milton, K. 2008. Introduction: Anthropological perspectives on climate change, Soapbox Forum. *The Australian Journal of Anthropology* 19(1): 57–58.

Montgomery, L. 2014. *In Norfolk, Evidence of Climate Change Is in the Streets at High Tide*, http://www.washingtonpost.com/business/economy/in-norfolk-evidence-of-climate-change-is-in-the-streets-at-high-tide/2014/05/31/fe3ae860-e71f-11e3-8f90-73e071f3d637_story.html.

Nesbit, J. 2015. *Climate Change Caused California Drought: The Science behind the Drought Is Unquestionable*, www.usnews.com/news/blogs/at-the-edge/2015/04/14/climate-change-and-the-california-drought.

Nuttall, M. 1998. *Protecting the Arctic: Indigenous Peoples and Cultural Survival*. London: Routledge.

———. 2009. Living in a world of movement: Human resilience to environmental instability in Greenland, in S. A. Crate and M. Nuttall (Eds.), *Anthropology and Climate Change: From Encounters to Actions*. Walnut Creek, CA: Left Coast Press, Inc.: 292–310.

———. 2010. Anticipation, climate change, and movement in Greenland. *Études/Inuit/Studies* 34(1): 21–37.

———. 2013. Human security and climate change in the Arctic, in M. Redclift and M. Grasso (Eds.), *2013 Handbook on Climate Change and Human Security*. Cheltenham: Edward Elgar.

Nuttall, M., Berkes, F., Forbes, B., Kofinas, G., Vlassova, T., and Wenzel, G. 2005. Hunting, herding, fishing, and gathering, in ACIA, *Scientific Report*. Cambridge: Cambridge University Press, pp. 650–90.

Oliver-Smith, A. 2009. Climate change and population displacement: Disasters and diasporas in the twenty-first century, in S. A. Crate and M. Nuttall (Eds.), *Anthropology and Climate Change: From Encounters to Actions*. Walnut Creek, CA: Left Coast Press Inc.: 113–36.

Orlove, B. S, Chiang, J. C. H., and Cane, M. A. 2002. Ethnoclimatology in the Andes. *American Scientist* 90: 428–35.

Orlove. B. S., Wiegandt, K., and Luckman, B. (Eds.). 2008. *Darkening Peaks: Glacier Retreat, Science and Society*. Berkeley and Los Angeles: University of California Press.

Overland, J. E., and Wang, M. 2013. When will the summer Arctic be nearly sea ice free? *Geophysical Research Letters* 40: 2097–2101.

Pettenger, M. E. (Ed.). 2007. *The Social Construction of Climate Change: Power, Knowledge, Norms, Discourses*. Aldershot, Ashgate.

Price, S. J., Ford, J. R., Cooper, A H., and Neal, C. 2011. Humans as major geological and geomorphological agents in the Anthropocene: The significance of artificial ground in Great Britain. *Philosophical Transactions of the Royal Society* 369(1938): 1056–84.

Rayner, S. 1989. Fiddling while the globe warms? *Anthropology Today* (6): 1–2.

Rayner, S., and Malone, E. L. (Eds.). 1998. *Human Choice and Climate Change: An International Assessment*. Columbus, OH: Battelle Press (four volumes).

Redclift, M., and Grasso, M. (Eds.). 2013. *Handbook on Climate Change and Human Security*. Cheltenham: Edward Elgar.

Roscoe, P. 2014. A changing climate for anthropological and arcaheological research? Improving the climate-change models. *American Anthropologist* 116(3): 535–48.

Rudiak-Gould, P. 2011. Climate change and anthropology: The importance of reception studies. *Anthropology Today* 27 (2): 9–12.

Rudiak-Gould, P. 2013. *Climate Change and Tradition in a Small Island State: The Rising Tide*. New York: Routledge.
Sandweiss, D. H., and Kelley, A. R. 2012. Archaeological contributions to climate change research: The archaeological record as a paleoclimatic and paleoenvironmental archive. *Annual Review of Anthropology* 41: 371–91.
Skrydstrup, M. 2009. Planetary resilience: Codes, climates, and cosmoscience in Copenhagen, in K. Hastrup (Ed.), *The Question of Resilience: Social Responses to Climate Change*. Copenhagen: The Royal Danish Academy of Science and Letters.
Steffen, W., Crutzen, P. J., and McNeill, J. R. 2007. The Anthropocene: Are humans now overwhelming the great forces of nature? *Ambio* 36(8): 614–21.
Steffen, W., Grinevald, J., Crutzen, P., and McNeill, J. 2011. The Anthropocene: Conceptual and historical perspectives. *Philosophical Transactions of the Royal Society A* 369: 842–67.
Steffen, W., Richardson, K., Rockström, J., Cornell, S. E., Fetzer, I., Bennett, E. M., Biggs, R., Carpenter, S. R., de Vries, W., de Wit, C. A., Folke, C., Gerten, D., Heinke, J., Mace, G. M., Persson, L. M., Ramanathan, V., Reyers, B., and Sörlin, S. 2015. Planetary boundaries: Guiding human development on a changing planet. *Science*. DOI: 10.1126/science.1259855, 15th January.
Strauss S., and Orlove, B. S. (Eds.). 2003. *Weather, Culture, Climate*. London: Berg.
Urry, J. 2011. *Climate Change and Society*. Cambridge: Polity Press.
Van de Noort, R. 2008. Conceptualising climate change archaeology. *Antiquity* 85(329): 1039–48.
———. 2013. *Climate Change Archaeology: Building Resilience from Research in the World's Coastal Wetlands*. Oxford: Oxford University Press.

Part 1: Building Foundations of Anthropology and Climate Change

Chapter 1

Climate Knowledge: Assemblage, Anticipation, Action

Kirsten Hastrup

Climate is not new on the anthropological agenda; it has been immanent in ethnographic descriptions since the early days of anthropology. *Immanence* is a key word here, because, until fairly recently, climate was seen mainly as a basic condition of social life (Hastrup 2013a). In contrast, the contemporary phenomenon of *climate change* is a relatively new item on the human agenda, reflecting the fact that it became prominent on the global, political agenda only a couple of decades ago. The Intergovernmental Panel on Climate Change (IPCC) reports have been instrumental in this, with five reports to date—the first in 1990, the second in 1995, the third in 2001, the fourth in 2007, and the fifth in 2013. The messages from these reports have gained momentum over the years and are now seen as (more or less) incontrovertible within an otherwise very diversified field of climate research. Among the findings are far-reaching environmental changes around the globe, projected for both a near and a more distant future.

Against this background it comes as no surprise that anthropologists have taken an increasing interest in contemporary climate change and that a number of significant volumes on this issue have seen light since the beginning of this century, seeking to find a distinct anthropological voice that may contribute to the discussion that has so far been dominated by other sciences (for example, Strauss and Orlove 2003; Crate and Nuttall 2009; Hastrup 2009a; Hastrup and Olwig 2012; Greschke and Tischler 2015). The historical dominance of the natural sciences in climate research has induced anthropologists to take a submissive attitude, not being immediately able to match the level of generality or the use of statistical averages (Krauss 2015: 69–71). This situation has by no means restrained anthropologists from contributing their own analyses of particular regional or local developments in works that substantiate the diverse and often momentous implications of climate change for social life across the globe. However, such separate action has drawbacks that are twofold; first, the anthropological engagement with climate change has tended to feature marginal or indigenous communities as (isolated) victims of the massive industrial impact on climate more or less uniformly, and, second, it has isolated anthropology from other kinds of scientific understanding of climate change (Bravo 2009; Pálsson et al. 2012). It is of course right that anthropologists have published their analyses of particular communities, providing evidence for the destabilization not only of global climate but also of social life. In this process itself the entwinement of natural and social histories has become firmly established. The time is now ripe to take up the greater challenge of dealing with the boundless processes of the unknown and even partly unknowable impingement on human life that climate change signals and that questions anthropology's time-honored focus on local communities—without undermining the key methods of fieldwork. We must learn to theorize across ethnographic fields and offer our theories to the wider community of scholars and scientists for inspection and inclusion in the general field of climate change research (Hastrup 2015).

Anthropology can and must contribute significantly also to the current discussion of the Anthropocene, now a familiar term, dating back only to 2002 (Crutzen 2002). With the human footprint all over the globe, the human science of anthropology must pull its weight. It is no longer possible to entertain a notion of a self-regenerating nature, beyond the human domain. Humans are everywhere, not only as destroyers of nature but also as providers of collective solutions. We need to understand better the human responses to climate change, unfolding at the interface between natural and social histories that always outstretch a particular moment or place.

Thus anthropologists must turn their attention to the processes by which climate change is configured and how a motley combination of knowledge forms enter into the phenomenon's making, which I discuss in the first section on assemblage. In the subsequent section I address the issue of anticipation being different from prediction in my argument and signposting a way of being. In the third section, I discuss action, with a focus on social action as based both on past experiences and anticipated futures. The key example throughout relates to the Arctic, where people live with rapid changes in their environment and from where many scientific climate models draw some of their main observations.

Assemblage: The Making of Knowledge

In setting the frame for this section I want to stress that the identification of *climate* as a discrete phenomenon, and by implication of *climate change*, is an analytical human endeavor. Whether it is talked about by scientists, discussed by fishermen experiencing turbulent seas, or pondered by urbanites witnessing wild cloudbursts, climate knowledge is assembled from many sources. It is not enough simply to designate the diversity of understanding to different scientific scales; scale itself is a product of particular knowledge practices, not a solid ground on which its stands.

In short, the field of climate knowledge remains wide open, even as we try to stabilize it, making the potential outcome of anthropological analyses to be of the same generality as results of general circulation models (GCMs), now seen as authoritative models of climate prediction. The authority of GCMs is vested in the general agreement of the complexity of the global climate system, which functions to transport heat from the equator to the poles and determines how it is retained, distributed, and circulated in the oceans, the atmosphere, and land surfaces. "Owing to its immense size and long-time scales, the climate system cannot be studied by experimental methods. Therefore, scientists have relied on climate models—theory-based representations that characterize or simulate essential features and mechanisms—to explore how Earth's climate works" (Edwards 2011: 128). Thus simulations are the bases of the computerized general circulation models, which are at the core of present-day climate change scenarios, seductively depicting "time-dependent three-dimensional flows of mass, heat, and other fluid properties" (Lahsen 2005: 898). Closer to the ground, anthropologists have documented recent flows of people across the continents in response to fluid political, economic, and climatic conditions (Hastrup and Olwig 2012)—a phenomenon also known from the past (Orlove 2005). These flows are not computerized but can easily be foreshadowed in imagination, fueled by images of increasing numbers of climate refugees seeking to cross the Mediterranean in sinking vessels or to transgress high-security borders such as the one between Mexico and the United States—often to their peril.

As I have dealt with in more depth elsewhere (Hastrup 2013b), the technological feat inhering in the simulation of the complex interactions of various elements and processes in the earth system, coupling atmospheric, oceanic, and land-surface processes, is not free of subjectivity or untouched by humans. Yet these simulations have great power over the human mind, in part because of their technological sophistication:

> In recent decades, our understanding of the climate has been revolutionized by the development of sophisticated computer models, known as general circulation models (GCMs). GCMs are a representation of the physical laws . . . expressed in such form that they are suitable for solution on fast super-computers. (Williams 2005: 2932–33)

Note the authoritative framing used—"representation of the physical laws" and "suitable for solution," acknowledging a somewhat selective editing process. Such framing is not to denigrate the value of the climate models but simply to humanize them. They are results of hard work and sophisticated reflections made by many people, agreeing (or possibly disagreeing) on their meaning. In short, they are the outcome of a particular "knowledge space" in the sense suggested by David Turnbull, as an "interactive, contingent assemblage of space and knowledge, sustained and created by social labour" (Turnbull 2003: 4). Looking back on the history of climate ideas shows one how these were formed under particular circumstances and by people who were situated in particular places (Heymann 2010; Carey 2012). Thus we must integrate the history of climate ideas into any current discussion of climate.

In the ancient world, astronomer-geographers such as Ptolemy (2nd century C.E.) connected climate to the inclination of the sun and identified 15 climatic zones on the basis of their longest day, equally an expression of latitude (Edwards 2011: 128). This idea informed the European perception of climate until the 19th century, when climate became seen as the condition of a particular region pertaining to temperature and dryness; climate was now local and static, and on the scale of human lifetimes (Weart 2010: 67). Seeing climate as the stable, long-term average of local weather became untenable during the 19th century, as natural history broke new ground:

> Three discoveries changed the general view of a steady world. One was the discovery by Richard Owen (1841) of the fossilized remains of the no-longer-existing giant dinosaurs; the second and most important was Charles Darwin's expedition on the Beagle in 1839 leading to the discovery of evolution of the species. The third was Louis Agassiz' discovery of the ice ages (1837). (Ditlevsen 2013: 183)

That the theory of an ice age was first formulated by a Swiss scientist, Agassiz, is no coincidence; he observed Alpine glaciers and formations of big boulders, which were far away from current glacier tongues, and eventually suggested that the entire northern hemisphere had once been covered in ice, of which the Alpine glaciers were but remnants. At the same time, Swedish expeditions to Svalbard were able to ascertain that shellfish fossils found in mainland Scandinavia were still present as living species in the icy ocean around these far northern and uninhabited islands. The idea of an ice age was further solidified by Hinrich Rink's observations of the huge and very productive glaciers in Greenland and his suggestion that they were outlets from ancient ice on the ice cap (Rink 1877). Ice gradually became recognized as a repository of climate histories in the depth of time. With the high modern ice core research on the Greenlandic ice cap (and in Antarctica), the value of the ice archive of shifting climates over the past 100,000 years, also in relation to future scenarios, cannot be overestimated (for example, Mason-Delmotte et al. 2012).

When Greenland was gradually being mapped in the early 20th century, the ice age had become an established truth. In 1916–1918 a cartographic expedition was made to northern Greenland, where geologist Lauge Koch made important geomorphological observations. He also made the following note on the ice age on his way north: "As will be known, almost all of North- and Middle Europe was covered by one continuous mass of ice, which arched up as a shield from Scandinavia and across the neighbouring countries. A similar case obtained for Canada and the northern parts of the United States. In Greenland, [where] the ice has remained, one is still in the middle of the Ice Age, and travelling from South to North Greenland, is to experience the return of the ice age" (Koch 1919: 565). Thus the ancient climate history was readable in space; cartography and the long story of the ice came together. In the process, it became established that climate could change radically on a planetary scale (Weart 2010: 67).

The main point of this historical review is to show how "evidence" for the ice age was found locally—in the Swiss Alps, in Svalbard, and in Greenland—and was assembled by people looking for boulders around earlier glacier edges or identifying living counterparts in the far north to fossils known from elsewhere. The discoverers were people, steeped in the actualities of the world, which is also true of present-day scientists. Assemblage is the key word in the process of knowledge making. Knowledge does not emerge out of the deep; it is made and authorized in a community of potential dissenters and only later becomes common knowledge. This assemblage pertains equally to scientific and nonscientific knowledge about climate and climate change.

Therefore, climate knowledge must always be contextualized historically and socially (Carey 2012: 238). How climate is understood has changed over time and has "depended not only on scientific achievements but also on broader technological, social, political, and cultural contexts" (Heymann 2010: 582). This need to contextualize becomes very obvious in the 20th century, where the idea that humans could influence the climate gradually took root, albeit unevenly (Weart 2010: 68ff). The process by which it changed from being merely a scientific puzzle to a more ominous and heavily politicized arena of discussion was a long one, heavily mixed up in international politics, warfare, demographic developments, and technological advancements. The establishment of the IPCC was instrumental in transforming the idea into a fact; in the IPCC's first report (1990), it remained mostly a scientific puzzle, but already in the second report (1995), a consensus was voiced: "not only was the world getting warmer, but 'the balance of evidence suggests' that humanity was exercising a 'discernible' influence of global climate" (Weart 2010: 75). Recognition of humanity's role led to the Kyoto Protocol (1997) and onward to ever more concerted efforts on the part of the scientific community to assemble evidence for reports that followed. The understanding of humans' role also induced the scientific community to integrate the models from different laboratories focusing on different elements—oceans, sea ice, atmospheric conditions—into a comprehensive Earth Modelling Framework in 2002, allowing the different component models to interoperate (Edwards 2011: 135). This surprisingly recent feat should not make us forget that the "fact" of anthropogenic climate change became mainstream only through the combined efforts of thousands of scientists, standing on the shoulders of pioneers who had voiced the hypothesis over almost a century (Weart 2010: 76). It is this human effort of assemblage that we should note.

Anthropologists need not presuppose total consistency with the field of climate research to seek to scale and contribute an anthropological model to this larger multidisciplinary effort for our work to be incorporated. Rather we should take comfort in the volatility of climate science and contribute our own theories in conjunction with other theories, since theories are of equal generality. We might think of David Turnbull's description of the field of turbulence research, balancing on an edge between stable and volatile knowledge. When entering the field of turbulence research as an anthropological researcher, Turnbull realized that there was no agreement on either the phenomenon or how to deal with it.

> Yet despite the lack of consensus there is sufficient coherence for the practitioners to act as if there is a field of turbulence research. Coherence in this case does not derive from a unifying paradigm or the adoption of an agreed set of instruments or methods. It derives from a very loose recognition that the phenomenon at issue is turbulence, even though its nature cannot be specified and even though it occurs in a very diverse set of flow situations from blood vessels to aircraft wings to the earth's atmosphere. But equally important, coherence results from the work of the researchers in the field trying to establish equivalences and connections in problem solving while also struggling for authority. (Turnbull 2003: 190)

The implicit recognition that scientific knowledge is not monistic and that there are many ways of contributing to a particular field is significant here. The field of climate change seems as inherently indeterminate as turbulence; perhaps paradoxically, I find this encouraging. Here, too, knowledge makers of all kinds now drift toward acknowledging the need to establish equivalences that highlight shared concerns, immediately opening up space for a dialogue across outmoded epistemological boundaries, between natural and social sciences on the one hand and between scientific and traditional knowledge on the other. In terms of climate change, everyone inhabits the same field of uncertainties, and nobody can escape the global terminology. Climate change is nothing if not global, even if backed by located observations and weather-events. Anthropology should contribute with substantial analyses of how knowledge is made in practice and how the human capacity for anticipation is shaped and stretched within such practices, however named. The time has come for anthropology to straddle the gap between the particular and the universal—in recognition of our shared humanity.

In the Arctic region the changes in weather and wind, and not least in the ice conditions, are remarkable, and their direct implications for social life are well documented (Huntington and Fox 2005). In the natural sciences the Arctic has played a crucial role in identifying both the long-term changes in global climate through ice-core research and the current, rapid changes seen to endanger the sea ice (Luedtke and Howkins 2012). The World Wide Fund for Nature has recently issued a position paper wherein the Last Ice Area is identified in Northwest Greenland; a small patch of ice-cover is all that will remain of the vast ice-cover of the Arctic Ocean within the next generations (WWF 2013). Anthropologists, too, are mapping the changing sea ice together with locals (for example, Krupnik et al. 2010). While such ethnographic effort is important in its own right, it makes no big difference in the wider field of climate change research; it may add valid observations at a level where no one is really in any doubt any more. It is well known that the Arctic is warming and that the sea ice is melting, and that we must understand it in context (Aporta 2010). What we now need is to acknowledge and harness the human power at theorizing also in the Arctic; we should leave behind the tendency to attribute only "observations" to people living outside academia.

All kinds of knowledge are equally located and equally based in observation and analysis. As I have related elsewhere (Hastrup 2013d), this fact was hammered home to me when I was conversing with a seasoned hunter in Northwest Greenland. He had contributed to a report from the Inuit Circumpolar Council (ICC) to which he referred me for further information about the actual changes in the ice conditions and so on over the past generations, which was based on anthropological interviews in the area. I expressed my excitement at the prospect of eventually being able to read it, to which he replied: "It is not exciting for me. They do not know our reality. They just want to become professors. Why would we want more professors with no clue about reality?" The anthropologists had collected their words, but the hunters did not need to have their observations simply repeated. They had their own theories and wanted a true conversation on future scenarios.

It is sometimes difficult to shed received visions of Arctic plights. The Arctic has experienced "one of the most rapid warming trends of any place on earth. Since the 1950s, average annual temperatures in the region have increased by 2 to 3°C, a rate more than twice the global mean" (Luedtke and Howkins 2012: 150). Given that the fluctuations in sea ice have had dramatic ramifications for humans and animals in the region, "perhaps most emotively by the plight of polar bears" (ibid.), it is no wonder that the Arctic has gained prominence as a canary in dominant climate change narratives, with the rhetoric matched by highly visible changes and by species that lend themselves easily to tales of extinction (Hamblyn 2009). In my perspective, the major drawback of this attention to the Arctic is a general victimization of people actually living there, on the one hand, and, on the other, a construction of any Arctic climate narrative as a crisis narrative, "where people are seen as lacking the agency to fight back and [as] the keepers of valuable traditional knowledge" (Bravo 2009: 258). It has been argued that "global change science is both shaping and being shaped by a new kind of citizen, namely, the Arctic citizen" (Martello 2004: 107).

The polar bear clinging to a minute ice floe is not only an icon of climate threats that extend to social communities but is also experienced directly by people living in proximity to accessible polar bear populations. Thus, each autumn the small town of Churchill on the western shore of Hudson Bay in Canada (800 inhabitants)—dubbed the Polar Bear Capital of the World—attracts thousands of tourists wanting to catch a glimpse of the ultimate victim of unsustainable lifestyles elsewhere in the world (Grill 2015). It is the time when bears congregate by the coastline awaiting the formation of the sea ice. Thus the iconic status of the bear is now a source of income for the inhabitants of Churchill, if also somewhat of a strain at least for the season. In a neighboring community, people have a similar experience, when the bear hunt starts and licenses are given away to trophy hunters—or rather sold (Tyrrell 2006). In both cases, anthropologists have shown how there is a clash of discourses at play, one of global climate change and threatened species, attracting outsiders wanting to see or to shoot the bears before it is too late (!), and another of seasonal cycles in the animal-human relationship wherein the bears are not icons of climate threats but coinhabitants of the social space, and not seen as particularly endangered.

There is not necessarily a conflict between these narratives; they are just different. They are not different in scale, however, as in global versus local, because both are equally based in knowledge that transcends the "local" by far. Thus it is untenable to see the situation in terms of global, scientific knowledge on the one hand, and local or indigenous knowledge on the other. Indigenousness may be of political importance in some cases, but knowledge is a dynamic domain, and—as suggested by Emilie Cameron—to equate indigeneity and the *human self* with the traditional and the local may actually extend colonial forms of knowledge and practice (Cameron 2012: 111). In the example here, people are simply placed differently in relation to ice and polar bears; their knowledge is in that sense equally *located*. All kinds of knowledge are both situated and situating (Turnbull 2003: 19).

The practices by which people deal with the challenge of projected changes in climate always depend on records, experiences, and observations. The general point is that the foundation of climate change knowledge is empirical across the board; all knowledge is based on empirical observations and experiences, which may be processed differently but which are equal in their being assembled and processed—before they are recognized as knowledge (Hastrup 2015). The empirical basis of climate change experiences and observations is where anthropology must take heart.

Anticipation: The Making of Scenarios

If assemblage is linked to the establishment of facts, anticipation is closely linked to making scenarios for the future, which contain realistic expectations based on the knowledge assembled. Scenarios in this sense are not predictive but akin to the imaginative horizons suggested by Vincent Crapanzano, which we construct wittingly or unwittingly, and "that determine what we experience and how we interpret what we experience" (Crapanzano 2004: 2). Within such a view of horizons there are undercurrents of both openings and closures that sit well with "anticipation" in the way I use it here.

If the concept of climate change is based on assembled knowledge from a variety of sources, each individual, group, culture, and society assembles it both from the immediate surroundings and from other people, television features, Internet communication, and scientific reports. The Arctic is like everywhere else in that sense, except that the actual emplacement of people in that particular landscape lends its own distinct tinge to their theories. The seasonal changes between light and darkness, between ice and water, and the presence of various species, bear all the marks of the specific latitude—the ancient denominator of the climate. Yet, while the inclination of the sun is stable, everything else is in flux. Like other people who live in close conversation with landscape and living resources, the hunters and families in Northwest Greenland whom I have come to know acutely sense that a new era is on their doorstep. The sea ice, which has provided the basic infrastructure for social life, allowing people to move between settlements and hunting grounds on dog sledge, is increasingly affected by the warming and the derived changes in oceanic conditions, among which unpredictable sea currents were often mentioned. While the ice is in some way a known territory, it is also a shifty partner in the social life as elsewhere in the Arctic (Henshaw 2009; Hastrup 2013e, 2014b).

Moving about in the Arctic landscape always demanded an astute sensitivity to cues in the environment and a capacity to respond appropriately to these cues (Ingold 2010: 134). When far northern hunters drive their dog sledges toward the ice edge, an important hunting ground for narwhal, walrus, and bearded seals, cues about the state of the ice and the passable routes are taken from the color of the ice, the slush on the surface, and the position of small icebergs that may have turned upside down because they had become top-heavy, thus indicating a melt-off from below, on the one hand, and, on the other, a sea ice so thinned out that it lost its grip. More long-distance cues are taken from seabirds known to congregate over open water, from distant sounds of tumbling icebergs, clouds, and winds. Such cues enable the travelers to foresee openings and closures in the ice, by way of a kind of diagrammatic reasoning (Hastrup 2013c).

Presently, this particular logic of spatial reasoning often fails; over the past 10 years the sea ice has become unpredictable, the new ice forming several months later than previously and being thinner. Thus old hunting grounds are becoming increasingly inaccessible, and communication between the settlements precarious. The animals act in erratic ways, too, owing to the changes in their habitat—for instance, in the balance between fresh and salt water, affected by the increasing velocity and break-off of the glaciers. In these circumstances social life has become strained; the recent arrival of "fishable" halibut from more southerly waters relieves the economic strain (for some, and in some years), but the halibut still depend on a stable sea ice, from where it is fished with long lines. Meanwhile, place-names referring to topographical features no longer fit the landscape, a phenomenon that applies all over the Arctic (for example, Henshaw 2009: 161; Nuttall 2010: 27).

The Northwest Greenland population is decreasing, now down to about 700 overall. Young men are no longer trained to become hunters, and young women are not properly taught to process skins and turn them into clothing and footwear, of the kind that is still absolutely necessary for hunting trips on the ice. All these changes are part of the same process of experiencing changes *and* anticipating their effects. What we see here is how the community's future is being formed along with the scenarios; clearly, the future is not envisioned as based on hunting in the old sense. The scenario is based on an analysis of assembled knowledge, not only from moving about in a changing landscape but also from notions of global climate change, incessantly filtering into their world from scientists visiting the place and measuring everything, including the thickness of the ice, the salinity of the open water, the amount of soot and algae on the icecap that precipitate the melt-off, and so on. Northwest Greenlandic communities are also well versed about the IPCC reports, and surveys made by the Greenland Institute of Natural Resources of their observations of the ice and the wildlife are the order of the day (for instance, Born et al. 2011).

Clearly, even in this remote corner of the world—which is of course at its own center—there is no such thing as local knowledge as opposed to scientific knowledge. All of it enters into one located knowledge space, producing its own *spatial vernacular,* to introduce a term suggested by Whatmore (2002: 6). The spatial vernacular is based on an engagement with nonhuman agents such as sea ice, walruses, glaciers, seals, and dogs, not to speak of whiteouts, unpredictable winds, and shifty sea currents, and incorporates knowledge from motley practices, instrumentations, theories, and people that pass by their world into age-old and trusted theories of the environment.

This kind of orientation challenges any notion of bounded and local social communities and forces us to think of social resilience as a feature of social flexibility and a human capacity for *re*orientation and anticipation of future opportunities (see Hastrup 2009b, c). Considering the almost unimaginable historical changes, technological advancements, colonial upheavals, and now climate change that Arctic peoples have faced since they first moved across America and eventually into Greenland, there is a strong case for the formidable human capability to identify and anticipate new possibilities. As Mark Nuttall has it:

> Inuit have not just adapted to the Arctic environment; they have anticipated the possibilities and conditions for successful engagement with it. In Greenlandic traditional communities, e.g., hunting and fishing involve not merely procurement, but also anticipating, waiting, hoping, pondering, and imagining the movements of seals, narwhals, fish and other animals to be caught, as well as anticipation and apprehension of the return home. (Nuttall 2010: 25)

We may see the age-old migrations across and between the North American Arctic and Greenland as part of this trend of anticipation but also, significantly, the social patterns of food distribution and sharing, which have leveled out the differences between people with shifting hunting luck in contemporary Arctic communities (Wenzel 2009).

At present, social life is circumscribed in the Greenlandic far north. It seems that the map of opportunities, within which people have been able to maneuver (more or less) successfully for generations is now shrinking to the point where the pillars of time-honored diagrammatic reasoning are undermined. The cues have become erratic, limiting—which implies that there is less freedom to move and to choose different steps toward the future; social flexibility is lost. I use *flexibility* as defined by Gregory Bateson as "uncommitted potential for change" (1972: 497). Bateson uses a parable of the acrobat on a high wire to illustrate the implications of this limitation of flexibility and freedom. To maintain his or her position on the wire, the acrobat must be free to move between one position of instability to another, and his arms must have maximum flexibility to secure the stability of more central parts. If the arms are locked, the acrobat falls. During the period when the acrobat is *learning* to walk on the wire, and thus learning to move his arms in an appropriate way, a safety net is necessary; this gives him the freedom to fall off the wire. "Freedom and flexibility in regard to the most basic variables may be necessary during the process of learning and creating a new system of social change" (ibid.: 498).

Presently, the people with whom I have worked have next to no freedom or flexibility. Their arms are tied, and their energy goes into short-term survival. They have increasingly gotten the ear of politicians in the south, but otherwise they are left to themselves. Considering that "to maintain the flexibility of a given variable either that flexibility must be exercised or the encroaching variables must be directly controlled" (Bateson 1972: 503), they fall short on both accounts. During conversations they express their concerns about the future, but these are far from exclusively related to climate. The uncertain political and economic development of Greenland also plays an important part in the uncertainty: even if the inhabitants agreed to move south, which most of the remaining people (the men in particular) refuse to seriously consider, there would be no work for them down there and very little opportunity to continue even as part-time hunters, and to keep dogs. There is no real safety net anywhere. They prefer to stay in their own changing landscape and sometimes they even hope for more rapid changes, because then—possibly—they could have a small port, bigger boats, and make a new way of living. Their view of the future thus lingers between doubt and hope, while their youth are encouraged to move south for more education than is available in the far north. The future is seeping south in spite of everything.

What I have provided here is not an account of local knowledge but the contours of a climate model, including scenarios for the future, based on located weather experiences. It operates on a generational time scale, and as such it matches the general view of a 30-year average of weather emerging as "climate," in terms of temperature—for instance, it also displays the same fault lines in predictive power. "The distinction between predictability of the first kind, the weather prediction, and prediction of the second kind, the climate projection, is well defined in a mathematical sense. But when it comes to Nature the situation is much more complex" (Ditlevsen 2013: 193). Here multitudes of temporal and spatial scales intermingle and make it difficult to properly predict even the near future. "There is thus a close connection between the spatial scales of variation that we want to predict and the time scale of predictability" (ibid.: 194). Ditlevsen continues:

> If we make a weather prediction of a day or more, which is an initial value problem, we already passed the time limit for some of the smaller scales, where we can thus only make predictions of the second kind. When the forecast says "showers" it is not predicting rain at a specific location at a specific time; it is predicting some probability of rain in a statistical sense. We thus have a mixture of predictions of the first kind and the second kind. This would also be the case had the models had such a high resolution that they actually resolved single clouds. (Ditlevsen 2013: 194)

Such a mixture or predictions is important in comparison to the anthropological observations from various places, where long- and short-term observations of weather phenomena combine into a sense of the "right" climate for social life to continue at particular places. Hildegard Diemberger has reported how in Tibet a particular mountain with a little snow on its summit was seen as a portent of scenarios of different time scales.

> The elders of Porong used to say that as long as this snow is there, the land, the people, and the rulers of Porong will prosper. If this disappears it will be the end.

> As well as being part of a local theory of political and social stability linked to the environment, mountains like this have been important indicators of weather trends affecting the region. It is even possible to suggest that they have been politically effective because their "readings" entail enough empirical knowledge and observation to ground them in people's experience. (Diemberger 2013: 108)

Natural scientists working in the same area have shown how important the cloud formation on the top of this and other snow-mountains is in terms also of the local moisture circulation ensuring the vegetation cover (ibid.: 108ff). Such downscaling was not possible to capture with a GCM, but the Porong alerted the scientists to the need for a higher resolution model that provided them an understanding also of the larger patterns in this highly mountainous area. Note Diemberger's use of the word *theory*; like other theories it emerges in a particular social and environmental context and aims at generalization.

On the Tibetan plateau as well as on the Mongolian desert steppe, the environment is vast and arid; for it to be livable at all, the recycling of moisture is essential. This fact is well known to the herders, and it is acknowledged that "by nature of their lifestyle pastoralists are in the position to gather environmental information (including climatic) over much larger areas and with much finer spatial resolution than weather stations which only register point occurrences" (Marin 2010: 74). It is not only pastoralists that may contribute vitally to domains that have escaped the GCMs; anthropologists, too, have privileged access to high-resolution models of climate at the level of human life—where many time scales and environmental features combine into emerging scenarios both for the immediate future (the day or the season) and for the more distant future (the next generation).

The challenge for people in the Arctic and on the Mongolian steppe, for instance, is that they (soon) have to make serious choices for their own future; models become all too real. In Northwest Greenland, moving south would fulfil the worst-case scenario, and, as in human life generally, it is hard to decide when there is no possible future left on the known premises. The question is when the weather development will undercut the social order. In terms of the environmental reality, this was not always orderly at all, as they well know, but there used to be a sense that a "bad year" or an "odd season" was the exception. Now it is the other way round, and evidently they reflect a lot on the situation, and thoughts about moving south in Greenland or even to Denmark, where many of them have family, are recurrent in the community. For some of these inhabitants, these thoughts transform into action. The general point is that people in the Arctic have not simply adapted to a rather harsh natural environment; they have also had to anticipate continuously the possibilities for successfully engaging with it (Nuttall 2010: 25). In North Greenland the world has been in constant flux, and to live there demands astute attention to possibilities, which again presupposes alertness and an ability to be open to surprise and uncertainty (ibid.: 26).

The question of when to move (or if) is fraught with theoretical difficulty, not unlike the scientific identification of tipping points. In ordinary forecasting practices scientists apply an implicitly linear thinking between cause and effect; if the temperature rise goes from 2 to 4 degrees, heat has doubled, and the melt-off of the ice is supposed to double as well; if the temperature rises by 6 degrees, the melt-off theoretically triples. If this situation were true, it would be easy to forecast its course, and, conversely, we might hope for restoration if the temperature curve turns round. However, both the climatic records from the ice-core and the current development in the Arctic exhibit a nonlinear dynamic, precisely where the system undergoes a major change—or reaches a tipping point.

> Tipping points in the climate may lead to dramatic changes, as seen in the palaeo-climatic record. If the Greenland ice sheet collapses the global sea level will raise seven metres. If this unfortunate situation should occur it will not happen overnight, but the actual time scale for this to happen is poorly understood. The recently observed speed of shrinking of the ice sheet has been surprisingly fast in comparison to the present understanding. (Ditlevsen 2013: 198)

The nonlinear response of actual processes cannot be captured in the GCMs, logically acting in linear fashion. While mathematical models are excellent for simulations at a low resolution, they can neither figure out the actualities on "the ground" so to speak, as we discussed, nor predict the actual points in time when a particular climate breaks completely down, and at which pace it will happen.

Exercising practical flexibility, as Arctic hunters must, is not necessarily to explode age-old knowledge, nor is it to think in linear terms. It could even be argued that disciplined and recognizable ways of nonlinear and spatial thinking are preconditions for a sustained effort to transcend the limits of the known; in other words, old knowledge provides the only safety net while people are still uncovering the basic variables of the emerging reality. It is the tipping point that neither they nor the scientists can predict—even if they may somehow *anticipate* it. I want to stress the nature of anticipation as distinct from prediction in the vein of Nuttall, whose concern is this:

> to focus on the place of anticipation in climate change studies and to point to the ethnographic possibilities in a multiplicity ways—as a form of knowledge, as ontology, as foresight or insight, as engagement, as orientation, as self-realisation, and as a consideration of potential. Anticipation concerns the future, or a range of possible futures—whether immediate near or distant—how people imagine themselves into it, and how they prepare for possibility and opportunity, for challenges, and for the effects of events and action. (Nuttall 2010: 33)

In the process of anticipation, nature is reconceptualized and the future reimagined. We may disagree on climate change (Hulme 2009), but we may still agree that all the anticipated scenarios, scientific or not, build on a particular empirical knowledge that is captured somewhere. For instance in ice-cores, atmospheric compositions, geological traces, place-names, memories, bodily sensations, stories, simulations, diagrams, or some other medium may close the gap between past experience and future expectation by incorporating them into a comprehensive model. The elements of the model may then be processed and transformed into realistic climate scenarios by imaginatively experimenting with the variables.

Anthropologists may contribute to climate change scenarios by actively unpacking the nature of anticipation as distinct from prediction. Yet possibly the greatest promise of anthropology is to offer up a comprehensive theory of the human potential in addressing the challenges posed by imagined tipping points. This action could enable us to get beyond the "Giddens paradox," in which "since the dangers posed by global warming aren't tangible, immediate, or visible in the course of day-to-day life, however awesome they appear, many will sit on their hands and do nothing of a concrete nature about them. Yet waiting until they become visible and acute before being stirred to serious action will, by definition, be too late" (Giddens 2009: 2). To resolve this paradox, we need more studies of human responses.

Action: The Making of Worlds

Human agency is based as much on future expectations as on past experiences (Hastrup 2007). To understand how climate change is incorporated into located theories and projected into action, we must first acknowledge that humans are *social* to the core. Sociality itself has wide ramifications in both time and space. In any community, including in the Arctic, both the temporal and spatial volatility of climate change as perceived and debated. Past, present, and future are linked in time-stories that are part of the baseline for action (Pahl et al. 2014). Spatially, climate concerns likewise contract and expand; however small and remote some of the Arctic communities are, they are tightly linked to the national (Greenlandic and Danish), and the international political order and to the global economy (Nuttall et al. 2005). Political interests, trade barriers, and conservation efforts affect and constrain the capacity for action in the Arctic communities. It has even been suggested that a new *environmentality*—that is, the way human subjects understand themselves in relation to their environment—will change with new technologies of governance, which is significant in the Arctic, where people have until recently lived on the margin of the state (Lovecraft 2008).

The complex social basis of action goes to substantiate initially why action is never simply a *re*action to what has already happened; it is also a mode of acting on anticipation, which again is embedded in particular visions of pertinent temporal and spatial frames. Agency is closely tied to the anticipation of a story, a line of future development within a relevant spatial framework. It is a profound matter of *responding*—response being made within a moral horizon and within a social context that we interpret and project forward as we go along. "Anticipation is also potentiation" (Strathern 1992: 178). The sense of history is what integrates individual actions into a larger vision of the world, filled out imaginatively and acted on. In that sense any social action is a creation, contributing to a history that outlasts (and outwits) our imagination.

We perform a world into being, and thus we meet the future half way toward it. We may even speak the world into occurring, because speech in itself is an act. Words are not simply expressions of the world; they are means of dealing with it—after which it is no longer the same. Once the Anthropocene has been declared, it frames and forms social actions within the world. Speaking—like seeing and thinking and other processes by which one appropriates the world—is bound up with time and history. To understand a mode of action or a form of life, we must bring figure and ground, or "the passing occasion and the long story," into coincident view (Geertz 1995: 51). The event of action and the long story belong together. It is not for individual agents to comprehend their actions fully, because the descriptions that are available to them from inside their experience cannot at the same time bring the complete story into view. Yet act they must, on the basis of anticipated futures that thereby are being shaped, making the present rather than the future the momentous unknown (Strathern 1992: 178). Sometimes there is too much future crammed into the present, preventing people to respond consistently—as Shakespeare showed in the case of Macbeth (Hastrup 2007: 195–56).

The general point is that people make the world as they imagine it; the Arctic hunters do not go out in search of game simply as matter of habit, but also because they imagine an outcome in the shape of a possible catch. Expectation and imagination play important parts in action. When the environment changes rapidly, the imagination is strained, and we must revisit the ways in which people seek to anticipate their world in view of the comprehensive uncertainties. Responding to the changing environment implies taking social responsibility, which is different from mere reaction or adaptation; social action is different from behavior. The Inuit have always been astute observers of their environment; through hunting and other activities they have assembled massive information on the weather and processed long-lasting theories about the connections between seasons, weather variability, and animal behavior. These theories are now under siege; responding to the present situation implies an acute awareness also of possible new openings and emerging resources.

Among the (literal) openings in Northwest Greenland is the revival of the global dream of a reliable Northwest Passage that would cut down transport time between East and West considerably; as ever, an open sea potentially forges new connections. Yet in the process, the dream of fossil fuel on the bottom of the Arctic seas has changed the context of the International Law of the Sea (UNCLOS), reshaping territorial interests in the process—quite apart from the territorial interests of narwhals and other species. While people in Northwest Greenland may eventually profit from the increasing intensity of passage and possibly find new avenues for making a living, they will also have to live with an ocean of disappearing game. Hunting may soon be a feature of the past. Already now, the hunters are all too aware that the intensifying traffic, also sometimes of fearless tourist cruise ships, in the region severely affects the animal resources—and amplify the effects of the break-down of the age-old exchange between the deep sea, ice, animals, and people in the Arctic world.

The capacity for responding to the current challenges implies a capacity to decide; no scenario can predict the tipping point that may make social life in a particular place impossible. Decisions cut out particular courses of action as "right" and discard others as "wrong"—under the circumstances. Neutrality in decision making is an illusion; facts, including the fact of action, are imbued with value (Putnam 1981: 127ff). Thus, the capacity to decide is deeply affected by standards outside the domain of climate change. With the massive changes in Northwest Greenland, decisions about future action, about whether to stay or to move south, to train the young to become hunters, or to send them away for alternative education not offered in the high north, is premised by a complex set of values that are not immediately reconcilable but all of which must be taken into account.

As pointed out by Nuttall, the current topographical reshaping of Greenland as the ice dwindles coincides with the emergence of a new Greenlandic nation that in itself redefines people's relationships to place and to their natural and social environment (Nuttall 2009: 297). Politics is thus very much a part of current changes in Greenland, and the more so because of the present Greenlandic government's wish to centralize habitation and thus to close some of the more peripheral settlements—if not immediately. In general, it seems clear that the uncertainties related to climate change in North Greenland are magnified, because they are underwritten by the "signature of the state," a notion as proposed by Veena Das, who also identifies the paradox of illegibility (Das 2004). The state apparently works by way of decrees, laws, and rational calculations on the basis of intentionality and direction; it works on behalf of the common good. With each regulation, however, the hunters have become more integrated into the state, Danish and Greenlandic, yet at the same time they have become marginalized from their own history, always centering on a degree of mobility and social flexibility that fitted their mode of living. Today also, the "illegible" aspects of the distant (if ever-present) state overdetermine an interpretation that feeds into the worst fear, expressed to me by an inhabitant of the far north: "in 10 years we are no longer here." This is certainly a possible scenario—although it may become true mainly by an altered sense of who "we" are.

The most valuable general insight produced by anthropologists studying the effects of climate change on particular societies is that people never see themselves as without some kind of the future, somewhere. Already now, the hunters in North Greenland are beginning to play mentally with new possibilities, affecting the "we"; I was told: "If the sea ice melts away completely, then maybe we can have a small harbour and real fishing vessels. We may even become the last port of call before the bigger ships go into the Northwest Passage or towards the North Pole." Although this idea still seems somewhat dreamy, it certainly testifies to the powers of anticipation. It also, inadvertently, shows how the inhabitants of the far north are truly cosmopolitan. In more general terms, and as noted by a scholar of the Arctic: "For most people in Arctic nations the phenomenon of climate change is now a piece of their lifeworlds, because it has entered the space of communicative action, the social space where people engage ideas and have meaningful debate" (Lovecraft 2008: 92).

To act meaningfully means to acknowledge the complexity of the situation, temporally and spatially, to assess the relative potential of various scenarios and to make the best possible decisions. Decision making incorporates ways of imagining oneself in social space in a constant process of reorientation. This claim does not to suggest that all actions are equally constructive in the long-term mitigation of the effects of climate change, only that they respond to particular scenarios for the future, assembled from diverse forms of knowledge.

An interesting case is provided by geoengineering, defined as "deliberately manipulating physical, chemical, or biological aspects of the Earth system" (Bonnheim 2010: 891). This case has a long pedigree in attempts at modifying regional weather events, through rain-making and cloud-seeding, having the first major breakthrough in mid-20th century, if imagined much earlier (ibid.: 892). In the Arctic (and elsewhere) geoengineering was closely tied to the military needs, such as the wish to control the ice and the weather-whiteouts in Northwest Greenland, where an American military base was established in 1953 as part of the cold war scenario (Martin-Nielsen 2012). Controlling ice and weather proved very difficult, and there were many sceptics and few results. Now, in the 21st century, the idea is back on the agenda, as one possible form of action in response to global warming, especially in the form of "stratospheric aerosol injection" (Bonnheim 2010: 894). Among the proposals were also suggestions "to reduce the amount of incoming sunlight and include ideas such as launching 55,000 mirrors into space, creating an orbiting 'space dust' cloud, and placing billions of small, hydrogen-filled balloons into the stratosphere" (ibid.).

Although such proposals may sound weird, and while most people probably would rather see a decrease in CO_2 emissions, they are still expressions of a surplus of ideas responding to the current challenges. Highlighting such surplus, also of less high-flying ideas, is one domain where anthropologists and other human or social scientists may contribute vitally to climate change research. For all the merits of "local" ethnography, anthropological studies of climate change of this kind have been sidelined as more or less irrelevant outside the discipline itself. This situation is a side effect of the sustained focus on the local particularities. To talk across disciplinary boundaries, anthropologists need to cultivate a more comprehensive interest in the interpenetration of different registers of knowledge.

As has recently been suggested, anthropology suffers from a specific kind of nostalgia: "nostalgia for disappearing worlds constitutes a foundational trope, lying at the heart of the major anthropological traditions" (Berliner 2014: 373). The common denominator for these observations is a sense of dealing with a vanishing world and being too late both to study it in its proper form and to prevent it from disappearing. Although these positions often show the heartfelt engagement with the worlds studied by the anthropologists, they are also basically adverse to an ethnographic realism.

With climate change equally shared by all, nostalgia for an unspoiled life has no place in the world. "Climate change effortlessly transcends boundaries and categories and is making cosmopolitans of us all" (Hulme 2010: 268). To discuss climate change in anthropology is to see even the remotest and smallest of social groups as global citizens, equally (if differently) affected by anthropogenic forces deriving from nowhere in particular—because it is all over the place at this point in time. This fact is a major inducement for anthropology to aim at delocalized theories of the human capacity for taking action in the face of new challenges. These theories may be based in particular ethnographic studies but must also transcend them.

Concluding: Back to the Anthropocene

Anthropology is a relatively young discipline. It is actually of the same age as the earliest suspicion about anthropogenic climate impact first hypothetically voiced by Arrhenius (1896). Looking back, we clearly see that the new "ethnographic subjects" peopling the earliest anthropological publications based on fieldwork emerged as entwined with particular landscapes, climates, and colonial relations—and with the dispositions of the ethnographers (Hastrup 2013f). This is still the case. Given that both climate and (post-)colonial relations seem to run their own course, we have little choice but to question the images of the ethnographic subjects that populate our books and articles—and thus to see ourselves in the mirror. We must cut ourselves loose from the nostalgia, on the one hand, and from the gut reaction of portraying people "out there" as inherently vulnerable, on the other. Instead we should acknowledge people's powers of theorizing, anticipating, and acting and to meet a future, where all humans are seen as potential contributors to sustainable solutions.

> There is more at stake in the Anthropocene than a simple addition of natural sciences and those concerned with *anthropos*. It is also not sufficient to identify planetary boundaries, tipping points, and limits of growth from a scientific perspective in order to successfully implement sustainable development or

> effective climate politics. We have to take into account the double challenge of global change, which affects our environment as well as our intellectual dispositions. The Anthropocene challenges the familiar distinction between nature and culture, which structures the order of knowledge and disciplines for such a long time. (Krauss 2015: 74)

I want to second the call for challenging our own intellectual dispositions; this is, I believe, the only way we can make sure that *people* are included in theories about Anthropocene realities—not only as victims but also most forcefully as responsible social agents. We should aim at contributing high-resolution theoretical knowledge to the global climate scenarios—not simply impart ethnographic knowledge—in the conviction that our theories about the human capacity for anticipation and action are of the same general order as are theories about ocean currents. In this way we may work toward an inclusion of people in future climate scenarios—so far portrayed primarily in terms of natural processes (Carey 2012: 243). In the Anthropocene it is no longer tenable to operate on the distinction between nature and society.

In responding to climate change, we recognize an increasing need for anthropology; at the core of its subject matter is human life itself, and the ways in which humans respond to anticipated futures. We may of course practice anthropology in numerous ways, and certainly not always with climate change as the main figure. But we need to acknowledge that the moment we engage with "climate change" we are already, implicitly at least, seeking to straddle the gap between the particular and the universal. This effort needs a new kind of awareness of the processes by which the world is theorized from below.

References

Arrhenius, S. 1896. On the influence of carbonic acid in the air upon the temperature on the ground. *The Philosophical Magazine* 41: 237–76.

Aporta, C. 2010. The sea, the land, the coast, and the winds: Understanding Inuit sea ice use in context, in I. Krupnik and C. Aporta, S. Gearhead, G. J. Laidler, and L. K. Holm (Eds.), *2010 SIKU: Knowing Our Ice*. New York: Springer Press, pp. 163–80.

Bateson, G. 1972. *Steps to and Ecology of Mind*. New York: Ballantine Books.

Berliner, D. 2014. On exonostalgia. *Anthropological Theory* 14(4): 373–86. DOI: 10.1177/1463499614554150.

Bonnheim, N. B. 2010. History of climate engineering. *Wires Climate Change* 1: 891–97. DOI: 10.1002/wcc.82.

Born, E. W., Heilmann, A., Kielsen Holm, L., and Lairdre, K. L. 2011. *Polar Bears in Northwest Greenland: An Interview Survey about the Catch and Climate*. Copenhagen: Museum Tusculanun Press. (MoG)

Bravo, M. 2009. Voices from the sea ice: The reception of climate impact narratives. *Journal of Historical Geography* 35: 256–78.

Cameron, E. S. 2012. Securing Indigenous politics: A critique of the vulnerability and adaptation approach to the human dimensions of climate change in the Canadian Arctic. *Global Environmental Change* 22: 103–14.

Carey, M. 2012. Climate and history: A critical review of historical climatology and climate change historiography. *Wires Climate Change* 3: 233–49. DOI: 10.1002/wcc.171.
Crapanzano, V. 2004. *Imaginative Horizons: An Essay in Literary-Philosophical Anthropology.* Chicago: University of Chicago Press.
Crate, S. A., and Nuttall, M. (Eds.). 2009. *Anthropology and Climate Change: From Encounters to Actions.* Walnut Creek, CA: Left Coast Press, Inc.
Crutzen, P. J. 2002. Geology of mankind: The Anthropocene. *Nature* 415: 25.
Das, V. 2004. The signature of the State: The paradox of illegibility, in V. Das and D. Poole (Eds.), *2004 Anthropology in the Margins of the State.* Santa Fe, NM: School of American Research.
Diemberger, H. 2013. Deciding the future in the land of snow: Tibet as an arena for conflicting forms of knowledge and policy, in K. Hastrup and M. Skrydstrup (Eds.), *The Social Life of Climate Models: Anticipating Nature.* London: Routledge, pp. 100–27.
Ditlevsen, P. D. 2013. Predictability in question: On climate modelleing in physics, in K. Hastrup and M. Skrydstrup (Eds.), *The Social Life of Climate Change Models: Anticipating Nature.* London: Routledge, pp. 183–02.
Edwards, P. N. 2011. History of climate modeling. *Wires Climate Change* 2: 128–39. DOI: 10.1002/wcc.95.
Fincher, R., Barnett, J., Graham, S., and Hurlimann, A. 2014. Time stories: Making sense of futures in anticipation of sea-level rise. *Geoforum* 56: 201–10.
Geertz, C. 1995. *After the Fact: Two Countries, Four Decades, One Anthropologist.* Cambridge, MA: Harvard University Press.
Giddens, A. 2009. *The Politics of Climate Change.* Cambridge: Polity Press.
Greschke, H., and Tischler, J. (Eds.). 2105. *Grounding Global Climate Change: Contributions from the Social and Cultural Sciences.* Dordrecht: Springer.
Grill, C. 2015. Animal belongings: Human-nonhuman interactions and climate change in the Canadian subarctic, in H. Greschke and J. Tischler (Eds.), *Grounding Global Climate Change: Contributions from the Social and Cultural Sciences.* Dordrecht: Springer, pp. 101–18.
Hamblyn, R. 2009. The whistleblower and the canary: Rhetorical constructions of climate change. *Journal of Historical Geography* 35: 223–36.
Hastrup, K. 2007. Performing the world: Agency, anticipation and creativity, in E. Hallam and T. Ingold (Eds.), *Creativity and Cultural Improvisation*, Oxford: Berg, pp. 193–206.
———. (Ed.). 2009a. *The Question of Resilience: Social Responses to Climate Change.* Copenhagen: The Royal Danish Academy of Sciences and Letters.
———. 2009b. *Waterworlds*: Framing the question of resilience, in K. Hastrup (Ed.), *The Question of Resilience: Social Responses to Climate Change.* Copenhagen: The Royal Danish Academy of Sciences and Letters, pp. 11–30.
———. 2009c. Arctic hunters: Climate variability and social flexibility, in K. Hastrup (Ed.), *The Question of Resilience: Social Responses to Climate Change.* Copenhagen: The Royal Danish Academy of Sciences and Letters, pp. 245–70.
———. 2013a. Anthropological contributions to the study of climate: Past, present, future. *Wire's Climate Change 2013.* DOI: 10.1002/wcc.219.
———. 2013b. Anticipating nature: The productive uncertainty of climate models, in K. Hastrup and M. Skrydstrup (Eds.), *The Social Life of Climate Change Models: Anticipating Nature.* London: Routledge, pp. 1–29.
———. 2013c. Anticipation on thin ice: Diagrammatic reasoning among Arctic hunters, in K. Hastrup and M. Skrydstrup (Eds.), *The Social Life of Climate Change Models: Anticipating Nature.* London: Routledge, pp. 77–99.
———. 2013d. Scales of attention in fieldwork. *Ethnography* 14(2): 145–64.
———. 2013e. The ice as argument: Topographical mementos in the high Arctic. *Cambridge Anthropology* 31(1): 52–68. DOI: 10.3167/ca.2013.310105.

Hastrup, K. 2013f. Andaman islanders and Polar Eskimos: Emergent ethnographic subjects c. 1900. *Journal of the British Academy* 1: 3–30. DOI: 10.5871/jba/001.003.

———. 2014a. Nature: Anthropology on the edge. In K. Hastrup (Ed.), *Anthropology and Nature*. London: Routledge, pp. 1–26.

———. 2014b. Of maps and men: Making places and people in the Arctic, in K. Hastrup (Ed.), *Anthropology and Nature*. London: Routledge, pp. 211–32.

———. 2015. Comparing climate worlds: Theorizing across ethnographic fields, in H. Greschke and J. Tischler (Eds.), *Grounding Global Climate Change: Contributions from the Social and Cultural Sciences*. Dordrecht: Springer, pp. 139–54). DOI: 10.1007/978-94-017-9322.3_8.

Hastrup, K., and Fog Olwig, K. (Eds.). 2012. *Climate Change and Human Mobility: Global Challenges to the Social Sciences*. Cambridge: Cambridge University Press.

Henshaw, A. 2009. Sea ice: The sociocultural dimensions of a melting environment in the Arctic. In S. A. Crate and M. Nuttall (Eds.), *Anthropology and Climate Change: From Encounters to Actions*. Walnut Creek, CA: Left Coast Press, Inc., pp. 153–65.

Heymann, M. 2010. The evolution of climate ideas and knowledge. *Wires Climate Change* 1: 581–97.

Hulme, M. 2009. Why we disagree about climate change: Understanding controversy, inaction, and opportunity. Cambridge: Cambridge University Press.

———. 2010. Cosmopolitan climates: Hybridity, foresight, and meaning. *Theory, Culture, & Society* 27(2-3): 267–76. DOI: 10.1177/0263276409358730.

Huntington, H., and Fox, S. 2005. The changing Arctic: Indigenous perspectives. In *Arctic Climate Impact Assessment*. Cambridge: Cambridge University Press, pp. 51–98.

Ingold, T. 2010. Footprints through the Weatherworld: Walking, breathing, knowing, in *The Objects of Evidence: Anthropological Approaches to the Production of Knowledge*. Special issue of the *Journal of the Royal Anthropological Institute*, 2008: 121–39.

Koch, L. 1919. Geologiske iagttagelser. In K. Rasmussen, *Grønland langs Polhavet*. Copenhagen: Gyldendal, pp. 553–76.

Krauss, W. 2015. Anthropology in the Anthropocene: Sustainable development, climate change, and interdisciplinary research, in H. Greschke and J. Tischler (Eds.), *Grounding Global Climate Change*. Dordrecht: Springer Press, pp. 59–76.

Krupnik, I., Aporta, C., Gearhead, S., Laidler, G. J., and Holm, L. K. (Eds.). 2010. *SIKU: Knowing Our Ice*. New York: Springer Press.

Lahsen, M. 2005. Seductive simulations? Uncertainty distribution around climate models. *Social Studies of Science* 35: 895–922.

Lovecraft, A. L. 2008. Climate change and Arctic cases: A normative exploration of social-ecological system analysis. In S. Vanderheiden (Ed.), *Political Theory and Global Climate Change*. Cambridge, MA: The MIT Press, pp. 91: 120.

Luedtke, B., and Howkins, A. 2012. Polarized climates: The distinctive histories of climate change and politics in the Arctic and Antarctica since the beginning of the Cold War. *Wires Climate Change* 3: 145–59. DOI: 10.1002/wwc.161.

Marin, A. 2010. Riders under Storms: Contributions of nomadic herders' observations to analysing climate change in Mongolia. *Global Environmental Change* 20: 162–76.

Martello, M. L. 2004. Global change science and the Arctic citizen. *Science and Public Policy* 31: 107–15.

Martin-Nielsen, J. 2012. The other cold war: The United States and Greenland's ice sheet environment, 1948–1966. *Journal of Historical Geography* 38: 69–80.

Mason-Delmotte, V., et al. 2012. Greenland climate change: From the past to the future. *Wires Climate Change* 3: 427–49. DOI: 10.1002/wcc.186.

Nuttall, M. 2009. Living in a world of movement: Human resilience to environmental stability in Greenland, in S. A. Crate and M. Nuttall (Eds.), *Anthropology and Climate Change: From Encounters to Actions*. Walnut Creek, CA: Left Coast Press, Inc., pp. 292–310.

Nuttall, M. 2010. Anticipation, climate change, and movement in Greenland. *Études/Inuit/Studies* 34(1): 21–37.

Nuttall, M., Berkes, F., Forbes, B., Kofinas, G., Vlassova, T., and Wenzel, G. 2005. Hunting, herding, fishing, and gathering: Indigenous peoples and renewable resource use in the Arctic. In *ACIA: Arctic Climate Impact Assessment.* Cambridge: Cambridge University Press, pp. 649–90.

Orlove, B. 2005. Human adaptation to climate change: A review of three historical cases and some general perspectives. *Environmental Science and Policy* 8: 589–600.

Pahl, S., Sheppard, S., Boomsma, C., and Groves, C. 2014. Perceptions of time in relation to climate change. *Wires Climate Change* 5: 375–88. DOI: 10.1002/wcc.272.

Pálsson, G., et al. 2012. Reconceptualizing the "Anthropos" in the Anthropocene: Integrating the social sciences and humanities in global environmental change research. *Environmental Science & Policy* 28: 3–13.

Putnam, H. 1981. Fact and value. In *Reason, Truth, and History*. Cambridge: Cambridge University Press, pp. 127–49.

Rink, H. 1877. On the glaciers and the origin of floating icebergs (Appendix I), in *Danish Greenland, Its People and Products* 1974: London: Hurst.

Strathern, M. 1992. Reproducing anthropology. In S. Wallman (Ed.), *Contemporary Futures: Perspectives from Anthropology*. London. Routledge, pp. 172–89.

Strauss, S., and Orlove, B. S. (Eds.). 2003. *Weather, Climate, Culture*. Oxford: Berg.

Turnbull, D. 2003. *Masons, Tricksters, and Cartographers*. London: Routledge.

Tyrrell, M. 2006. More bears, less bears: Inuit and scientific perceptions of polar bear populations on the west coast of Hudson Bay. *Études/Inuit/Studies* 30: 191–208.

Weart, S. R. 2010. The idea of anthropogenic global climate change in the 20th century. *Wires Climate Change* 1: 67–81. DOI: 10.1002/wwc.006.

Wenzel, G. W. 2009. Canadian Inuit subsistence and ecological instability—if the climate changes, must the Inuit? *Polar Research* 28: 89–99.

Whatmore, S. 2002. *Hybrid Geographies: Natures, Cultures, Spaces*. London: Sage.

Williams, P. D. 2005. Modelling climate change: The role of unresolved processes. *Philosophical Transactions: Mathematical, Physical, and Engineering Sciences* 2363 (1837): 2931–46. London: Royal Society.

WWF. 2013. *Greenland: The Last Ice Area. Scoping Study: Socio-Economic and Socio-Cultural Use of Greenland LIA*, Pelle Tejsner and Mette Frost. Copenhagen: WWF-DK.

Chapter 2

The Concepts of Adaptation, Vulnerability, and Resilience in the Anthropology of Climate Change: Considering the Case of Displacement and Migration

Anthony Oliver-Smith

To date, the local and regional effects of climate change are most significantly affecting and being responded to by vulnerable, exposed communities in various parts of the world. Our understanding of climate change effects for human communities depends on our ability to develop concepts that not only appropriately frame the nature of the risks and results those effects usher in but also that inform the development of strategies and practices that enable us to reduce those risks and results (Lavell 2011). Three concepts of varying longevity in anthropology have recently been deployed to help us to understand and respond to the effects of climate change: adaptation, vulnerability, and resilience. All three concepts have, in one guise or another, implicitly or explicitly, been the subject of anthropological enquiry throughout the history of the discipline.

Adaptation, originally developed in biology as an overarching concept that plays a central role in natural selection and evolution, has long been employed in anthropology to understand how cultures use their natural resources for social reproduction and long-term survival. The more recent, subsidiary concepts of vulnerability and resilience, emerging from disaster research and ecology, respectively, are framed as conditions that in some sense serve as measures or indices of the success or lack thereof of adaptive strategies. All three concepts involve a wide array of sociocultural variables in articulation with the uncertainties in evolving climate change scenarios. All three also, in various contexts and functions, address the question of risk defined as the combination of hazard, exposure, and vulnerability. Additionally, all three concepts often overlap, in some instances actually conflating with one another and reducing the possibility of conceptual, and indeed, practical clarity.

The social effects of climate change are always manifested in the context of local social and cultural organization in articulation with the

---

complexities of climate and weather phenomena. In other words, human engagement with the natural environment is neither a biological nor a technical response of an undifferentiated population to physical or material conditions. It is always internally complex, involving the diverse interests, forms, and varieties of knowledge and meanings of a differentiated population, interacting within both the material processes of a complex and dynamic physical setting and a set of social, political, economic, and ideological institutions and practices. More recently, human engagement with the natural environment also must include the interacting roles that international, national, regional, and other nonlocal institutions play in providing constraints and options for decision making. Last, this complex constellation of interacting variables on multiple spatial and temporal scales for a wide variety of adaptive agents (individuals, families, communities, and so on), each with differing values, priorities, and goals, are furthermore all transected by ethical considerations (Johnson 2013).

As the sociocultural context of climate change is thus complex, the physical phenomenon itself offers its own complexity. *Climate change* is defined by the United Nations Framework Convention on Climate Change (UNFCCC 1992) as "a change of climate which is attributed directly or indirectly to human activity that alters the composition of the global atmosphere and which is in addition to natural climate variability observed over comparable time periods." Two fundamental parameters have been used to characterize climate change: namely, averages or norms and extremes. Climate, in effect, is defined by those averages and norms, and human adaptive action is normally directed to them. Extreme events are fundamentally weather events, although extremes become factored into climate variability (Lavell 2011). Both climate and weather and their effects are influenced locally by such factors as topography, ground cover, and human action. According to the Intergovernmental Panel on Climate Change (IPCC), climate variability refers to variations in the mean state and other statistical measures of the climate on multiple temporal and spatial scales beyond that of individual weather events.

Climate change is an intensifying, cumulative, and compounding set of social-ecological feedbacks that will unfold over decades and centuries, establishing new averages and norms and in all likelihood new extremes (Wrathall et al. 2015). The adaptation (and mitigation), vulnerability, and resilience in human individuals, communities, and societies must be understood within these parameters of variability, norms and averages, and extremes. Under conditions of vulnerability, even relatively minor events in terms of the physical energy released can create extreme disasters. In fact, since many of the harshest effects of climate change will manifest as natural hazards (storms, floods, droughts, coastal and river bank erosion, and so on), insights and practices from the fields of disaster research and disaster risk reduction are highly relevant. Indeed, even a small increase in averages can push people living at the margins from chronic into acute risk. Here efforts to assist local adaptation are more appropriately focused on extreme vulnerability than extreme events. The responses to changes in norms and averages will largely guide new forms of production, energy uses, and resource (water) storage capacities and technologies. However, to the extent possible, extremes must be integrated into the overall adaptation framework, although it is generally impossible to mitigate irregular or unpredictable events completely (Lavell 2011). In effect, there will be new norms and averages and new extremes, both of which have important implications for understanding adaptation, vulnerability, and resilience for research and also for policy and practice.

Omnipresent in current research and policy dialogues, these three concepts have played a central if somewhat debated role in framing and developing responses to climate change. However, they also present complex problems for both analysis and application and have also proven to be imprecise and flawed in their usage. First of all, their definitions vary significantly, particularly regarding issues of temporal and spatial scales, cross-scale interactions and the fundamental units of analysis and application. These disparate definitions are due to different conceptual starting points and assumptions pursuing different inquiries into adaptation, vulnerability, and resilience, some taking a more focused view on human security issues and others a more holistic perspective that engages climate change in the context of other changes happening simultaneously (O'Brien et al. 2007). Therefore, in framing the challenges of climate change, each involves perspectives that have the potential to either reveal or obscure important questions. Despite this lack of coherence and consistency, all three concepts are used and are firmly situated in both scientific and policy discourses and have been deployed as policy instruments, serving a variety of purposes and agendas across a wide range of social and political interests. Contrary to this apparent dissonance, anthropological perspectives on adaptation, vulnerability, and resilience, grounded as they are in anthropology's long tradition of holistic research on local lifeways, potentially offer a consistent framework for climate change literature and policy.

Adaptation and Anthropology

In biological ecology, *adaptation* is defined as the process of developing or enhancing structural, physiological, and/or behavioral characteristics that improve chances for survival and reproduction in a given environment. Adaptation, or "survival of the fittest," is a process of natural selection in which those best equipped have the best probability of survival and also of passing on those characteristics to succeeding generations. From an anthropological perspective, adaptation is the fundamental conceptual nexus of a socioecological system. In the process of adaptation human and natural systems conjointly construct socioecological systems.

Adaptation has been one of the central concepts in anthropology since the 19th century emergence of the field and its focus on human biological and cultural evolution. To survive, including ensuring the maintenance, demographic replacement, and social reproduction of a population, as well as culturally meaningful lives, human beings interact with nature, both shaping and being shaped, through a set of material practices that are socially constituted and culturally coherent (Patterson 1994: 223). Socially constructed meanings create frameworks through which alternative material and social practices are analyzed, evaluated, and prioritized (Crane 2010). These material practices include food production, shelter, and, at the most fundamental level, security. All are accomplished through social arrangements, and all modify the natural and social world in ways that enable the persistence of the society through time. The goals of survival, maintenance, demographic replacement, social reproduction, and meaning are extremely variable culturally, which complicates interpretation and application in specific circumstances.

Certainly, among the factors that influence these decisions are the resources and hazards of an environment, generally as experienced under normal rather than extreme conditions. The premises on which humans make basic productive decisions are multiple. They emerge from direct environmental stimuli, social organizational forms, and ideological mandates. There is broad acceptance within anthropology that populations around the world have an intimate knowledge of the environments in which they live and possess a number of cultural constructs—technologies, forms of work organization, and the like—that allow them to make use of the available resources for social reproduction and sustainability and guard against the general level of the hazards in the environment through mitigation. In this context, mitigation, derived from disaster risk management, constitutes a form of proactive adaptation, designed to minimize the effects of normal hazards. In effect, mitigation increases the resilience of a society to the hazards in its surroundings.

When used by anthropologists, adaptation generally refers to changes in behavior and/or belief in response to altered circumstances to improve the conditions of existence, including a culturally meaningful life. Adaptation in human beings has a wider number of attendant features for adaptive capacity including complex human cognition, social organization, values, and meanings. Thus, human beings do not just adapt unconsciously as reactive organisms. Through cultural means humans perceive changes and consider their implications and possible responses through a grid of individually and collectively interpreted cultural knowledge and meaning, and they make decisions and elaborate responses, including the deployment of technology, that may reflect a variety of value positions enabling them to reconfigure themselves without losing functionality (Folke, Colding, and Berkes 2002). Further, the complexity of human societies ensures outcomes that are equally complex, often with unequal and/or unanticipated results.

In addition to the natural environment, human beings also must adapt to a social and cultural environment and its material manifestations (the "built" environment). In that sense, people also adapt to a set of socially constructed institutional circumstances; that is, they do not just adapt to natural features, land or water, for example, but also to human institutions such as labor, economics, markets, schools, governments, and churches and the resources and constraints they present. Indeed, local culture derived from lived experience with their surroundings is fundamental to understanding how environmental change is perceived, responded to, and adapted to (Enfield and Morris 2012). In that sense, our institutions are at once part of our adaptation but also part of what we must adapt to, in what Birkmann (2011) terms "second order adaptations." Moreover, human adaptation is not simply a function of technical adjustment; it also involves the need to frame responses that accord with social and cultural parameters (Crate 2008).

Therefore, human adaptation is constituted in forms of belief, behavior, or technology that have become part of culture, enabling its members to survive and reproduce in its total environment. An adaptation is part of a lifeway, in effect, an "evolved practice or perspective" constituting the way things are done that is both culturally sanctioned and socially enacted. Adaptation is thus different from coping, a concept and response often conflated with adaptation in policy circles but that refers to decision making in novel situations for which there is no ready culturally integrated institutionalized response. Coping behavior involves immediate problem solving and decision making, improvisation, and creativity. Adaptations, on the other hand, are part of a culture's general knowledge and practice, in effect, part of the overall "toolkit" for survival in a specific environment. However, coping strategies that effectively address challenges without longer term negative outcomes may be adopted as established practices and come to constitute culturally sanctioned and socially enacted adaptations or adaptive strategies in response to climate change effects.

In sum, adaptation implies social and cultural change that is usually complex and may involve change in belief or behavior with many, if not all, other dimensions of life. For example, because of the relatively integrated nature of culture, a change in livelihood strategies may reverberate in environmental relations, gender roles, religious belief, and other aspects of local life. Sometimes the occurrence of a severe disaster will stimulate rapid alterations in the way of doing things, but the tendency for the resilience of predisaster systems to reassert themselves in the aftermath is evidence that such rapid change is far from inevitable.

Adaptation in Climate Change Policy

Within climate change frameworks, *adaptation* is now a central and commonly used term. In effect, adaptation arose as a policy option when policies directed toward mitigation had not been successful. In short, the term *adaptation* has shifted from referring to a basic and omnipresent process of change in life to being a policy driven set of formal strategies and projects (Nelson et al. 2007).

In the IPCC's Third Assessment Report, adaptation is defined as an "adjustment in natural or human systems to a new or changing environment. Adaptation to climate change refers to adjustment in natural or human systems in response to actual or expected climatic stimuli or their effects, which moderates harm or exploits beneficial opportunities. Various types of adaptation can be distinguished, including anticipatory (otherwise known as mitigation) and reactive adaptation, private and public adaptation, and autonomous and planned adaptation" (2001). However, in the climate policy world, the concept of mitigation is constrained to refer primarily to the reduction of greenhouse gases, whereas adaptation refers to the process that enables a system at a variety of scales to manage or best adjust to changing conditions (Smit and Wandel 2006).

Concomitantly, in 2001 the UNFCCC developed programs and language to promote national-level adaptation planning, especially among the most vulnerable, the world's poorest countries. Currently adaptation projects cover a wide spectrum of issues defined by the multiple effects of climate change such as drought and desertification, water management, disease vector expansions (malaria, dengue, cholera), agricultural diversification, river basin management, seasonal forecasting, flooding, sea-level rise, and conservation, to mention a few. Supported by national governments and international organizations, such projects appear to miss the mark and more closely resemble externally generated sustainable development projects, perhaps a good thing to promote but not necessarily an adaptation. These projects may become subsumed into local culture as adaptations when they themselves are adapted to and integrated into cultural knowledge and practice, in essence becoming part of the "toolkit." However, a failure to frame interventions holistically, with a narrow focus on the biophysical or economic effects, precludes this and also accounts for many of the failures of development projects in general.

Many researchers, particularly geographers and anthropologists, question the way the term *adaptation* is used in climate change discourses. Most are concerned with how it is assumed to be a key element in climate change policy but fails both to capture the full impacts of climate change and to accurately represent the perceptions of the affected people and/or the range of alternatives open to them (Orlove 2009; Lavell 2011; Oliver-Smith 2013). Criticisms focus on the failure to address problems holistically, ignoring broader social and cultural ramifications of interventions strategically designed to address specific biophysical stressors which, in turn, are further complicated by issues of scale both in analysis and application. Natural and social processes unfold in different time frames and at different spatial scales. Our responses to current problems are usually short term and often come with unintended negative consequences. These immediate or short-term responses are coping strategies that can make our societies more vulnerable in the long run. Long-term responses, despite their resonance with local and regional contexts, are usually difficult to appreciate when societies (and politicians) want to see results quickly. Spatially, adaptations that may be highly effective in one context may prove to have extremely negative effects either up or down spatial and organizational scales, underscoring the importance of local institutions in the way adaptation programs and practices must be developed and applied (Agrawal 2008).

With literally hundreds of professional associations and nongovernmental organizations focused on climate change adaptation today, we must ask whether adaptation projects are being framed and implemented as truly adaptive or whether they are basically coping strategies that fail to address the fundamental problems. Are there temporal frames in which an adaptation can be judged successful or failed? Although clearly debated in some of the research literature (Pelling 2011), climate change policy generally does not fully engage the issue of systemically imposed vulnerability—that is, the socially constructed outcomes of the way resources, wealth, and security are distributed in a society. In effect, part of what people may be adapting to in climate change is precisely the systemic vulnerability imposed by society. Is adaptation, as it is being currently framed, about adjusting so the status quo can persist? Basically, the question becomes: What is being adapted to? Climate change or a system of structural disadvantage, perhaps made worse by climate change? In effect, the problems of the poor and vulnerable do not begin with climate change. They may be made substantially worse by climate change, but limiting interventions to dealing with climatic effects fails to systemically address imposed social vulnerability.

The Construction of Vulnerability

The concept of vulnerability emerged in the 1970s in the work of geographers and anthropologists researching hazards and disasters in the developing world. These researchers found little in mainstream disaster literature that helped them to understand why disasters were so much worse in the global south. Positing that it was necessary to take "the naturalness out of natural disasters" (O'Keefe, Westgate, and Wisner 1976), researchers began to focus on the multiple and variously scaled social causes in the normal order of things that imposed risk selectively and increased disaster significantly (Hewitt 1983). As such, vulnerability is a concept that encompasses many complex and interconnected social, economic, demographic, environmental, and political processes and parameters, making it problematic to define succinctly, in a large part because many of the social forces that drive it are qualitative and therefore harder to conceptualize and quantify (Thywissen 2006).

In disaster research and management, vulnerability came to describe the degree to which a community is either susceptible or resilient to the impact of the natural hazards of its particular socioecological system. It is the outcome of various factors, including awareness of hazards, settlement and infrastructural patterns, public policy and administration, the level of societal development, socioeconomic structure, and institutional capacities in disaster and risk management (Wisner et al. 2004). Vulnerability and risk refer to the relationships between people and the total environment, including the physical setting and the sociopolitical structures that frame the conditions in which people live.

There are many approaches and models for vulnerability, some focusing on broadening and deepening spatial and temporal scales to identify causal linkages between root causes, dynamic pressures, and unsafe conditions, which, when combined with a natural hazard, produce a disaster (Wisner et al. 2004; IRDR 2011). Others, for example, Turner and colleagues (2003), reduce the temporal dimension and focus on detailing exposure, susceptibility, coping impact response, capacity, adaptive capacity, and interactions with perturbations and stressors as elements of vulnerability. However, at the most basic level there is general agreement that vulnerability is a necessary precondition for a disaster to take place and refers to the social characteristics and conditions of a group that place people at risk in terms of their abilities to anticipate, respond to, and recover from a disaster impact (Wisner et al. 2004). Vulnerability links the relationship that people have with their environment with social forces and institutions and the cultural values that sustain or contest them to understand how basic conditions such as poverty or racism produce susceptibilities to specific environmental hazards. Because it is impossible to achieve complete environmental security, every society is characterized by some level of vulnerability. In that sense, vulnerability is an intrinsic, dynamic, multidimensional societal characteristic, independent of the magnitude of any particular hazard.

Much anthropological research on vulnerability has focused on the social construction of risk (see Oliver-Smith and Hoffman 1998; Hoffman and Oliver-Smith 2004; Button 2010). In so far as vulnerability is socially produced, risk therefore is not evenly distributed across the social spectrum, suggesting that everyone in a specific environment will not be equally vulnerable to the effects of climate change. High levels of vulnerability reflect either lack of, or inappropriate, adaptations or an unequal distribution of risk, both of which are socially constructed. Vulnerability thus explicitly ties environmental issues, such as hazards, with the structure and organization of society and the rights associated with membership.

While vulnerability has been adopted as a critical concept in the assessment of a variety of risks in a diversity of fields, it continues to draw criticism on a number of grounds. Among some hazard researchers the enthusiastic embrace of the vulnerability concept was seen to obscure the importance of exposure to specific hazards (Birkmann 2011). Others criticized the conflation of vulnerability with poverty, pointing out that vulnerability to environmental risk was not invariably a characteristic of poverty. For example, there are many communities that, although extremely poor, have high levels of resilience or the social-organizational adaptive capacity and effective strategies for dealing with environmental threats (Laska and Peterson 2013). Other critics have seen vulnerability as a profoundly disempowering concept for local populations (Cannon 2008; Cameron 2012). They argue that defining a community as vulnerable masks important forms of agency that exist in social organizational terms and results instead in characterizing the community as passive and helpless in the face of environmental forces. Moreover, the widespread acceptance and use of the concept of vulnerability has led to its dilution in many policy contexts, moving it from an inherently critical concept to being a term to describe a set of conditions—dehistoricized, lacking its former critical edge, defanged. Critical to our discussion here, this dilution of the concept would seem to be particularly true of the literature on adaptation and climate change.

Vulnerability and Climate Change

Because many of global climate change's effects will manifest as disasters such as drought, flood, storms, landslides, and so on, vulnerability now plays a significant role in framing both scientific analysis and policy options. Interestingly, the climate change literature in general takes a different approach to social vulnerability than does disaster research. The IPCC (2007) definition is concerned with "the degree to which a system is susceptible to and unable to cope with adverse effects of climate change, including climate variability and extremes. Vulnerability is a function of the character, magnitude, and rate of climate change and variation to which a system is exposed, its sensitivity, and its adaptive capacity." This approach is seen as constituting a return to an emphasis on exposure to physical processes rather than the social construction of risk (Kelman and Gaillard 2010).

The complexity of the interrelationships between ecological and social systems at multiple levels makes crafting a policy-relevant research agenda on vulnerability and the social effects of climate change a challenging task. It requires the integration of global projections, local and regional manifestations, and local patterns of vulnerability that are socially and economically constructed by local, regional, and global processes. To discern vulnerable communities, local level projections from both the ecological and the social domains must be addressed in ways that reflect their mutual constitution. This mutual constitution must inform framing a coherent research agenda in order to generate the information and perspectives necessary to inform policy that will provide adequate and durable solutions for vulnerable populations who face such prospects as loss of land, loss of ecosystem services, intensified storms, and other climate change effects.

Although the probability of more extreme events is high, it is likely that most of the effects of climate change will be gradual, incrementally affecting communities that are already dealing with high levels of social vulnerability, thus turning creeping, chronic risk or disaster into rapid-onset disaster (Lavell 2011). Therefore, climate change will tend to exaggerate the frequencies and effects of existing hazards that are the outcomes of preexisting social vulnerabilities. Indeed, by undermining local resilience, climate change effects may also increase the vulnerability of people to geological and other hazards not related to climate change. However, even in cases where the climate change driven hazard is novel, its effects will still be expressed and coped with through local patterns of social vulnerability. In the final analysis, the most effective interventions should adopt a sustainable development orientation to assist people primarily through reducing vulnerability and enhancing resilience in the face of changing environmental conditions. Nonetheless, efforts should not just be focused on adapting at the biophysical level alone but also on the social vulnerabilities that render people susceptible.

Since climate change will exacerbate existing hazards, adaptation should combine with and draw on disaster risk reduction to focus on processes that strain the capacity of communities to cope, perhaps made even more extreme by climate change processes. Indeed, the experience and tools that have been developed to promote disaster risk reduction should prove useful to the practice of climate change adaptation (Birkmann and von Teichen 2009). In effect, both climate change adaptation and disaster risk reduction can be complementary and should be framed and designed from a development perspective to address those social and economic features that render people vulnerable to environmental hazards in general. Climate change adaptation and disaster risk reduction are related undertakings, and both must address systemic vulnerabilities as well as risks exacerbated by specific climate change effects as fundamentally development related phenomena (Kelman and Gaillard 2010). In that sense, the most effective overall policy for both climate change adaptation and disaster risk reduction should focus on reducing development driven risk construction, addressing systemic vulnerabilities and reducing inequality, an approach entirely consistent with a model of development that directly addresses problems of poverty, malnutrition, health, education, and access to resources that can at the same time increase local resilience (Cannon and Muller-Mahn 2010).

Resilience and Adaptive Capacity

At roughly the same time that vulnerability emerged in the 1970s, ecological science, drawing on physics and engineering, employed the term *resilience* to refer to the capacity of an ecosystem to respond to disturbance by resisting damage and recovering quickly (Holling 1973). Within the ecosystem frame, disturbances of sufficient magnitude or duration can profoundly affect an ecosystem and result in a different state or regime, characterized by new or profoundly altered processes and structures. Originally derived from a framework developed by Gunderson and Holling (2002) and known as the adaptive cycle (exploitation, conservation, release, and reorganization), resilience has been incorporated into the larger discourse on global environmental issues to understand the character and dynamics of change at various stages. Systems ecologists began to explore this fundamental link using the term *socioecological system* (Folke, Colding, and Berkes 2002; Folke et al. 2004). By building systemwide resilience through "analysis, adaptive resource management, and adaptive governance" (Walker et al. 2004), ecologists hope to reverse the downward trajectory of biodiversity loss. Also in the context of long time frames, the resilience perspective proves useful at least at the heuristic level in conceptualizing socioecological processes. While the long-term human-environment dialectic is understandably beyond the scope of some strains of ecology, it is entirely germane in the context of climate change that landscapes are not seen as stable and timeless: impacts on land use, culture, politics, and environments do matter, at time scales, both short and long. Not yet incorporated into resilience thinking, for example, is the archaeological attention to scales of time and space. Historical ecology, with its deep-time perspective and regional context, enables archaeologists to take into account not only rapid variables but also slower, more obscure features of what would appear to be stable systems (Crumley 2013).

Resilience is now a central theme in both research and policy on climate change. However, it is a resilience that has largely dispensed with the adaptive cycle framework. The focus is on resilience as a condition that enables a society to absorb the impact of a stressor and retain functionality. Broadly speaking, resilience refers to the ability to prepare and plan for, absorb, recover from, or more successfully adapt to actual or potential adverse events (National Research Council 2012). From this perspective resilience in a community is framed largely as a function of the levels of internal coherence, social solidarity, and the capacity to organize and work on its own behalf in confronting the challenges of climate change. In this context it comes rather close to being synonymous with adaptation and is seen to be more a property of a social system and its members in their relationship to an environment than as a feature of a single socioecological system.

Despite the considerable reservations about the utility of the concept of resilience, many social scientists and policy makers apply the concept to society, defining resilience as the ability of social groups or individuals to bear or absorb sudden or slow changes and variation without collapsing (Holling and Meffe 1996). Increasingly, there is a focus on the examination of community resilience in understanding how communities can reduce risk and losses from climate change. Resilience in communities is embedded in the historical, social, and cultural constructions that govern social interactions and the material development of communities and the attendant institutions pertaining to management and growth. Resilience also involves continual feedbacks to preparedness and disaster mitigation activities, through social learning that may enhance adaptive capacity over longer periods of time (Cutter et al. 2008).

Approaches to adaptation and resilience building vary considerably, ranging from centralized, top-down efforts based on scientific management principles that focus on specific biophysical effects of climate change, to others, framed as adaptive governance, that also draw on science but that emphasize engagement with community building, multilevel politics, and ethical issues such as indigenous rights (Doubleday 2007; Nadasdy 2007; Brunner and Lynch 2010; Cote and Nightingale 2012; Hatt 2012; Welsh 2013). To be politically feasible and equitable, resilience projects must engage diverse stakeholders, and its implementation must adapt when conditions change. Robust—resilient—local and regional management requires understanding the region's past; that understanding requires a comprehensive grasp of the social, historical, cultural, and political aspects of humans in their environments than is apparent in resilience theory (Crumley 2013).

The tension between the ecological and the social in resilience theory and practice manifests itself in a general neglect of culture. There is a need to recognize both synergies and tensions between resilience as an analytic scientific lens and as a normative cultural process. The challenge is to articulate normative culturally bound positions constructively with empirically based analysis of socioecological resilience (Crane 2010). Resilience at the local level clearly is culturally defined and socially enacted. Social constructions of meaning (culture) need to be integrated with materialist analysis of adaptive socioecological processes. While climate change practitioners may promote adaptive strategies that impart resilience to the ecological material components of a socioecological system, the implications of such changes for local people, particularly if enacted without consideration for local views, may be profound if deeply held traditional values and beliefs are devalued or become untenable, triggering a transformation of the culture itself (Crane 2010). Recent research suggests that changes in climate and subsequent disruption to the social, economic, and environmental determinants of health may cause increased incidents and prevalence of mental health issues, such as family stress, drug and alcohol abuse, and potential for suicide ideation (Willox et al. 2013). Such findings are consistent with research from other fields such as disaster studies and development-forced displacement and resettlement (Turton 2006). The implications of top-down adaptation with or without a resilience framework carry a similar disruptive potential with concomitant second order adaptations necessary (Birkmann 2011). In effect, the cultural side of resilience requires that livelihoods fulfill both material and moral needs in the context of major environmental, social, cultural, economic, or political changes to maintain a sense of continuity of meaning and coherence (Marris 1975; Crane 2010).

However, there is a tension regarding both the time frame as well as the goals of resilience thinking; that is, is resilience about stability or change? Nelson and colleagues (2007) assert that adaptation from a resilience standpoint, rather than replicating the present, is seen to promote the capacity to deal with future change. Resilience is thus oriented toward maintaining flexibility and adaptive capacity to handle future stressors. However, by defining resilience as the ability to absorb disturbance, be reorganized, and retain the same basic structure and ways of functioning, resilience thinking encourages us to consider that it is possible to design a future not so different from the present, an assertion that has received considerable scrutiny by those concerned with issues of vulnerability, inequality, and exploitation.

Like vulnerability, resilience is embedded in social structures that govern the interaction of human systems and the natural world. However, resilience tends to be portrayed largely as a function of community social organization and adaptive capacity. In reference to climate change, resilience is not the opposite of vulnerability but a separate concept referring to the capacity of a society to withstand impact and recover with little disruption of normal function. Resilience to climate change, however, raises two important and fundamental questions that pertain as well to adaptation and vulnerability (Nelson, West, and Finan 2009): Resilience of what or whom? Resilience to what? The fact that answers to those questions are often ambiguous points to the need for further scrutiny of the concept. In resilience, the focus of study can be (1) on an individual or household, (2) on socially defined groups such as the elderly, the private sector, or infrastructure (such as levees); or (3) from a spatial perspective on communities, cities, or ecosystems (Cutter et al. 2008). And resilience to what? To the ongoing stressors of a social system that distributes risks and losses unequally? Or to a set of biophysical changes set in motion by climate change effects, a view that implies a return to a pre-1970s model of natural hazards as drivers of risk and disasters?

Climate Change and the Limits of Adaptation

When climate change was first recognized, a policy framework based on mitigation was developed under the Kyoto Protocol (2005) to reduce the amount of greenhouse gas (GHG) emissions. It was not long, however, before it became clear that it was too late to mitigate climate change fully. The amount of GHGs already in the atmosphere had entrained a set of cascading processes that cannot be completely mitigated but only reduced. Subsequently, the Cancun Adaptation Framework (2010) was developed to assist affected communities and regions in adapting to the changes that are and will be taking place.

However, adaptation processes are increasingly recognized as complex undertakings that face limitations in application and effectiveness. Such limits are discussed as a set of absolute thresholds in biological, economic, or technological parameters (Adger et al. 2009). Adaptation limits are defined and measured for ecological systems by rates of extinction. For human systems, allegedly absolute economic or technological limits are not generally recognized as culturally defined and socially enacted; they are in effect a matter of choice, framed by cultural values and priorities (Oliver-Smith 2013). Indeed, some limits on adaptation may take the form of strategies to preserve a status quo that favors the wealthy or the powerful, reflecting a conservative dimension of resilience. In short, the limits of adaptation for human systems involve far more than the existence of biophysical risks (or extinctions), opening up important questions about the goals of adaptation, acceptable and unacceptable risks, and the elasticity or malleability of culture and society. In effect, the limits of adaptation at least within a threshold of total destruction have been seen as endogenous to the adaptive unit (Adger et al. 2009).

If the human capacity to adapt is somehow considered as open-ended and infinitely elastic, the burden of responsibility is shifted away from mitigating greenhouse gases toward more cost-benefit analyses of strategies and resources for adaptation (Dow et al. 2013). Since human adaptation is largely sociocultural, and is therefore somewhat flexible, then it is difficult to define limits in some absolute sense, beyond those that endanger physical existence. However, if limits are so defined, they can then be subject to being overridden, thus opening the community to imposed forms of adaptation interventions and violations of human rights. Thus, if defining an adaptation limit ultimately as a function of local culture, then establishing such limits as inflexible becomes a conceptual, practical, and ethical problem.

Adaptive capacity varies greatly according to its actors, whether individuals, families, communities, nations, corporations, and other purposeful groups, all whom differ considerably in their risk perceptions and interpretations and in their resources, knowledge, access to power, and many other ways. At the community level, each case involves a complex interaction between endogenous factors—such as knowledge, resources, adaptive capacity, and cultural values—and the exogenous, the severity of the environmental stressor and the institutional resources and constraints of the larger society. People's capacity to adapt will be limited not only by the degree of severity of biophysical risk or impact but also by the characteristic features of the vulnerability or resilience of their circumstances. Since the circumstances of vulnerability and resilience are expressed in a multiplicity of ways and through differential access to a variety of resources, many of these "limits" are actually externally imposed or lifted. In addition, responses to environmental stress will involve a range of options such as changes in resource use, social organization, cultural values and, in some circumstances, migration.

Adaptation, Vulnerability, and Resilience in Migration

Migration, whether permanent or temporary, has long been a response or survival strategy for people experiencing environmental change (Hugo 1996). Nonetheless, the role of environment in migration, particularly in the modern era, cannot be reduced to a simple cause and effect relationship.

The linkages between environment and society have grown ever more complex, making it difficult to speak of direct environmental causality in human migration, except in the most extreme cases. The relationship between environmental change and migration is complicated by the complexity of both, and in the difficulty in establishing causality between such complex phenomena. And, as with all things human, culture and society play crucial mediating roles between a population and the environment it inhabits (or leaves) (Oliver-Smith 2009). In most circumstances, the combination of environmental, social, economic, and political forces may increase the risk of uprooting for many vulnerable populations in exposed regions. Black's critique (2001) that emphasizing "environmental" factors diminishes the role played by political and economic factors in migration is well taken and coincides with the viewpoint of most disaster researchers today that highlights the political or economic forces that together with natural agents produce disasters or, for that matter, any forced migration that might ensue. In this context the question of whether migration is a form of adaptation or evidence of a failure to adapt becomes a matter of practical and ethical significance.

The issue of environment and migration is among the most discussed and debated dimensions of the impact of global environmental change on human beings. The research and scholarship focusing on the relationship between environment and migration is shot through with controversy, centering largely on the issues of predicted numbers, appropriate terminology for people uprooted by environment and other causes, and the political implications of both research and policy pertaining to environmentally displaced people. The contingent nature of prediction of environmental impacts, the vast disparities in predictions of numbers of people to be affected, the elusive nature of definitional issues, the difficult question of causation, and the overall complexity of human-environment relations all present serious challenges to researchers attempting to analyze the relationship between environment and migration.

There is considerable debate about what exactly constitutes an environmentally induced move and how to measure and explain it. The actual processes through which major population dislocations might occur are still only partially understood (Adamo 2008: 2). The relationship between environment and migration is far from linear or straightforward, and understanding it presents a number of conceptual challenges. Arriving at a consensus definition of an environmentally displaced person has proven to be very difficult. The disparate causes of environmental migration, including disasters, environmental degradation, contamination, and climate change—as well as the different forms and trajectories the migratory process may take—have proved challenging to bracket within one overarching definition. The working definition of The International Organization for Migration for environmental migrant is "persons or groups of persons who, for compelling reasons of sudden or progressive changes in the environment that adversely affect their lives or living conditions, are obliged to leave their habitual homes, or choose to do so, either temporarily or permanently, and who move either within their country or abroad" (IOM 2013: 16). Nevertheless, there has been considerable resistance to the terms *environmental migrant* and *environmental refugee* on conceptual, legal, and political grounds (Oliver-Smith 2012).

One of the complications of the lack of a consensus definition is the enormous disparity in estimates of people who have been or will be displaced by the effects of environmental change. Estimates are at least in part contingent on how environmental migration is defined and who will fall under a given definition. The range of estimates is considerable, ranging from the tens to the hundreds of millions, but the failure of most of those who offered estimates to specify the methods by which they arrived at their numbers has generated significant scientific debate and presented policy makers with confusing data. There was also no commonly agreed-on methodology (Gemenne 2011). Some estimates are based on field reports from relief agencies. Some are based on population figures in areas that are experiencing environmental change. Few have solid bases for the numbers that are estimated. Among the more reliable figures at present are those produced by the International Displacement Monitoring Center (IDMC) and the Norwegian Refugee Council (NRC) that use a baseline of events from the EM-DAT database to produce a core data set for events where over 50,000 people were affected. Data on the displaced from each event are then sought from organizations involved in relief for those events (Yenotani 2011). The most recent IDMC estimates are reflected in Figure 2.1. Note that for all but one of the years listed the percentage of disasters associated with climate and weather phenomena has never been less than 90%.

Nonetheless, debate has been active regarding how extensive the displacement and subsequent migration will be Suhrke (1994) divided the controversy into two camps: the maximalists, who predict massive numbers into the hundreds of millions of people who will be displaced and forced to migrate, and the minimalists, who are unwilling to privilege the role of the environment in migration, citing the complexity of interaction between social and environmental systems. The definitional challenges and the unreliability of the numbers undermined some of the maximalist claims that were, in turn, reflected in the shift in the IPCC reports toward questions of adaptation, vulnerability and resilience as key in mediating the question of displacement and migration by climate change effects. However, as Morrissey (2012) points out, the maximalist-minimalist distinction is more about the legitimacy of the nomenclature and tends to mask a general agreement that the environment, and particularly climate change, will factor into decisions by affected people regarding migration, both sides in effect recognizing that other considerations may also play a role in those options.

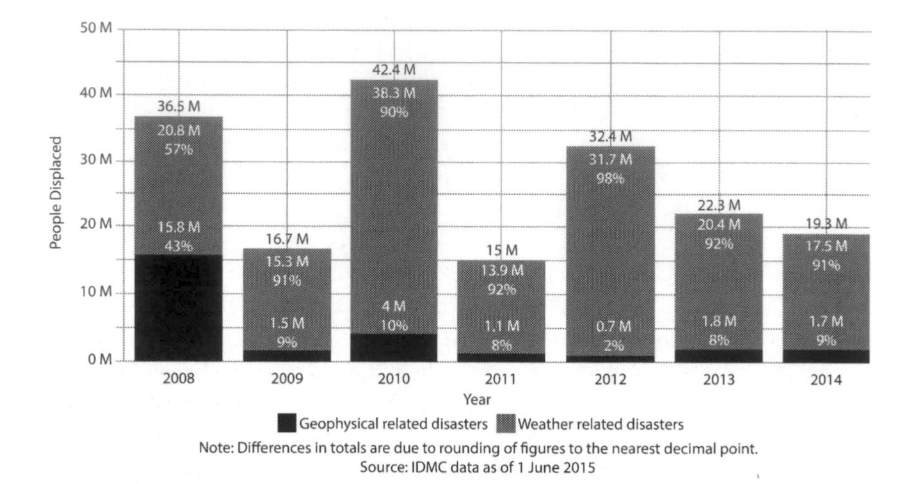

Figure 2.1 Global scale of displacement caused by disasters 2008–2014 (redrawn after *Global Estimates 2015: People Displaced by Disasters*, Norwegian Refugee Council and the Internal Displacement Monitoring Centre, 2015)

It is also clear that the "environmental migration" controversy is both highly charged and deeply embedded in the way complex human-environment relations are understood by scholars, politicians, and the general public. In both public policy and academic debate four distinct framings of environmental migrants are notable: as victims, security threats, adaptive agents, and political subjects (Ransan-Cooper et al. 2015). The UNHCR sees five displacement scenarios emerging in the near future: hydrometeorological disasters, population removal from high risk areas, environmental degradation, the submergence of small island states, and violent conflict (2009: 4). The question of whether migration is a form of adaptation or a coping response to the limits of adaptation hinges on the definition of adaptation. From the perspective advanced in this chapter, if migration is a strategy that is culturally sanctioned and socially enacted, then it can be considered an adaptation. The historical and ethnographic record abundantly shows that migration has been adopted as an adaptation, an evolved practice, to environmental and other forms of stress quite frequently. For example, Brazilian peasants of the arid northeast have engaged in migration as a standard adaptation to cyclical drought for generations (Kenny 2002).

In the context of climate change, one view is that migration is a form of adaptation. Migration allows people facing diminished livelihoods caused by climate change to diversify their income streams. The remittances provided by migrants allow families to remain in home regions despite environmental changes that have reduced livelihood security (Tacoli 2009; Warner 2009; Black et al. 2011). A contrasting view holds that if migration is a last resort for people living in conditions of extreme vulnerability, such that even slight changes in climate averages turns chronic risk into acute risk or who face sudden life-threatening climate extremes, migration is at best a coping mechanism whose outcomes may solve an immediate challenge but may not prove to be adaptive in the medium or long term (Lavell 2011). One telling dimension is that in much of the climate change and migration literature the focus is on the "adaptable" individual rather than on the sociocultural context or any possible collective responses by the state (Felli and Castree 2012).

What can we expect from future increasing effects of climate change for specific regions and communities? Answering that question is difficult for many reasons and most notably because of the numerous variables and the nonlinearity of their interactions. The challenge lies in determining not just absolute exposure and absolute exposed population but specific lands and populations in different socially configured conditions of resilience and vulnerability. In fact, conditions of vulnerability are accentuating rapidly owing to increasing human-induced pressures on ecosystems. Moreover, the vulnerability of a nation to environmental change effects is partially a function of its level of development and per capita income (Nicholls et al. 2007: 331). The lesser-developed countries have a significantly higher level of vulnerability to climate change effects and may generate significant displacement and migration. By the same token, these nations may also see populations, who for lack of resources, are "trapped" and unable to migrate from threatened environments (Black et al. 2011; Foresight 2011).

However, the problem with assessing the exposure of both lands and people to climate change is that we are dealing not only with projected environmental change effects but also with various future projections about various physical, societal, and infrastructural trajectories—including greenhouse gas emissions (in the case of climate change), demographic change, migration trends, infrastructural development, mitigation strategies, adaptive capacities, vulnerabilities, and patterns of economic change—all of which will play out in different ways, according to the political, economic, and sociocultural dispositions of national governments, international organizations, and general populations (Nakicenovic and Swart 2000). Now, however, it is becoming increasingly clear that there are situations in which peoples and places are experiencing devastating cumulative and compounding climate-driven stressors that will make livelihoods and continued habitation impossible. Basically, limits to adaptation have been reached, and no adaptation measures or strategies will either reduce the risks or safeguard the lives and welfare of the people of these regions. In other words, it is now recognized that the degree to which adaptation will help to ameliorate the effects of climate change is limited. Vulnerability assessments of specific regions such as small island states, the Arctic, high-altitude and glacial regions, and many coastal and river delta regions present conditions for which adaptation presents no viable means of continued habitation. Notwithstanding the difficulty in establishing limits or fixed thresholds beyond which there are no possible adaptive interventions that will enable continued habitation, today many people are facing precisely that circumstance in which they will lose habitat, livelihood, and culture. In effect, in specific locales productive livelihoods and continued habitation have reached their limits, engendering situations of forced displacement and "crisis migration" (Martin, Weerasinghe, and Taylor 2014).

It is projected that many of the emerging climate changes, such as increased risks of flooding, storms, deforestation, desertification, soil erosion, salt-water intrusion, and sea-level rise, will increase the number and scale of forced migrations in the near future, with all the consequent stresses and the need to adapt to new or radically changed environments. The displaced may experience privation, loss of homes, jobs, and the breakup of families and communities. They must mobilize social and cultural resources in their efforts to reestablish viable social groups and communities and to restore adequate levels of social and material life. Moreover, the combination of increasing population, population density, increasing poverty, and occupation of hazardous sites has accentuated vulnerability and exposure to natural hazards exacerbated by climate change effects and increases the probability of displacements and forced migrations. While many disasters trigger temporary displacements, the progressive nature of many climate change effects indicates that many of those migrations will become permanent as lands become permanently damaged and uninhabitable. To assist people threatened with major losses and damages from uprooting by climate change effects, the fields of Disaster Risk Reduction (Wisner et al. 2004) and Displacement and Resettlement research (Cernea 1997; Scudder 2009) have developed an array of tools and practices that will have immediate relevance to the tasks ahead.

With the recognition of these eventualities, the international climate change community has been faced with the necessity to address the challenges that cannot be ameliorated through adaptation and a policy framework to prepare for inevitable losses and damages entered negotiations. The primary outcome of negotiations in November 2013 was the Warsaw International Mechanism on Loss and Damage from Climate Change. In a few short years, climate negotiations have given birth to a new policy regime aimed at addressing losses and damages from climate change.

In broad terms, Loss and Damage constitutes the policy domain where existing mechanisms are not enough to prevent private and collective loss and damage, or to ensure human welfare. It assumes that climate change will generate conditions that can neither be mitigated, adapted to, nor insured against (Dow et al. 2013). In effect, loss and damage are the residual that exists between climate change adaptation (CCA), disaster risk reduction (DRR), and public/private risk transfer. Recent research documents losses that cannot be mitigated or adapted to, resulting in displacement (Lazrus 2012; Marino and Schweitzer 2009; Warner et al. 2010). The loss and damage policy asserts that climate change effects will overwhelm human systems and trigger permanent physical displacement (or resource displacement), migration, resettlement, and abandonment of systems that no longer sustain human life. Loss and damage further imply that DRR strategies are not adequate, because the changes involved are successive, progressive, accelerating, and permanent, at least in human timescales. Further, this policy assumes that insurance instruments, even where available, will not be able to cover the collective losses, such as cultures, languages, indigenous knowledge systems, livelihood practices, social networks, and statehood. Such losses will be particularly acute for indigenous and traditional peoples around the world, whose livelihoods and cultural worlds are identified with specific environments. However, the traumatic aspects of forced migration will be deeply felt by many more, including massive urban populations who face the possibility of displacement by storms and sea-level rise in Tokyo, Mumbai, Shanghai, Dhaka, and other megacities of coastal regions around the globe. Loss and Damage imply that for some people climate change will require new life systems, in all likelihood undertaken after a process of displacement and forced migration (Wrathall et al. 2015). The Loss and Damage initiative addresses the attendant losses that people will experience in these displacements by natural hazards made even more potent and complex by climate change and from the increasing vulnerability of growing populations, exposed in ever greater numbers and densities.

However, this initiative poses its own set of problems. It essentially limits analysis to surface conditions, taking us back to a focus on disasters or biophysical impacts in terms of outcomes rather than drivers and, in particular, responsibility for those drivers. The last 40 years have been spent broadening the temporal and spatial scales of analysis of disasters to include deeply embedded characteristics and processes that increase vulnerability, exposure, and impact. In effect, the Loss and Damage policy dispenses with that kind of analysis. It does not reference vulnerability. Moreover, it dispenses with the inevitable conclusions of that kind of analysis in terms of responses, because it contains no imperative for structural change. It's only dealing with the losses and damages being experienced and carries with it no transformational strategy.

Furthermore, this policy in effect substitutes a strategy aimed succinctly at compensation. However, we have learned from the work of Michael Cernea and Hari Mohan Mathur (2008) and Thayer Scudder (2009) in development-forced displacement and resettlement studies that compensation does not work well. It generally fails to restore livelihoods and leads to greater impoverishment. In the case of climate change, compensation, even in the guise of loss and damage, does not address systemic vulnerabilities. It basically focuses on surface conditions, which will no doubt be dire in many cases but with no imperative for structural change.

Moreover, there are as yet no comprehensive legal instruments for the protection and assistance of people affected and/or displaced by natural and technological disasters or the effects of climate change. Although not legally binding, the United Nations Guiding Principles on Internal Displacement pertaining to people forced to flee their homes by armed conflict, violence, violations of human rights, or natural or human made disasters—but who have not crossed an international border—are seen as a useful guide for developing appropriate policies. In addition, the Conference of the Parties to the UN Framework Convention on Climate Change has recently prioritized "measures to enhance understanding, coordination, and cooperation with regard to climate change induced displacement, migration, and planned relocation" (UNFCCC 1992). In 2012 the African Union Convention on the Protection and Assistance of Internally Displaced Persons in Africa, known as the Kampala Convention was ratified as the world's first binding agreement on internally displaced persons, including natural disasters and other causes of displacement. Researchers and activists are currently advocating for the construction of a new global policy for the protection and resettlement of people forced to leave their habitats because of sudden or gradual climate changes such as sea-level rise, extreme weather events, and drought and water scarcity (Peninsula Principles 2013; Brookings-Georgetown University-UNHCR 2015; Nansen Initiative 2015).

Humanitarian, development, and human rights actors are joining with climate change experts to develop internationally recognized general principles based on the definition of rights and responsibilities, protection and human rights, the prevention of risks of impoverishment, and monitoring mechanisms to protect the rights of people resettled because of climate change. The legal policy bases for planned resettlement and the land and property issues must also be specified. Basically, a consultative process should be organized to develop specific protection principles and concrete guidelines that will be useful to all stakeholders, including affected peoples, development and humanitarian actors, and governments who may be obligated to consider resettlement as an adaptation to climate change (Ferris 2012).

Conclusion: The Articulation of Adaptation, Vulnerability, and Resilience in Climate Change

Adaptation, vulnerability, and resilience are central concepts of climate change understandings in both academic research and policy frameworks. However, there are huge inconsistencies in definitions and applications. In anthropology they pertain to social and material change, a longstanding focus of anthropological research, particularly in the context of displacement and migration. All three concepts are articulated with and involved in the formulation and application of each. The concepts of vulnerability and resilience address the degree to which a society is adapted to the hazards and risks of its environment, including the effects of climate change. Both vulnerability and resilience, in referring to those conditions in which people live that render them either susceptible or resistant to environmental hazard, necessarily address the concept of adaptation. However, the relationship between vulnerability and resilience is not linear but rather dialectical (Aguirre 2007); that is, lowering vulnerability may or may not increase resilience, but it also may create other forms of vulnerability. For example, adaptations based on technology may create risks, exposure, and vulnerability through their malfunction or failure.

Adaptation requires change, but the direction and purpose of that change must be defined. In one context adaptation as currently deployed may foster approaches that more deeply embed neoliberal environmental relations and exploitation (Felli and Castree 2012). Interventions that do not challenge current practices may promote or exacerbate vulnerability. Vulnerability analysis therefore becomes a virtual requirement before any adaptation intervention if it is to have any transformational potential (Ribot 2011). Is transformational adaptation possible? As currently practiced, much climate change adaptation today does not address the real adaptive challenge, which requires questioning the beliefs, values, commitments, loyalties, and interests that have created and perpetuated the structures, systems, and behaviors that drive climate change (O'Brien 2012). Indeed, current definitions of climate change adaptation are positioned far more to accommodate change rather than to challenge the causes and drivers, leaving current development approaches essentially unchallenged (Pelling 2011). From their various point of genesis, the concepts of adaptation, vulnerability, and resilience have been brought to bear on climate change along similar, if not exactly parallel, policy-relevant paths. They foregrounded the social roots of environmental risk. They internalized and "routinized" risk and disturbance, emphasized the centrality of interactions across temporal and spatial scales, and underscored the importance of historical analysis in assessing the trends of current conditions. Finally, vulnerability and resilience required a new perspective on the relationship between society and environment. In effect, with the concepts of adaptation, vulnerability, and resilience, both the social and the natural sciences moved from understanding environment and society as a duality to a more mutually constitutive relationship. This mutual constitution becomes abundantly clear in the context of climate-change-associated migration (Oliver-Smith 2009, 2012).

Both vulnerability and resilience required greater scrutiny of the underlying causes of environmental change, whether associated with climate change effects or natural hazards, and are also complexly intertwined with the questions of adaptation and development. Indeed, both concepts have proved to be important advances in our understanding of hazards and climate change. Barring a few notable exceptions, such as the great reduction in mortality through early warning and shelter systems in flood prone Bangladesh (Hofer and Messerli 2006), neither has led to widely adopted policies or practices that have reduced risk, losses, or damages from either disasters or climate change in much of the world. The reasons for our lack of progress toward such goals lie at the core of the cultural and social organizational principles of the dominant states and societies of the world, which, moreover, have now influenced the process of societal development across much of the globe. In many senses, these same cultural and social forces also hinder the adoption and implementation of measures both to mitigate climate change drivers (that is, GHGs) and to reduce their effects as traditional models of development continue apace to inform energy use and natural resource exploitation around the globe. The disruptive effects of climate change oblige us to continue our efforts to mitigate by reducing GHGs, to adapt where possible by addressing both biophysical threats and local vulnerabilities, and to assist populations directly that have been uprooted and forced to migrate by both addressing their losses and damages and by facilitating socially informed planned resettlement. However, the degree to which any or all of these goals are possible is questionable without significant structural change in global and national organization and governance.

References

Adamo, S. B. 2008. Addressing environmentally induced population displacements: A delicate task. Background paper for the Population Environment Research Network Cyberseminar on Environmentally Induced Population Displacements, August: 18–29, www.populationenvironmentresearch.org/seminar082008.jsp, accessed Jan. 10, 2013.

Adger, W. N., Dessai, S., Goulden, M., Hulme, M., Lorenzoni, I., Nelson, D. R., Naess, L. O., Wolf, J., and Wreford, A. 2009. Are there social limits to adaptation to climate change? *Climatic Change* 93: 354.

Adger, N., and Vincent, K. 2005. Uncertainty in adaptive capacity. *CR Geoscience* 337: 339–410.

Agrawal, A. 2008. *The Role of Local Institutions in Adaptation to Climate Change.* Paper prepared for the Social Dimensions of Climate Change, Social Development Department, The World Bank, Washington, D.C. (March 5–6).

Aguirre, B. E. 2007. Dialectics of vulnerability and resilience. *Georgetown Journal of Poverty Law and Policy* 1(1), https://articleworks.cadmus.com/geolaw/z5800107.html

Almeria Statement. 1994. *2006 II International Symposium Desertification and Migrations,* http://www.sidym2006.com/eng/eng_doc_interes.asp

Bennett, J. 1996. *Human Ecology as Human Behavior.* New Brunswick: Transaction Publishers.

Birkmann, J. 2011. First and second order adaptation to natural hazards and extreme events in the context of climate change. *Natural Hazards* 58(2): 811–40.

Birkmann, J., and von Teichman, K. 2009. Addressing the challenge: Recommendations and quality criteria for linking disaster risk reduction and adaptation to climate change, in J. Birkmann, G. Tetzlaff, and K.-O. Zentel (Eds.), *DKKV Publication Series* 38: Bonn.

Black, R. 2001. Environmental refugees: Myth or reality? *UNHCR Working Papers* (34): 1–19.

Black, R., Adger, W. N., Arnell, N. W., Dercon, S., Geddes, A., and Thomas, D. S. G. 2011. The effect of environmental change on human migration. *Global Environmental Change* 21s: s3–s11.

Black, R., Bennett, S. R. B., Thomas, S. M., and Beddington, J. R. 2011. Migration as adaptation. *Nature* 478: 447–49.

Brookings-Georgetown University-UNHCR. 2015. *Guidance on Protecting People from Disasters and Environmental Change through Planned Relocation*, http://www.brookings.edu/research/papers/2015/10/07-planned-relocation-guidance

Brunner, R. D., and Lynch, A. H. 2010. *Adaptive Governance and Climate Change*. Boston: American Meteorological Society.

Button, G. 2010. *Disaster Culture: Knowledge and Uncertainty in the Wake of Environmental Catastrophe*. Walnut Creek, CA: Left Coast Press, Inc.

Cameron, E. S. 2012. Securing indigenous politics: A critique of the vulnerability and adaptation approach to the human dimensions of climate change in the Canadian Arctic. *Global Environmental Change* 22: 103–14.

Cannon, T. 2008. *Reducing People's Vulnerability to Natural Hazards: Communities and Resilience*. Research Paper No. 2008/34 UNU-WIDER. Bonn: United Nations University.

Cannon, T., and Müller-Mahn, D. 2010. Vulnerability, resilience, and development discourses in context of climate change. *Natural Hazards*. DOI: 10.1007/s11069-010-9499-4.

Cannon, T., Twigg, J., and Rowell, J. 2003. *Social Vulnerability, Sustainable Livelihoods, and Disasters, Report to DFID Conflict and Humanitarian Assistance Department (CHAD) and Sustainable Livelihoods Support Office*, http://www.benfieldhrc.org/disaster_studies/projects/soc_vuln_sust_live.pdf

Cernea, M. 1997. The risks and reconstruction model for resettling displaced populations. *World Development* 25(10): 1569–88.

Cernea, M., and Mathur, H. M. 2008. *Can Compensation Prevent Impoverishment: Reforming Resettlement through Investments and Benefit-Sharing*. New Delhi: Oxford University Press.

Christian Aid. 2007. *Human Tide: the Real Migration Crisis*. http://www.christianaid.org.uk/Images/human-tide.pdf

Cote, M., and Nightingale, A. J. 2012. Resilience thinking meets social theory: Situating social change in socio-ecological systems (SES) research. *Progress in Human Geography* 36(4): 475–89.

Crate, S. A. 2008. Gone the bull of winter: Grappling with the cultural implications of and anthropology's role(s) in global climate change. *Current Anthropology* 49(4): 569–95.

Crate, S. A., and Nuttall, M. 2009. *Anthropology and Climate Change: From Encounters to Actions*. Walnut Creek, CA: Left Coast Press, Inc.

Crane, T. A. 2010. Of models and meanings: Cultural resilience in socio-ecological systems. *Ecology and Society* 15(4): 19.

Crumley, C. L. 2013. New paths into the Anthropocene, in C. Isendahl and D. Stump, (Eds.), *Oxford Handbook of Historical Ecology and Applied Archaeology*. Oxford: Oxford University Press.

Cutter, S. L., Barnes, L., Berry, M., Burton, C., Evans, E., Tate, E., and Webb, J. 2008. A place-based model for understanding community resilience to natural disasters. *Global Environmental Change* 18(4): 598–606.

Doubleday, N. 2007. Culturing co-management: Finding "keys" to resilience in asymmetries of power, in D. Armitage, F. Berkes, and N. Doubleday (Eds.), *Adaptive Co-Management: Collaboration, Learning, and Multi-Level Governance*. Vancouver: University of British Columbia Press, pp. 228–48.

Dow, K., Berkhout, F., Preston, B. L., Klein, R. J. T., Midgley, G., and Shaw, M. R. 2013. Limits to adaptation. *Nature Climate Change* 3: 305–07.

Dow, K., Kasperson, R. E., and Bohn, M. 2006. Exploring the social justice implications of adaptation and vulnerability, in W. N. Adger, J. Paavola, S. Huq, and M. J. Mace (Eds.), *Fairness in Adaptation to Climate Change*. Cambridge, MA: The MIT Press, pp. 79–96.

El-Hinnawi, E. 1985. *Environmental Refugees*. Nairobi: United Nations Environmental Programme.

Felli, R., and Castree, N. 2012. Commentary. *Environment and Planning* 44: 1–4.

Ferris, E. 2012. *Protection and Planned Relocations in the Context of Climate Change*. UNHCR Legal and Protection Policy Research Series. Geneva, UNHCR, July.

Folke, C., Carpenter, S., Elmqvist, T., Gunderson, L., Holling, C. S., and Walker, B. 2002. Resilience and sustainable development: Building adaptive capacity in a world of transformations. *Ambio* 31(5): 437–40.

Folke, C., Carpenter, S., Walker, B., Scheffer, M., Elmqvist, T., Gunderson, L., and Holling, C. S. 2004. Regime shifts, resilience, and biodiversity in ecosystem management. *Annual Review of Ecology, Evolution, and Systematics* 35: 557–81.

Folke C., Colding, J., and Berkes, F. 2002. Building resilience for adaptive capacity in social-ecological systems, in F. Berkes, J. Colding, and C. Folke (Eds.), *Navigating Social-Ecological Systems: Building Resilience for Complexity and Change*. Cambridge: Cambridge University Press.

Foresight. 2011. *Migration and Global Environmental Change: Final Project Report*. London: The Government Office for Science.

Friends of the Earth. 2007. *A Citizen's Guide to Climate Refugees*. Australia: Friends of the Earth, http://www.safecom.org.au/pdfs/FOE_climate_citizens-guide.pdf

Gemenne, F. 2011. Why the numbers don't add up: A review of estimates and predictions of people displaced by environmental changes, *Global Environmental Change* 21s: s41–s49.

Global Humanitarian Forum. 2009. *The Anatomy of a Silent Crisi*. Geneva: Global Humanitarian Forum.

Gunderson, L., and Holling, C. S. 2002. *Panarchy: Understanding Transformations in Human and Natural Systems*. Washington, D.C.: Island Press.

Hamza, M. 2007. *Challenges in Measuring Vulnerability in Complex Environments: Environmental Transformations and Tipping Points of Population Displacement and Humanitarian Crises*, Expert Working Group "Measuring Vulnerability" United Nations University Institute for Environment and Human Security, Bonn, November 19–22.

Hatt, K. 2012. Social attractors: A proposal to enhance "resilience thinking" about the social. *Society and Natural Resources* 26(1): 30–43.

Hewitt, K. 1983. *Interpretations of Calamity*. Boston: Allen and Unwin.

Hofer, T., and Messerli, B. 2006. *Floods in Bangladesh*. Tokyo: United Nations University Press.

Hoffman, S., and Oliver-Smith, A. 2004. *Catastrophe and Culture: The Anthropology of Disaster*. Santa Fe, NM: School of American Research Press.

Holling, C. S. 1973. Resilience and stability of ecological systems. *Annual Review of Ecology and Systematics* 4: 1–23.

Holling, C. S., and Meffe, G. K. 1996. Command and control and the pathology of natural resource management. *Conservation Biology* 10(2): 328–37.

Hugo, G. 1996. Environmental concerns and international immigration. *International Migration Review* 30(1): 105–31.
International Displacement Monitoring Centre. 2015. *Global Estimates 2015*. IDMC/publications/2015/global-estimates-2015-people-displaced-by-disasters
Integrated Research on Disaster Risk (IRDR). 2011. *The FORIN Template*. Paris: ICSU.
Intergovernmental Panel on Climate Change (IPCC). 2007. Summary for policymakers, in S. Soomon, C. Qin, M. Manning, Z. Cher, M. Marquis, K. B. Avery, M. Tignor, and H. L. Miller (Eds.), *Climate Change 2007: The Physical Science Basis*. Contribution of Working Group I to the Fourth Assessment Report of the Intergovernmental Panel on Climate Change. Cambridge: Cambridge University Press.
International Organization for Migration (IOM). 2013. *Compendium of IOM Activities in Disaster Risk Reduction and Resilience*. Geneva: IOM.
Johnson, C. A. 2013. Governing climate displacement: The ethics and politics of human resettlement. *Environmental Politics* 21(2): 308–28.
Kelman, I., and Gaillard, J. C. 2010. Embedding climate change adaptation within disaster risk reduction, in R. Shaw, J. M. Pulhin, and J. J. Pereira (Eds.), *Climate Change Adaptation and Disaster Risk Reduction: Issues and Challenges*. Community, Environment, and Disaster Risk Management, Emerald Group Publishing Limited, Bingley, Chapter 2, pp. 23–46.
Kenny, M. L. 2002. Drought, clientilism, fatalism, and fear in northeast Brazil. *Ethics, Place, and Environment* 5(2): 123–34.
Laska, S., and Peterson, K. 2013. Between now and then: Tackling the conundrum of climate change. *Canadian Risk and Hazards Newsletter* 5(1): 5–8, http://207.23.111.231/sites/default/files/library/HazNet_2013-10_v5n1.pdf
Lavell, A. 2011. *Unpacking Climate Change Adaptation and Disaster Risk Management: Searching for the Links and the Differences: A Conceptual and Epistemological Critique and Proposal*, IUCN-FLACSO Project on Climate Change Adaptation and Disaster Risk Reduction.
Lazrus, H. 2012. Sea change: Climate change and island communities. *Annual Review of Anthropology* 41: 285–301.
Marino, E., and Schweitzer, P. 2009. Talking and not talking about climate change in northwestern Alaska, in S. A. Crate and M. Nuttall (Eds.), *Anthropology and Climate Change: From Encounters to Actions*. Walnut Creek, CA: Left Coast Press, Inc., pp. 209–17.
Marris, P. 1975. *Loss and Change*. Garden City, NY: Anchor Books.
Martin, S. F., Weerasinghe, S., and Taylor, A. (Eds.). 2014. *Humanitarian Crises and Migration: Causes, Consequences, and Responses*. London: Routledge.
Morrissey, J. 2012. Rethinking the "debate on environmental refugees": From "maximalists and minimalists" to "proponents and critics." *Journal of Political Ecology* 19: 36–49.
Myers, N. 2005. *Environmental Refugees: An Emergent Security Issue*. 13th Economic Forum, Prague (May 23–27), Session III Environment and Migration, http://www.osce.org/eea/1485
Nadasdy, P. 2007. Adaptive co-management and the gospel of resilience, in D. Armitage, F. Berkes, and N. Doubleday (Eds.), *Adaptive Co-Management: Collaboration, Learning, and Multilevel Governance*. Vancouver: University of British Columbia Press, pp. 208–27.
Nakicenovic, N., and Swart, R. 2000. Emissions scenarios. *Special Report of the Intergovernmental Panel on Climate Change*. Cambridge: Cambridge University Press.
Nansen Initiative. 2015. *Agenda for the Protection of Cross-Border Displaced Persons in the Context of Disasters and Climate Change-Final Draft*, https://www.nanseninitiative.org/global-consultations/
National Research Council. 2012. *Disaster Resilience: A National Imperative*. Washington D.C.: National Academies Press.

Nelson, D. R., Adger, W. N., and Brown, K. 2007. Adaptation to environmental change: Contributions of a resilience framework. *Annual Review of Environment and Resources*. DOI: 10.1146/annurev.energy.32.051807 (accessed 8/7/13).
Nelson, D. R., West, C. T., and Finan, T. J. 2009. Introduction to *In Focus*: Global change and adaptation in local places. *American Anthropologist* 11(3): 271–74
Nicholls, R. J., Wong, P. P., Burkett, V. R., Codignotto, J. O., Hay, J. E., McLean, R. F., Ragoonaden, S., and Woodroffe, C. D. 2007. Coastal systems and low-lying areas, in M. L. Parry, O. F. Canziani, J. P. Palutikof, P. J. van der Linden, and C. E. Hanson (Eds.), *Contribution of Working Group II to the Fourth Assessment Report of the Intergovernmental Panel on Climate Change*. Cambridge: Cambridge University Press.
O'Brien, K. 2012. Global environmental change II: From adaptation to deliberate transformation. *Progress in Human Geography* 36(5): 667–76.
O'Brien, K., Eriksen, S., Nygaard, L. P., and Schjolden, A. 2007. Why different interpretations of vulnerability matter in climate change discourses. *Climate Policy* 7(1): 73–88.
O'Keefe, P., Westgate, K., and Wisner, B. 1976. Taking the naturalness out of natural disasters. *Nature* 260(5552): 566–67.
Oliver-Smith, A. 2004. Theorizing vulnerability in a globalized world: A political ecological perspective, in G. Bankoff, G. Frerks, and D. Hilhorst (Eds.), *Mapping Vulnerability: Disasters, Development, and People*. London: Earthscan.
———. 2009. Nature, society, and population displacement: Toward an understanding of environmental migration and social vulnerability. *InterSecTions* 8. Bonn: United Nations University Institute for Environment and Human Security.
———. 2012. Debating environmental migration: Society, nature, and population displacement in climate change. *Journal of International Development* 24(8): 1058–70.
———. 2013. A matter of choice. *International Journal of Disaster Risk Reduction* 3: 1–3.
Oliver-Smith, A., and Hoffman, S. 1998. *The Angry Earth: Disaster in Anthropological Perspective*. London: Routledge.
Orlove, B. 2009. The past, the present, and some possible futures of adaptation, in W. Neil Adger, I. Lorenzoni, and K. O'Brien (Eds.), *Adapting to Climate Change: Thresholds, Values, Governance*. Cambridge: Cambridge University Press.
Patterson, T. 1994. Toward a Properly Historical Ecology, in C. L. Crumley (Ed.), *Historical Ecology: Cultural Knowledge and Changing Landscapes*. Santa Fe, NM: School of American Research.
Peninsula Principles on Climate Displacement Within States. 2013. *Displacement Solution*, http://displacementsolutions.org/ds-initiatives/the-peninsula-principles/
Ransan-Cooper, H., Farbotko, C., McNamara, K. E., Thornton, F., and Chevalier, E. 2015. Being framed: The means and ends of framing environmental migrants. *Global Environmental Change* 35: 106–15.
Pelling, M. 2011. *Adaptation to Climate Change: From Resilience to Transformation*. London: Routledge.
Ribot, J. 2011. Vulnerability before adaptation: Toward transformative climate action. *Global Environmental Change* 21: 1160–62.
Scudder, T. 2009. Resettlement theory and the Kariba case: An anthropology of resettlement, in A. Oliver-Smith (Ed.), *Development and Dispossession: The Anthropology of Development Forced Displacement and Resettlement*. Santa Fe, NM: SAR Press.
Smit, B., and Wandel, J. 2006. Adaptation, adaptive capacity, and vulnerability. *Global Environmental Change* 16: 282–92.
Stern, N. 2006. *The Stern Review: On the Economics of Climate Change*. Cambridge: Cambridge University Press.
Suhrke, A. 1994. Environmental degradation and population flows. *Journal of International Affairs* 47(2): 473–96.
Tacoli, C. 2009. Crisis or adaptation? Migration and climate change in a context of high mobility. *Environment and Urbanization* 21(2): 513–25.

Thywissen, K. 2006. Core terminology of disaster reduction: A comparative glossary, in J. Birkmann (Ed.), *Measuring Vulnerability to Natural Hazards: Towards Disaster Resilient Societies*. Tokyo: United Nations University Press, pp. 448–96.

Turner, B. L., Kasperson, R. E., Matson, P. A., McCarthy, J. J., Corell, R. W., Christenson, L., Eckley, N., Kasperson, J. X., Luers, A., Martello, M. L., Polsky, C., Pulsipher, A., and Schiller, A. 2003. A framework for vulnerability analysis in sustainability science, *Proceedings of the National Academy of Science* 100(14): 8074–79.

Turton, D. 2006. Who is a forced migrant? in C. de Wet (Ed.), *Development-Induced Displacement: Problems, Policies, and People*. Oxford: Berghahn Books.

UNFCCC. 1992. Definitions. UNFCCC.int/essential_background/convention/items/2536/php.

UNHCR. 2009. *Climate Change, Natural Disasters, and Human Displacement: A UNHCR Perspective*. United Nations High Commissioner for Refugees, http://www.unhcr.org/cgi-bin/texis/vtx/home/opendocPDFViewer.html?docid=4901e81a4&query=displacement%20scenarios

Walker, B., Holling, C. S., Carpenter, S. R., and Kinzig, A. 2004. Resilience, adaptability, and transformability in social-ecological systems. *Ecology and Society* 9(2): 5.

Warner, K. 2009. Migration: Climate adaptation or failure to adapt? Findings from a global comparative field study. *IOP Conference Series: Earth and Environmental Science* 6(56): DOI:10.1088/1755-1307/6/6/562006.

Warner, K., Hamza, M., Oliver-Smith, A., Renaud, F., and Julca, A. 2010. Climate Change, Environmental Degradation, & Migration. *Natural Hazards* 55: 689–15. DOI: 10.1007/s1069-009-9419-7.

Welsh, M. 2013. Resilience and responsibility: Governing uncertainty in a complex world. *The Geographical Journal*. DOI: 10.1111/geoj.12012.

Willox, A. C., Harper, S. L., Ford, J. D., Edge, V. L., Landman, K., Houle, K., Blake, S., and Wolfrey, C. 2013. Climate change and mental health: An exploratory case study from Rigolet, Nunatsiavut, Canada. *Climatic Change* 121: 255–70.

Wisner, B., Blaikie, P., Cannon, T., and Davis, I. 2004. *At Risk: Natural Hazards, People's Vulnerability, and Disasters* (2nd ed.). London: Routledge.

World Meteorological Organization. 2013. *Global Climate 2001–2010: A Decade of Climate Extremes-Summary Report. WMO 1119*. Geneva: World Meteorological Organization.

Wrathall, D., Oliver-Smith, A., Fekete, E., Gencer, E., Sakdapolrak, P., and Reyes, M. L. 2015. Problematizing loss and damage. *Journal of Global Warming* 8: 274–94.

Yenotani, M. 2011. *Displacement Due to Natural Hazard-Induced Disasters: Global Estimates for 2009 and 2010*. IDMC & NRC. Oslo.

Yonetani, M., Albuja, S., Bilak, A., Ginnetti, J., Glatz, A.-K., Howard, C., Kok, F., McCallin, B., Swain, M., Turner, W., and Walicki, N. 2015. *Global Estimates 2015: People Displaced by Disasters*. Geneva: Norwegian Refugee Council and the Internal Displacement Monitoring Centre.

Chapter 3

Apocalypse Nicked! Stolen Rhetoric in Early Geoengineering Advocacy

Clare Heyward and Steve Rayner

> We are on the precipice of climate system tipping points beyond which there is no redemption. (Hansen 2005: 8)

Talk of tipping points is now commonplace in climate science and policy and also in the media. Concerns that the prevailing linear models of climate change could be wrong were first voiced in the late 1980s. Scientists started to warn that environmental changes would not be smooth, gradual events but sudden "sharp jumps" (Broeker 1987: 123). Since then, warnings of abrupt, nonlinear climate change effects have been commonplace rhetoric for environmental campaigners and scientists alike (for example, Mastrandrea and Schneider 2001; Alley et al. 2005). Abrupt, nonlinear climate change effects are considered more pernicious, because they will shock human and environmental systems that will in turn struggle to adjust. As the concept of abrupt climate change effects became accepted, it was linked to the concept of tipping points, which had its origins in describing social change (for instance, Gladwell 1996). Tipping-point rhetoric has three elements, all captured by Hansen's much-quoted statement, at the beginning of this chapter. The first is irreversibility: once a tipping point is passed there is no return. The second is the abruptness of the change.

The third is catastrophe—the change will have dire consequences for human well-being. Examples of "tipping elements," features of the climate system that could pass a tipping point (Lenton et al. 2008), include loss of permafrost, loss of Arctic sea ice, the release of volumes of marine methane hydrates, decreases in the volume of the Greenland and West Antarctic ice sheets, Amazon and boreal forest die-back, and changes in the El Niño Southern Oscillation amplitude and frequency (Lenton et al. 2008).

The tipping-point trend in climate discourse commenced in earnest in 2001, when Hans Joachim Schellnhuber invoked the concept in a Linacre Lecture at Oxford University (Sciencewatch 2009). The term subsequently began to appear in the science policy literature as climate scientists used the concept when communicating climate change to policy makers and the public (see Russill and Nyssa 2009). Lindsay and Zhang (2005) made the first mention of tipping points in a scientific article, and subsequently the term became increasingly used in scientific literature and general climate discourse. For example, the 20-year anniversary of the Rio Earth Summit prompted reflection about the progress made and warnings about the possibility of passing climate tipping points were issued (Barnosky et al. 2012). Delegates at *Planet Under Pressure*, a major scientific conference in March 2012, were repeatedly told about the dire state of the global environment and the need for new thinking and urgent action (Rayner and Heyward 2013).

These recent developments are examples of an enduring trend of using claims about the natural world to justify political ones (ibid.). For example, the claim that "human societies must now change course and steer away from critical tipping points in the Earth system that might lead to rapid and irreversible change" (Biermann et al. 2012) is a contestable political claim about the appropriate response to climate change. Even granting that non-linear changes are involved, one could argue that scientific research and climate policy need not *necessarily* be focused on avoiding tipping points. For example, emphasizing the averting of possible physical catastrophe could obscure less dramatic but very real and severe problems that will occur in some of the world's poorest countries owing to "ordinary," linear, gradual anthropogenic climate change.

This is not to say that tipping points and abrupt changes have no basis in scientific fact. Geological records show that abrupt climate changes have happened (Alley et al. 2005), the greatest being the Younger Dryas, where temperatures dropped and glacial conditions returned in the northern hemisphere (Alley 2000). We merely suggest that increased concern with abrupt changes is not simply a response to improvements in scientific understanding of critical earth systems. The decision to highlight tipping points is an example of "stealth advocacy" (Pielke, Jr. 2007) by those concerned about lack of progress in curbing global greenhouse gas (GHG) emissions. The rise of tipping-point rhetoric has political roots. In this chapter we highlight some of the factors behind the success of tipping-point rhetoric in climate discourses and some of the political consequences of its success, using the interpretative framework of Cultural Theory.

Cultural Factors behind the Success of Tipping-Point Rhetoric

Cultural Theory is an anthropological framework that has been fruitfully applied to climate politics (see, for example, Rayner 1995a, b; Thompson and Rayner 1998; Verweij and Thompson 2006; Hulme 2009). The success of tipping-point rhetoric can be at least partly explained by its appeal to two enduring cultural tropes: millenarianism and a view of nature as being delicately balanced. Both are associated with an "egalitarian" worldview or "cosmology" (Rayner 1982; Thompson 1987). However, the tipping-point rhetoric and the rhetorical strategy associated with the egalitarian cosmology has been adopted by holders of a rival "hierarchical" cosmology but used to justify a very different kind of response to climate change. In this chapter we first present a brief introduction to Cultural Theory, largely following Thompson and colleagues (1990), and outline two long-standing cultural tropes on which tipping-point rhetoric is based. We then show how key elements of the rhetoric were adopted and deployed for very different ends.

A Brief Introduction to Cultural Theory

Cultural theory identifies four idealized ways of viewing the world and hypothesizes that individuals will draw on one of these four cosmologies when constructing arguments and advocating action. These four cosmologies are hierarchical, egalitarian, individualist, and fatalist. They are differentiated according to two factors: social network characteristics (*group*) and the extent of social differentiation (*grid*). The group factor is linked to the desire to maintain in-group/out-group boundaries. The grid factor is the degree of individual control (or conversely, external constraints) an individual has.

Each cosmology tells a different story about human nature, individual-group relations, the natural world, danger, blame, responsibility and distributive justice. For example, the archetypical hierarchical context is a group with high internal differentiation and strong collective identification. The hierarchical cosmology holds that human nature has the potential for evil and disorder and needs to be controlled by inculcation of strong social norms. It values conformity with group norms and respects authority and institutions. The egalitarian context also has a high group identification factor, but the internal character of the group is different, with a strong desire for social equality and distaste for internal differentiation. Each individual is regarded as having equal power to steer the group's actions, external attempts to impose norms are distrusted, and large institutions are regarded as corrupting. Like the egalitarian worldview, the individual worldview rejects attempts to constrain actions by recourse to social status (class, gender, age, and so on). However, unlike the egalitarian worldview, it sees humans as fundamentally rational and self-interested, with more interest in their own gain than in group membership. The individualist setting is typically a network in which competition between equals is rewarded. People operating in an individualist mode have a high sense of personal agency. This last element is absent for the fatalist context. As the name suggests, the fatalist worldview holds that individuals are constrained by forces external to them, but, unlike the hierarchical context, there is no sense of belonging to a collective. Because of the lack of agency the fatalist voice is not present in political debates, but its cause might be taken up by one of the other voices. Figure 3.1 depicts the three politically active voices.

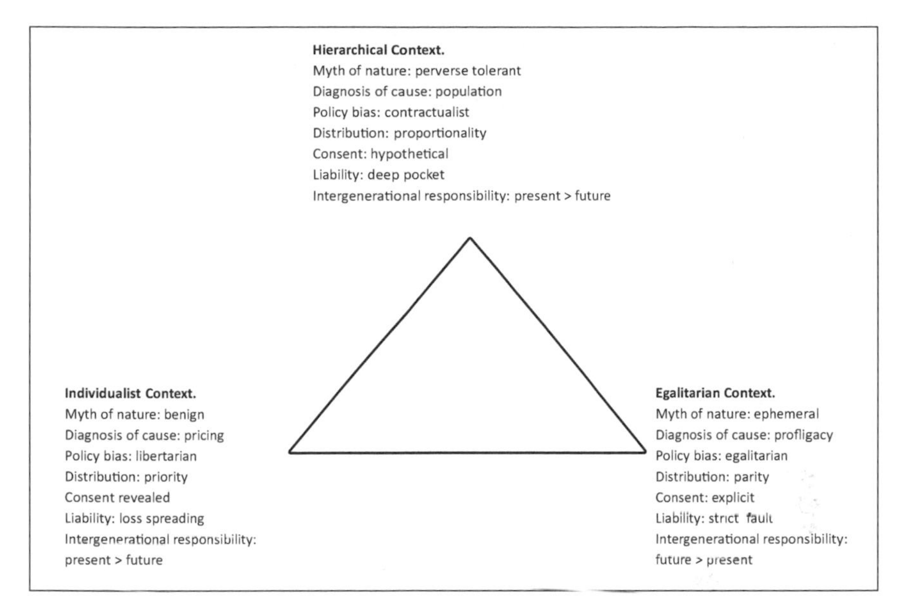

Figure 3.1 Map of the three active cosmologies (after Rayner 1995b)

Cultural Theory does not claim that an individual holds to one particular cosmology in all situations and at all times. In fact, as individuals move between different social contexts they are likely to adjust their rhetoric accordingly (Thompson et al. 1990). Accordingly, the cosmologies are best viewed as sets of argumentative resources that can be used to critique existing social arrangements. It is this power of critique and prescription that is of interest here. Initial uses of tipping-point rhetoric and political prescriptions drew on the enduring narrative of *millenarianism.*

Millenarian Thinking

Millenarian thinking has been found in cultures and societies across the world and throughout human history. Originally it was a feature of religious narratives, wherein a powerful supernatural force (God or ancestors) acts to reward the devout and dutiful and usher in a new age of peace and justice. *Secular apocalypticism* (Barkun 1986) is a more recent phenomenon, with *green millenarianism* being one of its more prominent manifestations. Owing an intellectual debt to Thomas Malthus (Linnér 2003), green millenarianism began in earnest with the rise of the environmental movement in the 1960s. The seminal text of the decade, *Silent Spring* (Carson 1962), told a story of humanity's use of pesticides destroying nature. Soon after, *The Closing Circle* (Commoner 1972), *The Population Bomb* (Ehrlich 1971), and *The Limits to Growth* (Meadows et al. 1972) all warned of impending crisis caused by toxic waste build-up, famine, and severe resource shortages, respectively. Proponents of these views appealed to the authority of modern science to justify a given course of action. It should come as no surprise then that, following in this tradition, apocalyptic framings are frequently encountered in discourses of climate change (Buell 2010).

General Features of Millenarianism

Millenarian narratives share four key structural features (Rayner 1982): (1) the *end of history* (2) brought about by an *external force*, and (3) *time compression*, meaning that the end is imminent. These three features are used to construct a story about (4) the need for *behavioral change*. A brief elaboration follows.

1. The End of History One feature of millenarian thinking is that the current era is transient. Millenarian thinking traditionally posits a final purpose. The current era is one (albeit the penultimate) of an unfolding series of steps to that end-state. In much religious apocalypticism, the final state will be one of guaranteed peace and prosperity (material and spiritual) for all (Cohn 1957). Humanity is urged to prepare physically, mentally, and/or spiritually for the new era. Another form of apocalyptic thinking is *cataclysmic forewarning* (Wojcik 1997). Most green millenarianism and some religious prophecies (for example, in the Old Testament) take this form. The end of the epoch threatens desolation and suffering on a vast scale; however, it can be avoided through human effort. Accordingly, the message of green millenarianism is that humanity (or large sections of it) must change its ways in order to avert catastrophe. However, the actions undertaken will ultimately bring about a new era for human societies with the promise of flourishing lives for all.

2. External force The second feature of millenarian thinking is the presence of a powerful external force that will bring the current epoch to its end by delivering justice. For example, in religious apocalypticism, the end of the epoch is to be brought about by God, either to punish humans for their sins or to redeem humans as part of the divine plan. Green millenarianism replaces God or spiritual powers with a quasidivine Nature. The appropriate attitude to nature is one of awe, wonder, and reverence at the powerful force capable of producing both sublime beauty and terrible destruction. If humanity adopts the wrong attitude toward this powerful but alien force, it will react by wreaking catastrophe on humankind.

3. Time compression A continual refrain in millenarianism is that the end of the epoch is imminent and that action in preparation for, or to avert, the catastrophe is urgently required. In religious apocalypticism, when the anticipated day of judgement passed, predictions were often quietly revised

to another date in the near future. In green millenarianism, humanity is exhorted to act before it is too late and the force of nature is set on a new path. The fact that environmental change is often slow and gradual does nothing to lessen the sense of urgency. Even if the catastrophe will occur decades or centuries in the future, action is needed now to avoid it.

4. BEHAVIORAL CHANGE Millenarian rhetoric ultimately aims to promote behavioral change. Humanity is admonished for its current failings and encouraged to pursue a different path. Most millenarian accounts prescribe that material goods must be redistributed or in some cases rejected outright. Conventional activities aimed at securing those goods must cease. In religious apocalypticism, this change is regarded as essential preparation for the new era of very different forms of social and spiritual relations. In narratives of cataclysmic forewarning, such as green millenarianism, rejection and redistribution of key material goods are necessary to avert the impending catastrophe. In both cases, the momentous nature of the changes required means that resistance is to be expected, but it is permissible to overcome it in order to achieve the desired ends.

The Egalitarian Influence

In Cultural Theory, millenarianism is associated with an egalitarian cosmology. It has been shown that a high-group, low-grid structure can be maintained by adopting the compressed conceptions of time and space, which are key structural features of millenarianism (Rayner 1982). The course of action advocated in the face of an apocalypse is thus the egalitarian demand to reduce and equalize consumption. "Small is beautiful" expresses the egalitarian ideal. A steady state economy, frugality, self-control (including in reproduction), and simplicity in an individual's personal life are favored. These factors help to maintain the lack of internal differentiation desirable according to the egalitarian cosmology: all group members should have the same command of resources. Moreover, scaling back human impact is the only way to avoid catastrophe, according to the egalitarian view of the balance of nature.

The idea of a balance of nature is traceable at least as far as the ancient Greeks. It posits an equilibrium point between natural factors. Despite academic criticism (for example, Kircher 2009), this age-old metaphor persists in the public imagination, although there is significant variance as to how the balance of nature is understood (Zimmerman and Cuddington 2007). Holling (1986) elaborated four conceptions of nature, which were incorporated into the four main cosmologies of Cultural Theory by Michael Thompson (1987). The myth of stability is associated with the individualist cosmology. Nature is robust, justifying a laissez-faire attitude to managing the natural world. There is no need to regulate human activity, because deviations from the equilibrium point will be temporary and natural systems will shift back. The egalitarian cosmology holds the opposing myth of instability. It holds that nature is ephemeral and that perturbations will trigger the collapse of a delicately poised balance. Meanwhile, in the hierarchical cosmology, nature is mostly capable of withstanding human activities, but there are limits, and it is important to know where they lie. Finally, the fatalist conception of nature is that it is capricious and entirely unpredictable (Thompson et al. 1990). These different conceptions of nature are depicted in Figure 3.2.

The similarities between this conception of nature and the idea of tipping points should be evident. Tipping-point rhetoric thus exhibits the main features of green millenarianism. A catastrophe is looming, and it is the end of the era of humanity's use of nature's resources carelessly and greedily, disturbing the delicate balance. The force of nature will react to centuries of mistreatment and react in ways that will make humanity quake unless immediate action is taken. For example, Europe's politicians were told by Tony Blair and Jan Peter Balkenende: "we have a window of only 10–15 years to take the steps we need to avoid crossing catastrophic tipping points" (quoted in Watt 2006), and the New Economics Foundation argued in 2012 that action is needed in the next 50 months (Simms 2012). Unless greenhouse gas (GHG) emissions are curbed, the positive feedback mechanisms will proceed, prompting the onset of one or more tipping points (Romm 2011a, b).

Statements such as these are not scientific predictions because of the uncertainties involved in forecasting what is likely to happen. Additionally, there are a vast number of factors that could affect eventual outcomes, prompting many scientists to talk only about "projections." Instead, the conditional aspects, plus the behavior changes advocated, makes these statements akin to cataclysmic forewarnings found in both religiously motivated and secular apocalypticism. Humanity is on the path to disaster

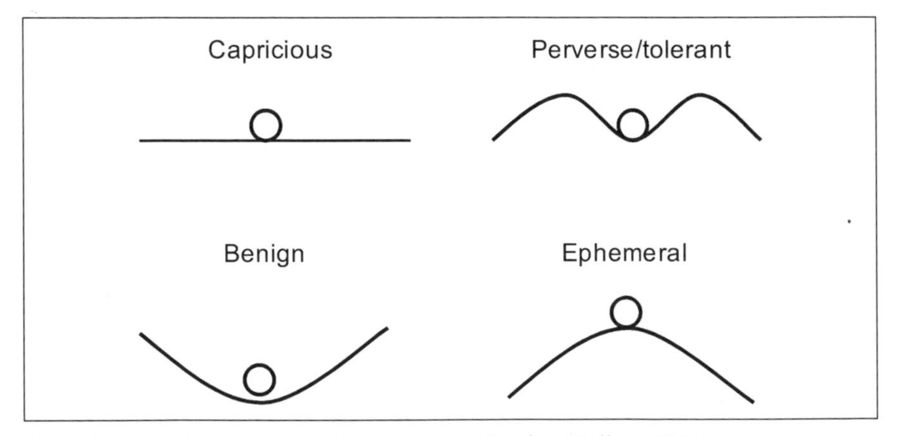

Figure 3.2 Myths of nature (Thompson 1987, after Holling 1986)

unless intended audiences make profound changes in the way they live. The egalitarian voice claims that humans must respect nature's fragility and make only modest demands on it. The most popularly advocated approach is to start dramatically cutting the use of fossil fuels, on which industrialized economies are based.

The Success of Tipping-Point Rhetoric

Tipping-point rhetoric is not only the latest variant of green millenarianism but also the most respectable. This form of apocalyptic rhetoric is not confined to isolated or oppressed groups but promulgated by eminent scientists and world leaders. What factors can account for this success? The following points offer at least a partial explanation. First, tipping-point rhetoric is emotive. Newspaper articles about environmental disaster, complete with awe-inspiring images of icebergs calving or a landscape of dense green forest, can do much to stir public imagination and make climate change more tangible. Tipping-point rhetoric also has a sense of drama, provided by the idea of the delicate balance of nature and the ancient narrative of millenarianism.

Second, this picture cannot easily be challenged; the global climate is extremely complex and significant uncertainties remain about key climate mechanisms. Even those who dislike the rhetoric acknowledge that the presence of tipping points cannot be ruled out. Indeed, as the next section explains, there is greater acceptance of tipping-point rhetoric than might be expected. Indeed, we might say that tipping-point rhetoric is the latest way of expressing the dominant *hegemonic myth* of climate change discourse—namely, that the planet is fragile, vulnerable, and alone (Rayner 1995a). A hegemonic myth is not a shared episteme or worldview but a rhetorical theme that sets the terms of debate within which even opposing arguments must be framed to be admissible to a discourse, despite how they seek to subvert it once admitted (Thompson and Rayner 1998).

According to Cultural Theory, a lack of challenge is to be expected from the politically inactive fatalist perspective. We might expect contestation from those in the individualist corner, who are quick to decry "scaremongering" when it means that they will no longer be allowed to behave as they choose. Indeed, individualism is the worldview most associated with climate skepticism. It is noteworthy, however, that there is little challenge to the hegemonic myth from the hierarchical cosmology. At first glance, tipping-point rhetoric seems to form a middle ground between the egalitarian and the hierarchical cosmologies, but this middle-ground is rather narrow. Those who hold a more hierarchical cosmology can accept the hegemonic myth but use it to promote their own preferred strategy. They can accept the basic concept of tipping points and elements of the millenarian narrative, but the ending of their story and their prescribed behavioral changes differ from those who hold an egalitarian view. Rather than rely solely on mitigation, holders of a hierarchical cosmology consider an additional response: research into geoengineering should be conducted.

Millenarianism in Pro-Geoengineering Rhetoric

Since the mid-2000s a growing number of scientists have been calling for concerted research into various techniques of *geoengineering*, the "deliberate large-scale manipulation of the planetary environment" (Shepherd et al. 2009: 1). Out of all the justifications of pursuing geoengineering research (see Rayner et al. 2013), one of the most prominent and powerful has been the "climate emergency" argument (see, for example, Crutzen 2006; Blackstock et al. 2009; Victor et al. 2009; Caldeira and Keith 2010; Long et al. 2011; Goldblatt and Watson 2012; Victor et al. 2013). A climate emergency is a rapid and drastic physical change in the Earth's climate, which could have extremely pernicious effects on human well-being. One particular form of geoengineering, sulphate aerosol injection into the stratosphere to diffuse sunlight, is thought to be fast-acting and therefore the means of averting a climate emergency. Hence, it is argued, there is a pressing need for research into that particular geoengineering technology.

The climate emergency argument borrows from the green millenarian narrative in three ways. First, the climate emergency narrative maintains the green millenarian feature of positing the possible end of an epoch and the dawn of a new era. Second, this era is brought about by the realization that external natural forces will react to humanity's behavior unless humankind adopts an appropriate relationship with the natural world. It is another cataclysmic forewarning: humanity can be saved from the crisis, providing appropriate action is undertaken. However, the concept of nature invoked and humanity's appropriate relationship to it is different. Instead of the egalitarian view of nature as ephemeral and delicately balanced, advocates of the climate emergency argument adopt the perverse/tolerant view associated with hierarchical context. In this view, tipping points indicate that nature is extremely complex, and mismanagement can lead to disaster. However, there is some scope for maneuver, and good management of natural processes is not impossible. The climate emergency argument thus coheres with the egalitarian worldview and challenges the individualist cosmology by postulating that there are limits to nature's resilience. Ultimately, however, it rejects both individualist and egalitarian cosmologies by postulating that Nature can (and should) be carefully managed and done so by suitably qualified experts. In this narrative, the control over the Earth's biological, chemical, and physical systems offered by geoengineering marks the completion of the transition to the Anthropocene. As in the green millenarian story, rightful order is achieved, but here humans achieve their destiny as the stewards of nature.

Second, the shift from talking of dangerous climate change to "climate emergency" immediately adds a sense of urgency characteristic of millenarian thinking. Proponents maintain that, whereas the occurrence of any possible "climate emergency" might be many decades away, action—in the form of research into stratospheric sulphate aerosols—is needed *now*. This could be because (1) a tipping point, which would trigger a climate emergency, could be passed in the near future; (2) the technologies needed, either to avoid the passing of a tipping point or to cope with the effects of doing so, will take years to develop (testimony from Ken Caldeira to the United States House of Representatives Committee on Science and Technology, quoted in Gordon 2010); (3) technical limitations must be investigated (Blackstock et al. 2009) either to guard against the temptation to deploy initially promising but ultimately unsafe and undeveloped technologies (Caldeira and Keith 2010) or (4) to avoid "moral hazard," the temptation to delay mitigation in the hope that this form of geoengineering will make it unnecessary (Keith et al. 2010).

Those who use climate emergency arguments in relation to geoengineering have thus adapted the green millenarian narrative, keeping key elements of its powerful rhetoric. There are some differences: climate scientists who use the emergency argument say that an emergency *could* happen, not that it *will* happen unless behavior is changed. They are also careful to use the narrative to propose another type of response to climate change, not to displace entirely the case for mitigation. Finally, in keeping with the different ending to the hierarchical narrative, a different redistribution of resources is called for. To safeguard humanity against a possible catastrophe, more research funds should be provided to the scientific elite.

Political Consequences

The framing of the natural world as fragile and vulnerable, based on the egalitarian view of nature, has long been one of the hegemonic myths in climate discourse (Rayner 1995a). As such, it is accepted by anyone who wishes to participate in the discourse. Thus "climate skeptics" or "climate deniers", by holding that nature is robust, reject this myth and are in turn cast out by those who accept it. Within the climate discourse the success of this myth is also evident in the sidelining of adaptation policies, because they were thought to go against the "tread lightly" prescription associated with the myth of fragility (Pielke et al. 2007). Tipping-point rhetoric is now one of the dominant ways in which this hegemonic myth is expressed. Hence, despite its egalitarian origins, it is accepted by those of a hierarchical persuasion. However, once agents' acceptance of any hegemonic myth confirms them to be legitimate participants, they begin to posit different elaborations and exceptions according to their own specific views and to advance their own interests and solutions (Thompson and Rayner 1998). Egalitarian and hierarchical voices contest not the existence of tipping points but rather their *significance*. For the egalitarian worldview, tipping points show that nature is ephemeral and that a collapse is on its way. For the hierarchical voice, tipping points are indicative of what *can* happen if nature is not properly managed. They are a pathological symptom, real and dangerous enough, but not an essential aspect of the world. Tipping-point rhetoric thus serves both a green millenarian position and an alternative millenarian account invoked in the rhetoric of advocates of geoengineering research. Both use millenarian rhetoric, promising a new era and justifying urgent action accordingly, and both use the idea of catastrophe to argue for their own worldview and preferred response.

The ascendency of tipping-point discourse has several political consequences. First, history suggests that apocalyptic rhetoric is often used as justification for more authoritarian rule, so we might expect those who invoke tipping points to argue for political change. There is some evidence for this fact, as we show shortly. Second, the time compression inherent in the millenarian discourse would lead us to expect that the claims of the necessity of urgent action will be contested. While, again, there is some evidence for this expectation, there is less contestation than might be expected when millenarian narrative is used by the hierarchical voice. The relative lack of contestation leads to the third consequence of the success of tipping points: this particular green millenarian rhetoric has been used by advocates of geoengineering research to advance solutions quite inimical to conventional environmentalist ideals and to the egalitarian cosmology on which they are based.

The Tendency toward Authoritarianism

Tipping-point rhetoric can potentially lead to more authoritarian modes of decision making. In green millenarianism, if very severe mitigation is required, then regulation of personal consumption might be necessary. The possibility of eco-authoritarianism has always loomed over green politics. However, given the general egalitarian commitment to participatory democracy plus the fact that strict egalitarianism is rarely part of the "establishment" but usually a subaltern voice opposing it (Thompson et al. 1990), it is perhaps more likely that a more authoritarian form of governance will result from the success of the hierarchical use of the rhetoric. As Verweij and Thompson (2006) have observed, this likelihood has been evident in calls for global systems of observation, coordination, and planning, such as the 2003 United Nations Human Development Report's call for a "Life Observatory" (United Nations Human Development Programme 2003). This trend continues. In 2012, the *State of the Planet Declaration* called for a "new contract between science and society in recognition that science must inform policy to make more wise and timely decisions" (Brito and Stafford-Smith 2012). Meanwhile, from their claim that human societies "must steer away from tipping points," Frank Biermann and his colleagues argue that to do so requires "fundamental reorientation and restructuring of national and international institutions toward more effective Earth system governance and planetary stewardship" (Biermann et al. 2012, 1306). They recommend the "upgrading" of the

UN Environmental Programme so that it becomes a specialist UN agency equivalent to the World Health Organisation, with "a sizable role in agenda-setting, norm development, compliance management, scientific assessment, and capacity building" (Bierman et al 2012, 1306). Also recommended are measures to integrate sustainable development policies at all levels and to close gaps in global regulation, especially of emerging technologies, including geoengineering. Obviously, this form of expert authoritarianism should not be regarded as equivalent to dictatorship. Most scientific experts in the Western world profess to be democrats and see their role as advisory. (Indeed, Biermann and colleagues do acknowledge issues of legitimacy [2012, 1307].) It remains the case, however, that a transfer of political power is called for and that expert contributions should be taken much more seriously in the policy-making process.

In the case of geoengineering as a response to a climate emergency, we might expect that similar governance solutions will be advocated as the field matures. The climate emergency narrative vindicates the hierarchical view of nature and the corresponding ideal of management by experts who can maintain the boundary between climate stability and catastrophe. Additionally, widespread acceptance of climate emergency rhetoric could effectively cede the power to decide about appropriate action from the rest of the population. An early report stated that "in a crisis, ideological objections to solar radiation management may be swept aside" (Lane et al. 1997: 12). Whether the greater fear is eco-authoritarianism or expert-authoritarianism, we should ask what transfers of political power are being advocated.

Pitfalls of Time-Compression

The second danger is that, as a form of millenarian rhetoric, the non-occurrence of anticipated events tends to be used by holders of a rival cosmology to discredit it. For thousands of years, the imminent end of the world has been proved to be not so imminent after all. (For a truncated list of millenarian visions in the United States alone, see Stewart and Harding [1999].) When prophesies fail, their proponents make revisions while their opponents take the opportunity to mock them for their credulity. Because tipping-point rhetoric is being used both by holders of egalitarian worldviews and by those who take a hierarchical position, the opponents to it will be holders of individualist worldviews. Indeed, those who downplay climate change generally often espouse individualist values (Leiserowitz 2006).

Holders of individualist worldviews point out that disaster has not happened yet. Nor, they argue, is it going to. Not only is nature robust but humans are capable of adapting to any natural changes. Whatever the position taken on the facts of climate change, the real object of concern in the individualist viewpoint is the attempt to limit individuals' rights to use, and to profit from the use, of the Earth's natural resources. The time-compression introduced in the global warming debate means that each year a disaster *does not* happen is taken as further confirmation that it *will not*, just as every cold snap is taken as evidence that global warming itself is a sham and disproof of the hegemonic myth. Those who "believe in anthropogenic global warming" are merely the next set of doom-mongers. However, there is one important twist.

As those of hierarchical and egalitarian persuasions both appeal to millenarian rhetoric and accordingly emphasize urgency, we might expect both to be equally susceptible to the pitfalls of time compression. Instead, there is less contestation from individualist voices when millenarian rhetoric is promulgated from a hierarchical standpoint. Indeed, Weitzman (2009) argues that the costs of a climate emergency make it prudent to invest in sulphate aerosol injection research. Why might that be the case? One reason is that geoengineering turns out to be consistent with the individualist cosmology, as it is with the hierarchical cosmology. Several participants in the climate change debates who have endorsed broadly individualist views in the past accept and endorse sulphate aerosol geoengineering research (for example, Bickel and Lane 2008; Lomborg 2010). From the individualist viewpoint, nature is robust, so large-scale interventions will not pose any major problems. Nor is there any problem in managing environmental systems for greater human benefit. For example, the individualist perspective might agree with statements such as "humans have long been co-creators of the environment they inhabit" (Shellenberger and Nordhaus 2012) or that gardening and geoengineering differ only in their scale (Keith 2000). The individualist voice does not object to the idea of geoengineering, so contesting the hierarchical use of tipping-point rhetoric is less of a priority than contesting the egalitarian calls for redistribution. For the egalitarian green millenarians, this is a bitter irony. The solution proposed by their former rhetorical allies, those holding the hierarchical worldview, is now garnering support from another of their political opponents, those espousing the individualist position. Tipping-point rhetoric, the latest variety of green millenarianism will have played an instrumental role in bringing about a solution that is anathema to the egalitarian cosmology. From the hierarchical point of view, we suggest, this has been a successful case of *stolen strategy* and one means of promoting a preferred response to climate change.

Stolen Strategies

Stolen strategy is a concept similar to that of the more familiar "stolen rhetoric." In Cultural Theory, a case of stolen rhetoric is when, discussing a specific issue, an individual uses the rhetoric, the key values, and concepts of one culture to support the position of another (Thompson 1990). Stealing rhetoric can be a deliberate maneuver: an individual makes shifts in discursive strategy to support his or her desired conclusion in a particular context (West et al. 2010). With this potential advantage comes the risk that using the language of a rival cosmology will undermine the individual's current cosmology if too many exceptions are made. For example, the anti-abortionist who uses the egalitarian language of "the rights of the fetus" effectively abandons the hierarchical commitment that the community can differentiate between its members (Thompson 1990: 263). In addition to stolen rhetoric, there is also the concept of *stolen strategy*, whereby the "*means* corresponding to one cultural bias are used to achieve aims belonging to another bias" (Mamadouh 1999: 404, our italics). Stealing strategy carries risks similar to those of stealing rhetoric. We suggest that the hierarchical use of climate emergency rhetoric is best understood as an example of a stolen *strategy,* in which case the hierarchical worldview remains the same but the speakers take elements of a narrative normally associated with the alternative cosmology of egalitarianism. Climate emergency rhetoric shares similar features with the apocalyptic narrative *structure* of green millenarianism, but, as we have seen, it does not share its *content*.

If climate emergency rhetoric is an example of a stolen strategy, we might expect the hierarchical justification of geoengineering research to fluctuate as the political context changes. For example, it will be interesting to see whether climate emergency rhetoric wanes if governments and funding bodies agree to make substantial investments in sulphate aerosol injection. In the short term, we might expect the hierarchical justifications to occupy the middle-ground between the technological optimism of the individualist view and the pessimism of the egalitarian. This expectation would partly account for the oft-heard refrains that scientists endorse research but are agnostic on deployment and that research into all geoengineering technologies must include investigation of social effects, including impacts on mitigation. In the longer term, if geoengineering becomes part of the mainstream discourse on climate change, arguments offered in its support that are more congruent with the two "establishment" cosmologies (that is, the individualist and the hierarchical cosmologies) might become more common, and the frequency of climate emergency arguments may decline. In fact, there are signs that a decline is beginning. A recent study about geoengineering framings in English-language newspapers suggests that there are diverse justifications for pursuing geoengineering (Scholte et al. 2013). Another study reports that the "metaphorical landscape" is changing (Nerlich and Jaspal 2012). Moreover, some advocates of geoengineering research have recently stopped using climate emergency arguments to make their case. The decline is not necessarily permanent: the climate emergency justification retains considerable potential to be used as a "trump card" should other arguments fail to be convincing. Rather, we can expect the use of climate emergency justification to fluctuate according to political factors—for example, the relative priority given to geoengineering in climate policy.

In light of the variety of justifications for geoengineering research consistent with the hierarchical cosmology, we might ask why the climate emergency rhetoric was adopted in the first place. Why not, for example, make an individualist appeal to the relative cheapness of sulphate aerosol injection—as others (for instance, Gingrich 2008; Levitt and Dunbar 2009) have done? We conjecture that climate emergency rhetoric was instrumental in persuading members of the scientific community to break their self-imposed taboo on geoengineering. In one of the most influential papers on geoengineering to date, Paul Crutzen stated that warming of Arctic regions might result in accelerated carbon dioxide and methane emissions, leading to positive feedbacks and warned that the "Earth system is increasingly in the non-analogue condition of the Anthropocene" (2006: 217). Crutzen's paper was published before the concepts of tipping points and climate emergency were in common parlance, but the nonlinear conditions and his example of methane release have since been discussed in those terms. Also significant is the invocation of the Anthropocene. A new era is dawning, and markedly different action is warranted. Both egalitarian and hierarchical worldviews take the concept of hubris seriously, leading to members of the scientific community being reluctant to recommend geoengineering research. Positing a new era and the threat of a possible emergency enabled the hierarchical voice to overcome that obstacle. In these new circumstances, sulphate aerosol geoengineering could (regrettably) be the best course of action. In that case, scientists had better know what they are doing. Research is therefore necessary (Lawrence 2006).

Conclusion

The invocation of climate tipping points is a political strategy used in the discourse on climate change. Originally a concept promoted by environmental activists and concerned scientists in an attempt to galvanize action on mitigation, tipping points are now a key element in a new variant of green millenarianism. The narrative of a climate emergency displays key features of millenarianism and prescribes a course of action for avoiding environmental catastrophe. The narrative, the view of nature it depends on, and the prescribed action are all consonant with the egalitarian worldview of Cultural Theory.

However, the concept of tipping points has also featured in another apocalyptic narrative, that of climate emergency rhetoric. Climate emergency rhetoric presumes a hierarchical view of nature and accordingly prescribes different actions. The foregrounding of catastrophe and urgency of the climate emergency argument gave it considerable rhetorical power compared to other justifications for geoengineering research. Climate emergency rhetoric put geoengineering on the climate policy agenda by suggesting that scientific investigation geared toward sufficiently high-impact events was justified, even imperative. The possibility of drastic climatic events was used to break the scientific community's own taboo in order to advocate geoengineering research as part of the global response to climate change. Once sufficient momentum is created, the push toward geoengineering research can then be supported by both hierarchical and individualist viewpoints. The use of apocalyptic climate emergency rhetoric by climate scientists to justify research into geoengineering is seemingly a case of a successful stolen strategy.

Acknowledgments

This work was conducted as part of the Climate Geoengineering Governance Project (http://geoengineeringgovernanceresearch.org) funded by the Economic and Social Research Council (ESRC) and the Arts and Humanities Research Council (AHRC)—grant ES/J007730/1. Clare Heyward additionally acknowledges financial support from the Leverhulme Trust and, before that, the Oxford Martin School.

References

Alley, R. B. 2000. The Younger Dryas cold interval as viewed from central Greenland. *Quaternary Science Reviews* 19: 213–26.

Alley, R. B., Marotzke, J., Nordhaus, W. D., Overpeck, J. T., Peteet, D. M., Pielke, R. A., Pierrehumbert, R. T., Rhines, P. B., Stocker, T. F., Talley, L. D., and Wallace, J. M. 2005. Abrupt climate change. *Science* 299: 2005–10.

Barkun, M. 1986. *Crucible of the Millennium: The Burned-Over District of New York in the 1840s*. Syracuse, NY: Syracuse University Press.

Barnosky, A. D., Hadly, E. A., Bascompte, J., Berlow, E. L., Brown, J. H., Fortelius, M., Getz, W. M., Harte, J., Hastings, A., Marquet, P. A., Martinez, N. D., Mooers, A., Roopnarine, P., Vermeij, G., Williams, J. W., Gillespie, R., Kitzes, J., Marshall, C., Matzke, N., Mindell, D. P., Revilla, E., and Smith, A. B. 2012. Approaching a state shift in Earth's biosphere. *Nature* 486: 52–58.

Bickel, E., and Lane, L. 2008. *The Copenhagen Consensus: An Analysis of Climate Engineering as a Response to Climate Change*, http://faculty.engr.utexas.edu/bickel/Papers/AP_Climate%20Engineering_Bickel_Lane_v%205%200.pdf

Biermann, F., Abbott, K., Andresen, S., Bäckstrand, K., Bernstein, S., Betsill, M. M., Bulkeley, H., Cashore, B., Clapp, J., Folke, C., Gupta, A., Gupta, J., Haas, P. M., Jordan, A., Kanie, N., Kluvánková-Oravská, T., Lebel, L., Liverman, D., Meadowcroft, J., Mitchell, R. B., Newell, P., Oberthür, S., Olsson, L., Pattberg, P., Sánchez-Rodríguez, R., Schroeder, H., Underdal, A., Vieira, S. C., Vogel, C., Young, O. R., Brock, A., and Zondervan, R. 2012. Navigating the Anthropocene: Improving Earth System Governance. *Science* 335: 1306–07.

Blackstock, J., Battisti, D., Caldeira, K., Eardley, D., Katz, J., Keith, D., Patrinos, A., Schrag, D., Socolow, and R., Koonin, S. 2009. *Climate Engineering Responses to Climate Emergencies*. Santa Barbara, CA: Novim.

Brito, L., and Stafford-Smith, M. 2012. *State of the Planet Declaration*. London: Planet Under Pressure, www.planetunderpressure2012.net/pdf/state_of_planet_declaration.pdf

Broeker, W. S. 1987. Unpleasant surprises in the greenhouse? *Nature* 328: 12–126.

Buell, F., 2010. A short history of environmental apocalypse, in S. Skrimshire (Ed.), *Future Ethics: Climate Change and the Apocalyptic Imagination*. London: Continuum Publishing, pp. 13–36.

Caldeira, K., and Keith, D. 2010. The need for geoengineering research. *Issues in Science and Technology* 10: 57–62.

Carson, R. 1962. *Silent Spring*. Boston and Cambridge, MA: Houghton Miffin; Riverside Press.

Cohn, N. 1957. *The Pursuit of the Millennium*. London: Secker and Warburg.

Commoner, B. 1972. *The Closing Circle: Confronting the Environmental Crisis*. London: Cape.

Crutzen, P. 2006. Albedo enhancement by stratospheric sulfur injections: A contribution to resolve a policy dilemma? *Climatic Change* 77: 211–20.

Ehrlich, P. 1971. *The Population Bomb*. Cutchogue, NY: Buccaneer Books.

Gingrich, N. 2008. *Stop the Green Pig: Defeat the Boxer-Warner Lieberman Green Pork Bill-Capping American Jobs and Trading America's Future*, www.humanevents.com/2008/06/03/stop-the-green-pig-defeat-the-boxerwarnerlieberman-green-pork-bill-capping-american-jobs-and-trading-americas-future/

Gladwell, M. 1996. The pipping point. *The New Yorker*, June 3.

Goldblatt, C., and Watson, A. J. 2012. The runaway greenhouse: Implications for future climate change, geoengineering, and planetary atmospheres. *Philosophical Transactions of the Royal Society A: Mathematical, Physical, and Engineering Sciences* 370: 4197–216.

Gordon, B., 2010. *Engineering the Climate: Research Needs and Strategies for International Coordination*. Washington D.C.: House of Representatives Committee on Science and Technology.

Hansen, J., 2005. Is there still time to avoid "dangerous anthropogenic interference" with the global climate? *A Tribute to Charles David Keeling*. NASA Goddard Institute for Space Studies.

Holling, C. S. 1986. The resilience of terrestrial ecosystems: Local surprises and global change, in W. C. Clark and R. E. Munn (Eds.), *Sustainable Development of the Biosphere*. Cambridge: Cambridge University Press, pp. 292–320.

Hulme, M. 2009. *Why We Disagree about Climate Change: Understanding Controversy, Inaction, and Opportunity*. Cambridge: Cambridge University Press.

Keith D. W. 2000. Geoengineering the climate: History and prospect. *Annual Review of Energy and the Environment* 25: 245–84.

Keith, D. W., Parson, E., and Morgan, M. G. 2010. Research on global sun block needed now. *Nature* 463: 426–27.

Kircher, J. C. 2009. *The Balance of Nature: Ecology's Enduring Myth*. Princeton, NJ: Princeton University Press.

Lane, L., Caldeira, K., Chatfield, R., and Langhoff, S. 1997. *Workshop Report on Managing Solar Radiation*. Hannover, MA: NASA.

Lawrence, M. 2006. The geoengineering dilemma: To speak or not to speak. *Climatic Change* 77: 245–48.

Leiserowitz, A. 2006. Climate change risk perception and policy preferences: The role of affect, imagery, and values. *Climatic Change* 77: 45–72.

Lenton, T. M., Held, H., Kriegler, E., Hall, J. W., Lucht, W., Rahmstorf, S., and Schellnhuber, H. J. 2008. Tipping elements in the Earth's climate system. *Proceedings of the National Academy of Sciences* 105: 1786–93.

Levitt, S. D., and Dubner, S. W. 2009. *Superfreakonomics: Global Cooling, Patriotic Prostitutes, and Why Suicide Bombers Should Buy Life Insurance*. London: Allen Lane.

Lindsay, R. M., and Zhang, J. 2005. The thinning of Arctic sea ice 1988–2003: Have we passed a tipping point? *Journal of Climate* 18: 4879–94.

Linnér, B. O. 2003. *The Return of Malthus: Environmentalism and Post-War Population-Resource Crises*. Ilse of Harris: White Horse Press.

Lomborg, B. 2010. Geoengineering: A quick, clean fix? *Time Magazine*, www.time.com/time/magazine/article/0,9171,2030804,00.html

Long, J., Raddekmaker, S., Anderson, J., Benedick, R. E., Caldeira, K., Chaisson, J., Goldston, D., Hamburg, S., Keith, D., Lehman, R., Loy, F., Morgan, G., Sarewitz, D., Schelling, T., Shepherd, J., Victor, D., Whelan, D., and Winickoff, D. 2011. *Geoengineering: A National Strategic Plan for Research on the Potential Effectiveness, Feasibility, and Consequences of Climate Remediation Technologies.* Washington D.C.: Bipartisan Policy Center.

Mamadouh, V. 1999. Grid-group cultural theory: An introduction. *GeoJournal* 47: 395–409.

Mastrandrea, M. D., and Schneider, S. H. 2001. Integrated assessment of abrupt climatic changes. *Climate Policy* 1(4): 433–49.

Meadows, D., Meadows, D., Randers, J., and Behrens, W. 1972. *The Limits to Growth: A Report for the Club of Rome's Project on the Predicament of Mankind.* London: Earth Island.

Nerlich, B., and Jaspal, R. 2012. Metaphors we die by? Geoengineering, metaphors, and the argument from catastrophe. *Metaphor and Symbol* 27:131–47.

Pielke, Jr., R. 2007. *The Honest Broker: Making Sense of Science in Policy and Politics.* Cambridge: Cambridge University Press.

Pielke, Jr., R., Prins, G., Rayner, S., and Sarewitz, D. 2007. Lifting the taboo on adaptation. *Nature* 445: 597–98.

Rayner, S., 1982. The perception of time and space in egalitarian sects: A millenarian cosmology, in M. Douglas (Ed.), *Essays in the Sociology of Perception.* London: Routledge and Kegan Paul, pp. 247–74.

———. 1995a. Governance and the global commons. In M. Desai and P. Redfern (Eds.), *Global Governance.* London: Pinter, pp. 60–93.

———. 1995b. A conceptual map of human values for climate change decision making, in A. Katama (Ed.), *Equity and Social Considerations Related to Climate Change.* Nairobi: ICPE Science Press, pp. 57–73.

Rayner, S., Heyward, C., Kruger, T., Pidgeon, N., Redgwell, C., and Savulesuc, J. 2013. The Oxford Principles. *Climatic Change* 121(3): 499–512.

Rayner, S., and Heyward, C. 2013. The inevitability of nature as a rhetorical resource. In K. Hastrup (Ed.), *Anthropology and Nature.* London: Routledge, pp. 123–46.

Rockstrom, J., Steffen, W., Noone, K., Persson, A., Chapin, F. S., Lambin, E. F., Lenton, T. M., Scheffer, M., Folke, C., Schellnhuber, H. J., Nykvist, B., de Wit, C. A., Hughes, T., van der Leeuw, S., Rodhe, H., Sorlin, S., Snyder, P. K., Costanza, R., Svedin, U., Falkenmark, M., Karlberg, L., Corell, R. W., Fabry, V. J., Hansen, J., Walker, B., Liverman, D., Richardson, K., Crutzen, P., and Foley, J. A. 2009. A safe operating space for humanity. *Nature* 461: 472–75.

Romm, J. 2011a. Must-read Hansen and Sato paper: We are at a climate tipping point that, once crossed, enables multi-meter sea level rise this century. *Climateprogress,* http://thinkprogress.org/climate/2011/01/20/207376/hansen-sato-climate-tipping-point-multi-meter-sea-level-rise/

———. 2011b. NSIDC bombshell: Thawing permafrost feedback will turn Arctic from carbon sink to source in the 2020s, releasing 100 billion tons of carbon by 2100. *Climateprogress,* http://thinkprogress.org/climate/2011/02/17/207552/nsidc-thawing-permafrost-will-turn-from-carbon-sink-to-source-in-mid-2020s-releasing-100-billion-tons-of-carbon-by-2100/

Russill, C., and Nyssa, Z. 2009. The tipping-point trend in climate change communication. *Global Environmental Change* 19: 336–44.

Scholte, S., Vasileiadou, E., and Petersen, A. C. 2013. Opening up the societal debate on climate engineering: How newspaper frames are changing. *Journal of Integrative Environmental Sciences* 10: 1–16.

Sciencewatch. 2009. New hot papers [interview with Timothy Lenton and Hans Joachim Schellnhuber], http://archive.sciencewatch.com/dr/nhp/2009/09julnhp/09julnhpLentET/

Shellenberger, M., and Nordhaus, T. 2012. Evolve: The case for modernization as the road to salvation. *The Breakthrough Journal*, http://thebreakthrough.org/index.php/journal/past-issues/issue-2/evolve/

Shepherd, J., Caldeira, K., Cox, P., Haigh, J., Keith, D., Launder, B., Mace, G., MacKerron, G., Pyle, J., Rayner, S., Redgwell, C., and Watson, A., 2009. *Geoengineering the Climate: Science, Governance, and Uncertainty*. London: The Royal Society.

Simms, A. 2012. 50 months to avoid disaster—and a change is in the air. *New Economics Foundation*, www.neweconomics.org/blog/entry/50-months-to-avoid-climate-disaster-and-a-change-is-in-the-air

Stewart, K., and Harding, S. 1999. Bad endings: American apocalypsis. *Annual Review of Anthropology*: 285–310.

Thompson, M. 1987. Welche Gesellschaftsklassen sind potent genug, anderen ihre Zukunft aufzuoktryieren? in L. Burchardt (Ed.), *Design der Zukunft*. Cologne: Dumont, pp. 58–87.

Thompson, M., Ellis, R. J., and Wildavsky, A. 1990. *Cultural Theory*. Boulder, CO: Westview Press.

Thompson, M., and Rayner, S. 1998. Cultural discourses. *Human choice and climate change* 1: 265–343.

United Nations Development Programme. 2003. *Millennium Development Goals: A Compact among Nations to End Global Poverty*. Oxford: Oxford University Press.

Verweij, M., and Thompson, M. 2006. *Clumsy Solutions for a Complex World: Governance, Politics, and Plural Perceptions*. Basingstoke: Palgrave Macmillan.

Victor, D., Morgan, M. G., Apt, J., Steinbruner, J., and Ricke, K. 2009. The geoengineering option: A last resort against global warming? *Foreign Affairs* 88: 64–76.

Victor, D., Morgan, M. G., Apt, J., Steinbruner, J., and Ricke, K. 2013. The truth about geoengineering: Science fiction and science fact. *Foreign Affairs*, March27, www.foreignaffairs.com/articles/139084/david-g-victor-m-granger-morgan-jay-apt-john-steinbruner-kathari/the-truth-about-geoengineering

Watt, N. 2006. Blair warns against climate change tipping points. *The Guardian* October, 20, www.guardian.co.uk/world/2006/oct/20/greenpolitics.politics

Weitzman, M. 2009. On modelling and interpreting the economics of catastrophic climate change. *The Review of Economics and Statistics* 91: 1–19.

West, J., Bailey, I., and Winter, M. 2010. Renewable energy policy and public perceptions of renewable energy: A cultural theory approach. *Energy Policy* 38: 5739–48.

Wojcik, D. 1997. *The End of the World as We Know It: Faith, Fatalism, and Apocalypse in America*. New York: New York University Press.

Zimmerman, C., and Cuddington, K. 2007. Ambiguous, circular, and polysemous: students' definitions of the "balance of nature" metaphor. *Public Understanding of Science* 16: 393–406.

Chapter 4

Complex Systems and Multiple Crises of Energy

John Urry

This chapter examines multiple interdependent processes in environment, economy, climate, food, water, and energy (for more detail, see Urry 2011, 2013b). In particular it considers how late 20th-century neoliberalism, enabled by oil, ratcheted up the global scale of movement within the global North. This travel of people and goods became essential to most social practices that depend on and reinforce a high-carbon society, but it came to a shuddering halt when oil prices increased in the early years of the 21st century. Suburban houses could not be sold, especially those that were in far-flung oil-dependent locations. Financial products and institutions were rendered worthless. This story of oil in the United States portrays a bleak future as "easy oil" is scheduled to run out by the middle of this century. Herein I first develop a broad analysis of these issues to argue how complex energy systems and energy crises can bite back with interest and may portend some bleak futures. I do so by exploring a new trend in thinking about the future of societies—the "new catastrophism"—and also by considering how various analysts, scientists, and commentators are examining and advocating accounts of various futures characterized by an astonishing array of emergent risks. I then highlight how 20th-century social sciences mostly ignored issues of resource use and overuse, thereby replicating and reproducing a world where resources were unimportant for deciphering change and development, whereas now, in an era of resource depletion, contestation and collapse call the social sciences to attention in the context of interconnecting systems of oil, money, and property.

Societal Collapse

Systems set in place during the industrial period may contain the seeds of their own demise. There are no guarantees that increasing prosperity, wealth, movement, and connectivity will continue. Some argue that the North's rich 20th century was a short and finite period in human history. Its legacy could come to a sudden halt, with the societies of the rich North "reversing" or "collapsing." Joseph Tainter explored this idea of collapse in the late 1980s in *The Collapse of Complex Societies*. He examined how societies become more complex often as a response to short-term problems—complexity demands ever greater high-quality energy, increased energy produces diminishing returns, and there is growing concatenation of problems that can reinforce each other unexpectedly and unpredictably across domains. Tainter writes that "however much we like to think of ourselves as something special in world history, in fact industrial societies are subject to the same principles that caused earlier societies to collapse" (1988: 216). This statement echoes Marx and Engels's famous claim as to how modern bourgeois society "is like the sorcerer, who is no longer able to control the powers of the nether world whom he has called up by his spells" (1888: 58).

There are many new catastrophist social and scientific texts that examine the multiple interdependent processes in environment, economy, climate, food, water, and energy to analyze emergent risks. To name a few:

Our Final Century (Rees 2003)
The Party's Over: Oil, War and the Fate of Industrial Society (Heinberg 2005)
The Next World War: Tribes, Cities, Nations, and Ecological Decline (Woodbridge 2005)
The Revenge of Gaia (Lovelock 2006)
The Upside of Down: Catastrophe, Creativity, and the Renewal of Civilization (Homer-Dixon 2006)
The Long Emergency: Surviving the Converging Catastrophes of the 21st Century (Kunstler 2006)
The Next Catastrophe (Perrow 2007)
Field Notes from a Catastrophe (Kolbert 2007)
With Speed and Violence: Why Scientists fear Tipping Points in Climate Change (Pearce 2007)
Winds of Change: Climate, Weather and the Destruction of Civilizations (Linden 2007)
Meltdown (Haseler 2008)
Global Catastrophes and Trends: The Next Fifty Years (Smil 2008)
Down to the Wire: Confronting Climate Collapse (Orr 2009)
World at Risk (Beck 2009)
Requiem for a Species (Hamilton 2010)
The Vanishing Face of Gaia (Lovelock 2010)
Storms of my Grandchildren (Hansen 2011)
Living in the End Times (Žižek 2011)
Tropic of Chaos (Parenti 2011)
The Burning Question: We can't burn half the world's oil, coal, and gas. So how do we quit? (Berners-Lee and Clark 2013).

In examining one of the best-known examples of catastrophism, Jared Diamond in his book *Collapse* argues that in the past environmental problems brought about the collapse of societies (2011). He distinguishes eight environmental factors as being responsible: deforestation and habitat destruction; soil problems; water management problems; over-hunting; over-fishing; effects of introduced species on native species; human population growth; and the increased per-capita impact of people. Thus populations grew and exploited natural resources, particularly energy resources, to a breaking point, especially when such societies were at the very height of their powers. He uses such historical material to suggest that, in the 21st century, human-caused climate change, the build-up of toxic chemicals in the environment, and energy shortages will similarly produce abrupt declines. This situation involves the running out of oil and gas, the increased lack of resilience of many societies, a global failure of economy and finance, population collapse, increasing resource wars, huge food shortages, and increases of global temperatures that will make most plant, animal, and human life impossible. These factors might also constitute a "perfect storm" analogous to the societal collapse experienced by the Roman Empire and the Mayan civilization. Although historical time frames are divergent, the tendency of internal contradictions working slowly and imperceptibly over time and bringing down apparently dominant systems based on inexhaustible supplies of energy are in common.

Both Marx and Weber were aware of the significance of energy resources. Marx wrote of the "subjection of nature's forces to man, machinery, application of chemistry to agriculture and industry, steam navigation, railway, electric telegraphs, clearing of whole continents for cultivation, canalization of rivers" (quoted in Berman 1983: 93), while Weber recognized the finitude of the world's fossil resources (1939: 181).

But during the long 20th century this resource-dependence came to be forgotten. It seemed that societies had been able to spin off and break free from their resources. It was thought there were no finite limits and significant perverse consequences that flowed from the extensive and increasing burning of fossil fuels (Berners-Lee and Clark 2013). When Bauman, for example, writes of "liquid" everything, he seems to regard the social as infinite, without limits and without costs or consequences, with no resources that are finite and whose effects mean that they might bite back with interest (2000). Bauman does not notice that "liquid modernity" actually depends on burning the literal liquid of oil. Almost all social scientists, since perhaps Marx and Weber, have remained, we might say, energy-blind.

Systems Thinking

This increasing interest in how previous civilizations collapsed has stimulated many analyses using complex-systems thinking, in which small changes can tip large systems over the edge, beyond a threshold to trigger runaway changes beyond an equilibrium threshold. The starting notion here is that physical and social worlds are full of change, paradox, and contradiction. There are no simple unchanging stable states or states to which there is equilibrium-establishing movement. Physical and social worlds can be characterized through "the strange combination of the unpredictable and the rule-bound that governs so much of our lives" (Ball 2004: 283). So there are patterned, regular, and rule-bound systems; these rule-bound workings can come to generate various unintended effects; and unpredictable events can disrupt and abruptly transform what appear to be rule-bound and enduring patterns.

The "normal" state then is not one of balance and equilibrium. For example, populations of species can demonstrate extreme unevenness, with populations rapidly rising when introduced into a given area and then almost as rapidly collapsing. Any system is thus complex. Policies never straightforwardly restore equilibrium, unlike the claims of policy makers. Indeed, actions often generate the opposite or almost the opposite from what is intended. So many decisions intended to generate one outcome, because of the operation of a complex system, generate multiple unintended effects very different from what is planned (Budiansky 1995).

Systems thus generally do not move toward equilibrium (see Urry 2003). Much of the literature in economics presumes that feedbacks will be only negative and equilibrium-restoring (Arthur 1994; Beinhocker 2006). The equilibrium models dominant in most economic system analyses, especially general equilibrium models, ignore the huge array of positive feedbacks. Positive feedback mechanisms take systems away from equilibrium states. Thus systems should be viewed as dynamic, processual, and demonstrating the power of the second law of thermodynamics—that physical and social systems move toward entropy. Systems can be broadly viewed as unpredictable, open rather than closed with energy and matter flowing in and out. Systems are characterized by a lack of proportionality or nonlinearity between apparent causes and effects. There *can* be small changes that do bring about big, nonlinear system shifts, as well as the converse.

But not all aspects of a society change. Certain systems can be stabilized for long periods. Causation flows from contingent events to general processes, from small causes to large system effects, from historically or geographically remote locations to the general. Systems thus develop through lock-in but with only certain small causes being necessary to prompt or tip the initiation of each path. Systems once established are locked in and their patterns and rules survive for long periods even though there appear to be strong forces that should have undermined them. Institutions matter a great deal as to how systems develop, a fateful example of this being the steel-and-petroleum car system dating from the late 19th century but that has so far resisted all attempts to reverse its system dominance (Dennis and Urry 2009). Systems adapt and co-evolve in relationship to each other, and hence possible futures are irreducible to single structures, events, or processes, with futures being messy and complicated and difficult simply to reengineer.

Moreover, movement from one state to another can be rapid, with almost no stage in between. These strikingly abrupt transformations are normally known as *phase transitions*. As a gas is cooled it remains a gas until it suddenly turns into a liquid. This rapidity can also be seen in transformations between each ice age and the periods of relative warmth that occurred in-between. There are only two states, the glacial and the interglacial, with no third way in between. Ice-core research shows that there was abrupt movement from one state of the earth's system to the other, sometimes characterized as "punctuated equilibria" (Pearce 2007).

The importance of small but potentially fateful changes is increasingly described through black swans, which are those rare, unexpected, and highly improbable events that can have huge impacts. They are outliers not averages. Black swans are responsible for much economic, social, and political change in the world. The most important events are those least predictable (Taleb 2007). Thus when change happens it may not be gradual but may occur dramatically, at a moment, in a kind of rush. If a system passes a particular threshold, switches or tipping points occur through positive feedback. The system turns over as many climate scientists now fear with minor increases in global temperature provoking out-of-control "global heating" that could occur with "speed and violence" (Pearce 2007).

Along with global climate change various other system failures are likely. One is the probable decline in the availability of energy, especially oil, over the next few decades (Urry 2013b). Further, the world's population is growing by about 900 million per decade, the largest absolute increases ever recorded in human history. Rapid population growth in developing countries especially exposes populations to many hazards. The world went urban on May 23rd 2007 (http://www.prb.org/Publications/Articles/2007/623Urbanization.aspx). Cities consume three-quarters of the world's energy and are responsible for at least three-quarters of global pollution (Rogers 1997). Much food production depends on hydrocarbon fuels to seed and maintain crops, to harvest and process them, and to transport them to market, partly because of the exceptional food miles that have come to be involved in diets in the rich North (Pfeiffer 2006: 2). There may well be food protests as a result of likely flooding, desertification, and generally rising costs, as well as the tendency for rich societies to engage in land grabs. There are also growing insecurities in supplying clean usable water with huge demands from growing populations, especially those in mega-cities that have to both buy and transport, using carbon-based systems, their water from outside. A global temperature increase of 2.1° would expose up to 3 billion people to water shortages. Some commentators now refer to "peak ecological water," only 0.007% of water on Earth being available for human use (Pfeiffer 2006: 15). Already severe water shortages face one-third of the world's population.

Moreover, these systems are increasingly seen as interacting with each other. Homer-Dixon notes: "I think the kind of crisis we might see would be a result of systems that are kind of stressed to the max already . . . societies face crisis when they're hit by multiple shocks simultaneously or they're affected by multiple stresses simultaneously" (2006: 1). Human and physical systems exist in states of dynamic tension and are especially vulnerable to dynamic instabilities. Systems may well come to reverberate against each other and generate effects on larger systemic changes. It is the simultaneity of converging shifts that creates significant changes. Thus resource depletion, climate change, and other processes may come to overload a fragile global system, creating the possibility of catastrophic failure (Parenti 2011; Urry 2011).

Moreover, regions even in rich societies are often unable to cope with crises (New Orleans, Fukushima, New Jersey, the Somerset Levels), with potential oil shortages, droughts, heat waves, extreme weather events, flooding, desertification, highly mobile diseases, and the potential forced movement of millions of environmental refugees seeking security (Abbott 2008). States have to deal with such concatenating processes, although their tax revenues have been decreasing because of the proliferation of 60 to 70 offshored tax havens (Shaxson 2011; Urry 2014).

The social sciences have mostly ignored issues of resources and especially energy, and indeed of the perverse contradictory consequences of use and overuse. In a way, the social sciences replicated and reproduced a world where resources and especially energy were more or less free goods and unimportant for deciphering the lineaments of change and development. But that was the 20th century. Now, in the new century, the world looks different indeed, and issues of resource depletion, contestation, and collapse will be haunting it in some potentially catastrophic decades to come. In the rest of this chapter, I illustrate these points through the interconnecting systems of oil, money, and property.

Oiling the Wheels of Society

It is often said that money makes the world go round as trillions of dollars, euros, yen, yuan, and major currencies circulate around and across the globe at dizzying speed. The scale of this predominantly digital movement dwarfs the income and resources available to all individuals, most companies, and many countries.

But there is another powerful mobile force in the world, and that is oil. Žižek writes about its origins: "Nature is one big catastrophe. Oil, our main source of energy—can you even imagine what kind of ultra-unthinkable ecological catastrophe must have happened on earth in order that we have these reserves of oil?" (www.democracynow.org/2008/5/12/world_renowned_philosopher_slavoj_zizek_on; accessed May 20, 2010). Ancient fossilized organic materials settled on the bottom of seas and lakes and were buried. As more layers were laid on top, intense heat and pressure built up and caused the organic matter ultimately to change into liquid and gaseous hydrocarbons. The long hydrocarbon chains of oil are in effect "preserved sun," and only the sun exceeds oil in its energy. This was a unique gift, but like most gifts it is finite and irreplaceable.

This oil is central to Western civilization. Much like money, oil makes the world go round. It is remarkably versatile, convenient, mobile, and for most of the 20th century exceptionally cheap, since it first gushed out of the ground at Spindletop in Texas in 1901. When oil reaches $100 a barrel (despite current falls in price, oil is likely to return to this figure) the world's known oil reserves are worth $104 trillion, or 40 times the value of the United Kingdom's economy (Singh 2011).

Oil makes fast movement possible (see Urry 2013b). It provides almost all transportation energy in the modern world (at least 95%), powering cars, trucks, planes, ships, and some trains. It thus makes possible the mobile lives of friendship, business life, professions, and families. Oil also moves components, commodities, and food around the world in trucks, planes, and vast container ships, as well as oil itself in large tankers. Oil is also an element of most manufactured goods and much packaging and bottling worldwide (95%). It is crucial to at least 95% of food production and distribution for a rising world population through providing power for irrigation, moving food, and providing pesticides and fertilizers (Harvey and Pilgrim 2011: S42). Also, oil is used for much domestic and office heating, especially within oil-rich societies, and is crucial in providing back-up power and lighting. Overall oil generates around one-third of all greenhouse gas emissions and is a major category of GHG emissions currently increasing across the world. Besides being itself mobile, oil is energy-dense, storageable, and nonrenewable.

It was less the Enlightenment or Western science or liberalism that secured Western civilization but more contingent carbon resources and especially oil that enabled the West to dominate the long, highly mobile 20th century. Its carbon resources and especially the mobile energy resource of oil generated an exceptional modern, mobile civilization based on fast movement. Owen writes how "oil is liquid civilization" (Owen 2011: 77). The West consolidated its power and influence in the world beginning in the United States. Cheap plentiful oil was central to American economic, cultural, and military power (Heinberg 2005; Burman 2007). If the American Dream was to be experienced by all the world's population, it would take five planets to support it (Hulme 2009: 261). American production and consumption got there first, because the United States initiated and monopolized the manufacture of Large Independent Mobile Machines (LIMMs). LIMMs carry their own energy source, and, given the denseness and historic cheapness of oil, were highly efficient generating the addiction to oil (Rutledge 2005). Globally there are nearly 1 billion cars and trucks, and almost 1 billion international air journeys occur each year.

As the geologist M. King Hubbert, who discovered the problem of "American peak oil" in the 1950s, noted: "You can only use oil once. . . . Soon all the oil is going to be burned" (quoted in Heinberg 2005: 100). Many suggestions and prototypes are being developed to produce alternative energy—especially nuclear, solar, and wind-generated. But these do not replace oil and "oil civilization" (Owen 2011). The global economy-and-society became utterly locked into and dependent on this one source of mobile power. Moreover, all alternative fuels have a much poorer ratio of energy returned on energy invested, or EROEI. Furthermore, as oil's price rises there is not necessarily more produced and delivered to the market place. Currently the oil supply seems relatively fixed, and intermittent rapid increases in price are certain. Moreover, most new sociotechnical systems take a very long time to develop, often decades. An energy transition from one type of energy system to another happens only once a century or so but with utterly momentous consequences (Smil 2010).

It is a huge problem for one resource to be the basis of these modern systems and societies. The future of oil is the future of modern societies set on their pathway during the 20th century. This fact had a dramatic effect in exploiting the earth's resources. The 20th century used as much energy as the previous hundreds of centuries of recorded human history. And in using those resources to engender fast movement, that century generated carbon dioxide and other emissions that remain in the atmosphere for hundreds of years.

The oil-exploring, -producing and -using industries are a huge economic enterprise comprising many of the world's largest and richest companies. This "carbon capital" consists of Western-based oil "super majors" (ExxonMobil), state oil companies mainly found in producing countries (Saudi Aramco, the world's most valuable company), car- and truck-producing corporations (Toyota as the world's largest), huge engineering and road construction companies (Bechtel), and many corporations providing services to car drivers and passengers (Holiday Inn, McDonalds). This capital made the mobile 20th century.

Furthermore, finance and oil are fundamentally intertwined. Oil is speculated on in financial markets. Its price movements now stem from this speculation as well as from changes in the supply of and demand for oil for transportation and manufacturing. A major report for Lloyds of London shows how speculation destabilizes supply and price and further reduces energy security (Froggatt and Lahn 2010). There are now over 75 crude-oil financial derivatives where 15 years ago there was 1. This Lloyds Report maintains that extensive growth in financial trading in oil makes oil prices higher and more unstable (Dicker 2011); oil prices are exceptionally sensitive to even small changes in demand.

And according to the Chief Economist of the International Energy Authority the peaking of global oil supplies is not for the future; it already occurred in 2006 (www.good.is/post/international-energy-agency-s-top-economist-says-oil-peaked-in-2006; accessed December 27, 2011). I now show how that peaking has already had major economic and social consequences. The world economic crash of 2007–2008 onward was partly brought about by oil shortages and price increases.

Neoliberalism and Indebtedness

Crucial here is neoliberalism, which in about 1980 became the dominant global orthodoxy of economic and social policy. It moved from its birthplace within the economics department at the University of Chicago (Harvey 2005; Klein 2007; Crouch 2011). Even by 1999 Chicago School alumni included 25 government ministers and over a dozen central bank presidents throughout the world (Klein 2007: 166).

Neoliberalism is a doctrine and set of practices that assert the power and importance of private entrepreneurship, private property rights, the freeing of markets, and the freeing of trade. These objectives are brought about by deregulating private activities and companies and privatizing previously state or collective services—especially through low taxes, undermining collective powers of workers and professionals, and providing conditions for the private sector to find ever-new sources of profitable activity. Neoliberalism especially minimizes the role of the state to readdress the balance it perceives between the "bad" state and "good" markets. Neoliberals hold that states are always inferior to markets in "guessing" what should be done. States are thought of as inherently inefficient and easily corrupted by private-interest groups. Markets are presumed to be "natural" and so will move to equilibrium if unnatural forces or elements do not get in the way.

However, states are often important in eliminating "unnatural" forces, by destroying sets of rules, regulations, and forms of life that slow down economic growth and constrain the private sector. Sometimes that destruction is exercised through violence and attacks on democratic procedures. The "freedom of the market" is brought about through "shock treatment," the creation of an "emergency" that enables the state to wipe the slate clean and impose free-market solutions (Friedman 2002). Klein describes how the disaster of Hurricane Katrina in New Orleans in 2005 provided conditions for the large-scale privatization of the New Orleans school system (2007: 3–21). "Never let a good crisis go to waste" is a neoliberal mantra.

Harvey summarizes neoliberalism as involving "accumulation by dispossession" (Harvey 2005: 159–61). Peasants are thrown off their land, collective property rights are made private, indigenous rights are stolen and turned into private opportunities, rents are extracted from patents, general knowledge is turned into intellectual property, the state forces itself to sell off or outsource its collective activities, trade unions are undermined, and financial instruments and flows redistribute income and rights toward finance and away from productive activities.

From 1980 on, neoliberalism became the dominant global discourse. It especially involves the light regulation of banks and financial institutions and the resulting trillion-dollar growth in financial securitization. The distinction between commercial and investment banking has been undermined with the lowering of lending standards and the proliferation of innovative business models. There was the "macho" domination of financial services and a privileging of competitive individualism implemented through a bonus culture rewarding indebtedness and dangerous risk taking.

There has also been the extensive movement offshore of revenues that ought to be taxed and used for providing collective services and benefits onshore. There are about 70 secrecy jurisdictions. "Offshore is how the world of power now works" (Shaxson 2011: 7–8, 26–7). More than half of world trade, including much oil revenue, now passes through tax havens. A quarter of all global wealth is held offshore (Kochan 2006; Urry 2014). Such offshoring is central to the enormous shadow banking system and the overwhelming imbalance between "financialization" and the "real economy" (see the view from the inside in Soros 2008). By September 2008 the global value of financial assets was $160 trillion, more than three times the world GDP (Sassen 2009).

Finance in the 1990s and early 2000s especially flowed not into manufacturing industry but into many kinds of property. For property to get built the private sector often cajoles national or regional states to create and pay for—often through borrowing—related mobility infrastructures such as motorways, high-speed rail links, and airports. Property development is then undertaken by firms who borrow finance. The properties built are then purchased by buyers who also borrow to make the purchase. Loans are made to property purchasers who could not otherwise be able to buy. Thus there is a speculative funding of highly leveraged new developments that are leased or sold with indebtedness on all sides. Property developments here include suburbs, apartments, second homes, hotels, leisure complexes, gated communities, sports stadia, office blocks, universities, shopping centers, and casinos, all built on the indebtedness of developers and purchasers as well as states.

Central here is the speculative intertwining of property and finance involving new forms of finance. The debts incurred are turned into commodities, or "securitized." They are parceled up, sliced and diced, into financial packages that are sold on, with huge markets developing throughout the world for many of these "products," which presupposes that property can be worth only more. This process creates a financially complex house of cards resting on the bet that property rises in value. During the 2000s an unsustainable bubble of private, corporate, and national indebtedness developed especially within the United States and more recently in China (Stiglitz 2010).

Moreover, many new suburbs in the United States from the 1980s onward were built distant from city centers, and they were not connected to city centers by mass or public transit. Such *Sprawltowns* depended on car travel and hence plentiful cheap oil so that newly arriving residents could commute to work and drive about for leisure and social life (Ingersoll 2006; Owen 2011). Only one-half of U.S. suburbs have any access to public transport, and residents are wholly dependent on car travel and thus on the price of oil (Rubin 2009: 123).

Much of this suburban housing was sold to people with subprime employment, credit, and housing histories, involving financial innovations. Although subprime housing is known to be central in the events that triggered the crash of 2007–2008, it has not been recognized how subprime suburbs were driven to the brink by a problem of mobility, by oil dependence *and* oil price spikes in the few years beforehand. Even Stiglitz's dissection of the American "mortgage scam" does not grasp how energy resources can bite back and reverse what seemed at the time irreversible (2010). How did oil reverse what appeared inevitable: that property could keep going only upward in value?

There was cheap oil for much of the period since the early 1980s. The U.S. index of petrol prices was in money terms 134 in 1990 and more or less the same in 2000 (138). The roaring 1990s, especially of house price inflation, which ran two and a half times the increase in per capita income for Americans, was based on the falling real price of petrol (Cortright 2008: 3). Such petrol prices remained more or less constant in money terms until 2003 (145).

Indebtedness in the United States in turn fueled the huge growth in consumption as house prices rose. It became possible to cash in the rising values of property, especially during 2002–2004. That money was used to fund further consumer purchases especially of goods manufactured in China and transported to the United States in vast container ships. This situation generated a huge U.S. current account deficit (Brenner 2006: 318–19). Many Americans believed that they were really richer as house prices continued to rise and private debt skyrocketed. Also, many people were drawn into purchasing housing through very low rates in the first few years of their mortgage—but much higher rates were charged in later years (variable rate mortgages, or adjustable rate mortgages [ARMs]).

Writing during 2006, Brenner describes how financial speculation was producing a real estate mania. The total value of residential property in developed economies rose by more than 30 trillion U.S. dollars between 2000 and 2005. This staggering increase was equivalent to 100% of those countries' combined GDPs at the time (Brenner 2006: 342–43). Indebtedness was thus built on indebtedness. Such intersecting bubbles of asset prices seemed not to be understood by anyone, nor could they be "governed." George Soros, one of the financial masters of this universe, reports that no one understood the global system that they themselves were creating (2008; and see Tett's anthropological analysis of J. P. Morgan 2009).

Bubbles Burst

Bubbles do eventually burst, and they often burst most dramatically and painfully as they fill with more and more hot air. The bursting of this bubble first in the United States from 2005 onward and then worldwide had dramatic consequences for the real economy. This is where oil is central. Throughout the 20th century soaring oil prices generated economic crises. All recent major economic crises, bar one, have rising oil prices at their core. Murray and former U.K. Government Chief Scientist King maintain that the events of 2007–2008 were not just a credit crunch but an "oil-price crunch" (2012).

In the middle years of the 2000s global shortages of oil led to a rapid rise in petrol prices worldwide, including the United States. At that time the United States was importing two-thirds of its oil (Hofmeister 2010). There was a five-fold increase in real oil prices between 2002 and 2008 (Rubin 2009: 183). Over a longer period the price per barrel of oil increased over 14 times between 1990 and 2007, partly because the world output of oil and related products could not be raised beyond around 85 million barrels a day. Labban maintains that "the oil crisis arrives as a financial crisis" (2010: 551; Yergin 2011).

Moreover, hurricanes Katrina and Rita hit the Louisiana coastline during 2005. These extreme weather events destroyed billions of dollars of gas and oil infrastructures through flooding the Mississippi delta. Other refineries around the world were working to full capacity and could not increase production when these Mississippi refineries were shut down. Rita led to the capsizing of a production platform. Such extreme weather events intersected with the crisis of oil supply. This situation illustrates the argument that societies face crisis when they are hit by multiple shocks at the same time (Homer-Dixon 2006: 1).

These hurricanes were the extreme events that showed the vulnerability of the world's oil civilization. Because of the inability to replace Gulf of Mexico supplies, the price of oil skyrocketed. There will be many other occasions when floods or hurricanes or blowouts or revolutions reduce supply, dramatically increase prices, and devastate patterns of life—including of people living in suburbs built on "easy oil," easy movement, and the presumed rising price of property (Strahan 2007: 57–60). These oil shortages were reflected in the U.S. petrol price spiking in the 2000s. It reached 302 in 2007 and a peak of 405 in July 2008. The petrol price peaked in 2007–2008. By February 2009 the petrol index was still higher at 186 than it had been in 2000 (Urry 2013b: 32).

But many people living in new American suburbs were completely locked into cheap petrol and cheap movement. Without it they could have no life. As petrol prices increased property owners were forced to reduce their expenditure on housing and other goods and services, including on cars. Hamilton argues that "the oil price increase was one factor pushing home sales and house prices down very rapidly" (2009). In the heated atmosphere of the bubble, "gas [petrol] price increases may have been the trigger that broke the expectations of continued growth" (Cortright 2008: 5; Dodson and Sipe 2008). Suddenly, it cost more to fill up an SUV's petrol tank than it cost to buy a week's groceries.

The American housing boom was thus brought to a shuddering halt through the escalating price of petrol in the middle years of the last decade. This increase tipped financially weak households over the financial brink. They could no longer afford the mortgage payments that they already could only just manage to pay, given their subprime standing. Ten thousand homeowners lost their home to foreclosures every day. Millions of Americans received foreclosure notices, and tens of billions in real-estate assets were written off as losses by banks. This was a vicious circle. Foreclosures helped accelerate the fall of property values, helping to spur more foreclosures. The losses they created brought the financial system to the brink of collapse in the fall of 2008. The steep recession that followed led to greater homeowner foreclosures, as people who lost their jobs often also lost their homes.

Many households thus defaulted on their mortgages, suburbs collapsed with much property for sale, financial institutions around the world were left holding huge amounts of bad debt (albeit rated AAA by the major ratings agencies), some banks went to the wall or were "nationalized," and there was a global recession on an unprecedented scale that continued into 2013 at least.

There was a geographical distribution to these patterns. House price falls were the most marked in suburbs rather than in metropolitan cores, and they were especially steep in those distant oil-dependent suburbs. House price reductions were greatest where there were no alternatives to the car and the highest dependence on the price *and* availability of petrol for almost all aspects of life. Households were spending up to 30% of their income on travel. Cheap oil was a necessity (Cortright 2008).

Because of increases in the cost of petrol the value of housing in commuter belts dropped very steeply. Many such suburbs turned into "ghostburbs," full of foreclosures, for-sale signs, and empty housing. This situation meant that "households are being made to rethink another cherished American institution—the white picket-fenced suburban dream home" (Ghazi 2008; Rubin 2009).

These processes all reduced U.S. consumer spending, similar to what happened in 1990–1991 during the first Gulf War. This reduction led to multiple defaults as banks realized that they held huge amounts of bad, or toxic, debt. There was an escalating collapse of, especially, investment banks in the United States and then around the world, which had also invested in many American mortgages. This house of financial cards came tumbling down, beginning in oil-dependent suburbs full of American households who had been sold subprime mortgages through a vast "scam," according to Stiglitz (2010).

There were many consequences of this collapse, including a reduction in the distances Americans (and others) now travel; this was the first downward shift of U.S. mileage for 30 years or so (Urry 2013a). Such reductions produced a marked decline in car sales. And some commentators speculate that this slowdown marks the beginning of the end for oil addiction, as oil consumption collapsed first in these American subprime, oil-dependent suburbs that had been locked into cheap car-based movement.

Conclusion

In the final years of the 20th century neoliberalism ratcheted up the global scale of movement within the global North. Oil enabled this. This travel of people and goods is essential to most social practices that depend on and reinforce a high-carbon society. Much of this high-carbon production and consumption was based on indebtedness and on a greatly increasing significance of finance within modern economies. Indeed, money was increasingly borrowed from the rest of the world, especially from China, to finance this carbon extravaganza for the global rich.

But this extravaganza ended suddenly when oil prices increased in the early years of the 21st century. Suburban houses could not be sold—especially those located in far-flung oil-dependent places. Financial products and institutions were found to be worthless. Easy money, easy credit, and easy oil had gone together. And when oil prices hit the roof in these U.S. suburbs then easy money and credit also came to a halt, and the presumed upward shift in property prices was shown to be a false dream. The financial house of cards had been built on cheap oil and cheap mobility.

As oil got prohibitively expensive the house of cards collapsed. This story of oil in the United States provides a bleak vision of the future as easy oil more generally runs out by the middle of this century. The Chief Economist at HSBC Bank maintains that "even if demand doesn't increase, there could be as little as 49 years of oil left" (Ward 2011). And, of course, demand is increasing very rapidly, especially outside the West. Elsewhere, I show how we have not seen anything yet: collapsing supplies and dramatically rising prices will wreak future devastation around the world in decades to come. Bill Spence, former vice president of Shell, apocalyptically writes how in the next few decades "twice the energy with half the carbon dioxide is a challenge for us all" (cited in Lovell 2010: 171). The 20th century has left us with an unsustainable civilization based on the mobilities of people and goods requiring this one resource of cheap, plentiful easy oil. We might conclude: cannot live with oil, cannot live without it.

More generally we can note how complex energy systems and energy crises can bite back with interest. Systems set in place during the industrial period contain the seeds of their own destruction. There are thus no guarantees and but grave doubts that increasing prosperity, wealth, movement, and connectivity will continue. Some argue that the 20th century in the rich North was a short and finite period in human history that, if it were to end, could reverse or collapse and drag down the rest of the world.

References

Abbott, C. 2008. *An Uncertain Future: Law Enforcement, National Security and Climate Change*. Oxford: Oxford Research Group.
Arthur, B. 1994. *Increasing Returns and Path Dependence in the Economy*. Ann Arbor, MI: University of Michigan Press.
Ball, P. 2004. *Critical Mass*. London: William Heinemann.
Bauman, Z. 2000. *Liquid Modernity*. Cambridge: Polity.
Beck, U. 2009. *World at Risk*. Cambridge: Polity.
Beinhocker, E. 2006. *The Origin of Wealth*. London: Random House.
Berman, M. 1983. *All that Is Solid Melts into Air*. London: Verso.
Berners-Lee, M., and Clark, D. 2013. *The Burning Question*. London: Profile.
Brenner, R. 2006. *The Economics of Global Turbulence*. London: Verso.
Budiansky, S. 1995. *Nature's Keepers*. London: Weidenfeld and Nicholson.
Burman, S. 2007. *The State of the American Empire*. London: Earthscan.
Cortright, J. 2008. *Driven to the Brink*. Chicago: CEOs for Cities.
Crouch, C. 2011. *The Strange Non-Death of Neo-Liberalism*. Cambridge: Polity.
Dennis, K., and Urry, J. 2009. *After the Car*. Cambridge: Polity.
Diamond, J. 2011. *Collapse: How Societies Choose to Fail or Survive*. London: Penguin, 2011.
Dicker, D. 2011. *Oil's Endless Bid*. New York: Wiley.
Dodson, J., and Sipe, N. 2008. *Shocking the Suburbs*. Sydney: UNSW Press.
Friedman, M. 2002. *Capitalism and Freedom*. Chicago: University of Chicago Press.
Froggatt, A., and Lahn, G. 2010. *Sustainable Energy Security*. London: Lloyd's and Chatham House
Ghazi, P. 2008. Gas guzzlers and "ghostburbs." *The Guardian*, July 2nd.
Hamilton, C. 2010. *Requiem for a Species*. London: Earthscan.
Hamilton, J. 2009. The oil shock and recession of 2008: Part 2. *2009 Econbrowser*, www.econbrowser.com/archives/2009/01/the_oil_shock_a_1.html, accessed March 10, 2009.
Hansen, J. 2011. *Storms of my Grandchildren*. London: Bloomsbury.
Harvey, D. 2005. *A Brief History of Neo-Liberalism*. Oxford: Oxford University Press.
Harvey, M., and Pilgrim, S. 2011. The new competition for land: Food, energy, and climate change, *Food Policy*, 36: S40–51, p. S42.
Haseler, S. 2008. *Meltdown*. London: Forumpress.
Heinberg, R. 2005. *The Party's Over: Oil, War, and the Fate of Industrial Society*. New York: Clearview Books.
Hofmeister, J. 2010. *Why We Hate the Oil Companies*. New York: Palgrave Macmillan.
Homer-Dixon, T. 2006. *The Upside of Down*. New York: Island Press.
Hulme, M. 2009. *Why We Disagree about Climate Change*. Cambridge: Cambridge University Press.
Ingersoll, R. 2006. *Sprawltown*. New York: Princeton University Press.
Klein, N. 2007. *The Shock Doctrine*. London: Penguin Allen Lane.
Kochan, N. 2006. *The Washing Machine*. London: Duckworth.
Kolbert, E. 2007. *Field Notes from a Catastrophe*. London: Bloomsbury.
Krugman, P. 2008. *The Return of Depression Economics*. Harmondsworth: Penguin.
Kunstler, J. 2006. *The Long Emergency: Surviving the Converging Catastrophes of the 21st Century*. London: Atlantic Books.

Labban, M. 2010. Oil in parallax: Scarcity, markets, and the financialization of accumulation, *Geoforum* 41: 541–52.
Linden, E. 2007. *Winds of Change. Climate, Weather, and the Destruction of Civilizations*. New York: Simon and Schuster.
Lovell, B. 2010. *Challenged by Carbon*. Cambridge: Cambridge University Press.
Lovelock, J. 2006. *The Revenge of Gaia*. London: Allen Lane.
———. 2010. *The Vanishing Face of Gaia*. London: Penguin.
Marx, K., and Engels, F. 1888 [1848]. *The Manifesto of the Communist Party*. Moscow: Foreign Languages.
Murray, J., and King, D. 2012. Climate policy: Oil's tipping point has passed. *Nature* 481: 433–35.
Orr, D. 2009. *Down to the Wire: Confronting Climate Collapse*. New York: Oxford University Press.
Owen, D. 2011. *Green Metropolis*. London: Penguin, 2011.
Parenti, C. 2011. 2011. *Tropic of Chaos*. New York: Nation Books.
Pearce, F. 2007. *With Speed and Violence*. Boston: Beacon Press.
Perrow, C. 2007. *The Next Catastrophe*. Princeton, NJ: Princeton University Press.
Pfeiffer, D. 2006. *Eating Fossil Fuels*. Gabriola Island, BC: New Society Publishers.
Rees, M. 2003. *Our Final Century*. London: Arrow Books.
Richard, R. 1997. *Cities for a Small Planet*. London: Faber and Faber.
Rogers R. 1997. *Cities for a small planet*. London: Faber and Faber.
Rubin, J. 2009. *Why Your World Is about to Get a Whole Lot Smaller*. London: Virgin.
Rutledge, I. 2005. *Addicted to Oil*. London: I. B. Tauris.
Sassen, S. 2009. Too big to save: The end of financial capitalism. *Open Democracy News Analysis*, January 1.
Shaxson, N. 2011. *Treasure Islands*. London: Bodley Head.
Singh, M. 2011. What's all the oil in the world worth? www.fleetstreetinvest.co.uk/oil/oil-outlook/oil-world-worth-00027.html, accessed June 23, 2011.
Smil, V. 2008. *Global Catastrophes and Trends*. Cambridge, MA: MIT Press.
———. 2010. *Energy Transitions*. Santa Barbara, CA: Praeger.
Soros, G. 2008. *The New Paradigm for Financial Markets*. London: Public Affairs Ltd.
Stiglitz, J. 2010. *Freefall*. London: Penguin.
Strahan, D. 2007. *The Last Oil Shock*. London: John Murray.
Tainter, J. 1988. *The Collapse of Complex Societies*. Cambridge: Cambridge University Press.
Taleb, N. 2007. *The Black Swan*. London: Penguin, 2007.
Tett G. 2009. *Fool's Gold*. London: Little, Brown.
Urry, J. 2003. *Global Complexity*. Cambridge: Polity.
———. 2011. *Climate Change and Society*. Cambridge: Polity.
———. 2013a. A low carbon economy and society. *Philosophical Transactions of the Royal Society A*. DOI: 10.1098/rsta.2011.0566.
———. 2013b. *Societies beyond Oil*. London: Zed.
———. 2014. *Offshoring*. Cambridge: Polity.
Ward, K. 2011. HSBC, http://investmentwatchblog.com/oil-will-be-gone-in-50-years-hsbc, accessed December 28, 2011.
Weber, M. 1939. *The Protestant Ethic and the Spirit of Capitalism*. London: Unwin.
Woodbridge, R. 2005. *The Next World War: Tribes, Cities, Nations, and Ecological Decline*. Toronto: Toronto University Press.
Yergin, D. 2011. *The Quest*. London: Allen Lane.
Žižek, S. 2008. World-renowned philosopher Slavoj Zizek on the Iraq War, the Bush presidency, the War on Terror, and more, *Democracy Now!*, http://www.democracynow.org/2008/5/12/world_renowned_philosopher_slavoj_zizek_on
———. 2011. *Living in the End Times*. London: Verso.

Chapter 5

Entangled Futures: Anthropology's Engagement with Global Change Research

Eduardo S. Brondizio

Let me begin with a provocation. The irony of a world experiencing accelerated and interconnected changes—social, climatic, and environmental—is that we are forced, as scholars, to confront the fact that our disciplinary approaches are often unable to follow the pace and complexity of the transformations we are addressing; our toolkits and narrative devices are often too limited to examine cross-scale processes, and our interpretations are often overridden by the pace of change in the very system/situation we aim to understand and, intentionally or not, inform policy changes. Further, our institutional structures and cultures are often too rigid and conservative to adapt to new realities. In other words, if we accept the premise that we live in a time of social and environmental acceleration and interconnectedness, where social-environmental processes create continuously emergent features and new conditions, we should also accept that, although they provide fundamental expertise, disciplinary approaches are limited to explain the mechanisms and directions of change that transcend their specific domains or level of analysis; recognizing these limitations and imagining new ways of working together is no longer an option but a necessity and an opportunity.

This dilemma is not anthropology's alone, of course, nor is it specific to the anthropology of climate change for that matter, but it speaks to a challenge that Edgar Morin summarized as that of engendering "complex thinking," that is, "a thinking that is capable of unifying concepts which repel one another and are otherwise catalogued and isolated in separate compartments" (Morin 2008: 81). Can we consider the challenges posed by climate and global change, and, more broadly, sustainability, as opportunities for anthropology to rethink and reenergize a broader disciplinary debate, that is, to unite fragmented specialties and theoretical models that, more often than not, have limited the discipline to "be seated at the table" and bring expertise in support of a meaningful voice to issues of broader societal interest?

In the Epilogue of the first edition of this book, Crate and Nuttall (2009) addressed these issues head-on by urging the anthropology community (and more broadly the social sciences) to step up their contribution to climate change research and policy processes. By recognizing the long-standing division of work that has put the physical sciences at the helm of research of climate change dynamics, modeling, and scenario building and the social sciences at the supporting role of portraying local phenomena, they forwarded a challenge that still stands: how to bring the rich contributions and lived experiences that anthropologists examine around the world to the core of climate change research, policy making, and media-public interfaces? In many ways, this challenge implies that taking "engaged anthropology" in earnest means framing questions and working on research problems that are much broader than specific disciplinary interests and subcultures.

Crate and Nuttall challenged the anthropology community by calling anthropologists to advance climate change models and conceptual frameworks coupling biophysical, social, political, and ecological dynamics and to improve the reliability of prognostic scenarios that inform narratives of climate change and policy discussions. They also urged anthropologists to "study up" climate change dynamics by engaging with larger scale national and international policy-making processes, the media and the construction of climate change narratives, and the role of consumers, environmental groups, and power brokers in developed countries. These topics still stand at the forefront of climate change research.

Anthropologists participating in climate and global change research and policy forums recognize the need to engage with research questions that resonate beyond anthropology but within which anthropology can also find its own methodological applications, areas of interest, and productive research agendas (Brondizio and Moran 2012; Ogden et al. 2013). This is not an easy challenge by any means but a needed reflection at a time when the discipline struggles both to find its place within academic institutions and to [re]gain a larger voice in society.

Anthropology's role is further compounded by the fact that the disciplinary division of labor that marked the development of climate change scholarship and international global environmental change research more broadly is rapidly changing. Atmospheric sciences, oceanographers, and ecologists will continue to study aerosols, ocean currents, and biogeochemical cycles as anthropologists will continue to study how local communities experience, respond, and adapt to environmental change, yet these and other specialized forms of expertise are pieces of larger interpretations, models, and scenarios of climate change, which at the core involve learning and sharing disciplinary expertise and language and developing new forms of knowledge needed to conceptualize and analyze human-environment interactions in its evolving space-time complexity. Whether one decides to engage or not, climate change and global change research more broadly have become inherently collaborative in nature and scope. This call for working beyond disciplinary lines can be seen as a challenge for anthropology, which often finds itself in self-defense of its disciplinary identity and/or specific conceptual and methodological approaches, instead of leading the formulation of new conceptual frames and realizing the interdisciplinary and holistic potential often touted as inherent to the discipline. The opportunity for provoking "creative tensions" around complex questions concerning climate and global change and sustainability, within and beyond the discipline, can help to revitalize anthropological engagement with larger research and policy debates; collaborative engagement in global change research is an opportunity to add value to anthropological expertise so needed to advance modeling, prognostic scenarios, vulnerability analysis, and conversations about development and societal change.

Underlying my arguments in this chapter is the assumption that climate change research should be framed within the broader scope of global change (inclusive of environmental, social, economic, and climate change) and sustainability.[1] As one considers the multiple dimensions, historical contingencies, and compounding stressors associated with social change and environmental transformation, a broader frame of reference is not only instrumental but also critical if one intends to move toward high-level integration of social theory and Earth system science.

Here I provide a brief overview of the evolution of international Global Environmental Change research programs (herein GEC) and introduce an emerging research agenda framed under the new Future Earth initiative aiming to link global climate and environmental change research, development, and sustainability agendas more broadly. Since the 1980s climate change and environmental change more broadly have mobilized an unprecedented global network of scholars, both under the auspices of international institutions and resulting from individual and institutional networks of research collaboration. The history of GEC research programs, from predominantly disciplinary to interdisciplinary communities, serves as a thread for anthropology to reflect on the challenges already outlined and on its own approaches to develop conceptual models and research questions contributing to the next decade of research. During the last decades of climate and global change research, anthropology's participation and contribution has been both influential and diluted (Crate 2011; Barnes et al. 2013). An emerging new phase of climate and global environmental change research poses exciting opportunities for anthropologists to engage with broader questions, to challenge theoretical and methodological boundaries, and to contribute to knowledge and practices of broader relevance to society.

A [Very] Brief Overview of the Evolution of International GEC Programs

By the beginning of 2014, a major transition process of international GEC research programs was in full motion under the auspices of a new international initiative called Future Earth.[2] Bringing together three of the four existing GEC programs—IGBP, DIVERSITAS, and IHDP[3]—Future Earth has been charged by a global alliance of international organizations and national funders (Science and Technology Alliance for Global Sustainability[4]) with charting and coordinating a new 10-year vision and agenda for global environmental change and sustainability research. From the onset, Future Earth represents a constellation of 30 Core Projects originally functioning under GEC programs and representing a research community estimated to involve 60,000 scientists from around the world.[5]

Understanding this emerging agenda of global change, climate change, and sustainability research, however, requires attention, at least briefly, to the evolution of international research programs during the last 30-plus years and the gradual process of integrating the social sciences as an equal player in earth systems research.

Since the early 1980s a series of international efforts and programs began to be developed as a way of responding to concerns about global-level effects of human activities on the environment.[6] On the one hand, these efforts were concerned with developing a framework to understand the fundamental functioning of the earth as a system, particularly from a physical standpoint. On the other hand, concerns about understanding the social drivers behind global environmental change existed from the beginning. In just over a decade, this process led to the development of four main research programs (WCRP, IGBP, DIVERSITAS, and IHDP) and, later, one integrated program (ESSP) (Figure 5.1).

From their inception, international GEC research programs emerged largely as a combination of a bottom-up self-organization of scientists, the vision of particular leaders, the efforts of national research agencies concerned with Earth observation, national science academies, and networks of international organizations.[7] The initial development of these programs was clearly along disciplinary lines.[8] The integration of the social sciences was relatively slow and gradual but nonetheless incremental. Mooney and colleagues (2013) argue that in spite of early efforts, the initial design of GEC programs such as the WCRP and IGBP intentionally focused on maintaining a physical sciences approach and research program with a clear aim of developing the foundations of what became established as Earth Systems Science. While the conceptual frameworks developed at the time recognized the human dimensions of the Earth's systems, the implementation of the program and its research agenda focused on the physical-biological dimensions. This focus soon sharpened to make the way clearer for the social sciences community to see a way of being involved, particularly those working on land-related issues.

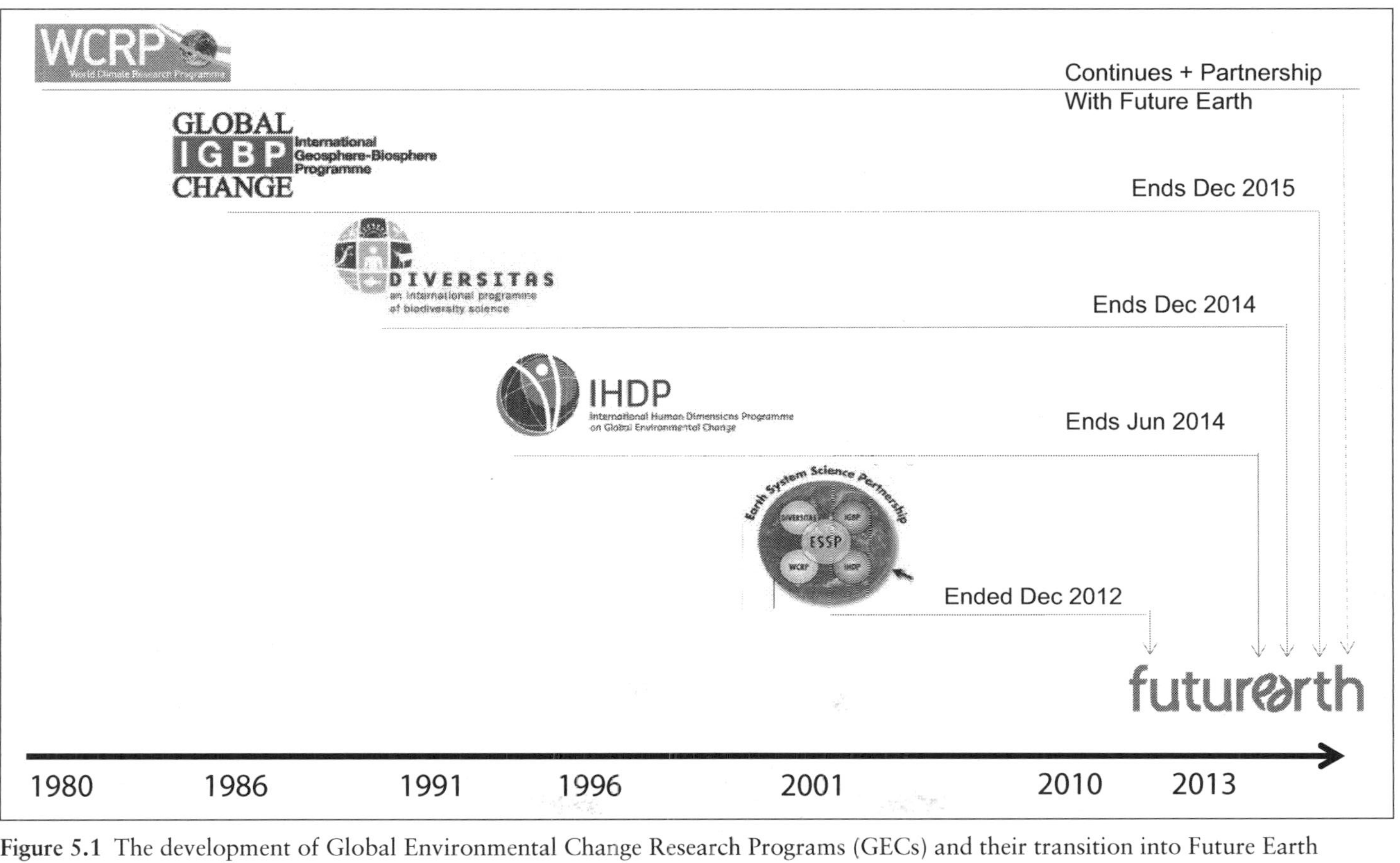

Figure 5.1 The development of Global Environmental Change Research Programs (GECs) and their transition into Future Earth

Several previous and parallel efforts contributed to the integration of the social sciences. During the late 1970s programs such as the World Climate Impact Program explicitly recognized the social-economic dimensions of climate research. In the United States, following the establishment of the U.S. Global Change Research Program, joint efforts by the National Research Council (NRC) and the Social Science Research Council (SSRC) led to the development of a foundational agenda that strongly influenced global environmental and climate change research nationally and internationally. NRC's Committee on the Human Dimensions of Global Change (HDGC) founded in 1989 contributed to the landmark volume *Global Environmental Change: Understanding the Human Dimensions* (Stern et al. 1992), which was instrumental in setting the research agenda for the next decade.

Internationally, the Human Dimensions Program initiative (HDP) of the International Social Science Council (ISSC) began a collaborative project with IGBP's Global Change and Terrestrial Ecosystems initiative, which led to the establishment of the Land Use and Cover Change (LUCC) initiative and a comprehensive research agenda to understand drivers, dynamics, modeling, and scenarios associated with land-change processes (Moran et al. 2004). The LUCC initiative and the ensuing decades of international research collaboration that it initiated are, arguably, the most important step toward the development of a social science presence and research agenda in global change programs. During the early 1990s the research agenda and conceptual framework proposed by LUCC provided an entry point for social scientists to "see" their intensive research cases contributing to a larger analysis of global change. Significant methodological development took place along the integration of social sciences data, geospatial technologies, modeling, and ecological analysis. The emphasis on drivers of change, underlying and proximate causes of change, and concerns with scaling up case studies, in spite of limitations, provided a bridge of dialogue with the physical sciences community examining and modeling scenarios of climate change. It is difficult to imagine today's advances in conceptual, methodological, and collaborative aspects of global change research without considering these early efforts.

By 1996 HDP had evolved into IHDP and was formally sponsored by a collaboration between ISSC, ICSU, and later UNU and was leading independently or in collaboration with IGBP and DIVERSITAS an expanded "human dimensions" agenda. Supported by a large international network of researchers and institutions from various specialties of the social sciences, several projects were put in motion in topics ranging from governance, urbanization, and industrial transformation to heath and coastal vulnerability.[9] These efforts had a direct influence on broadening the social science research agenda and in informing the scope of the agenda currently emerging under the Future Earth program.

At the turn of the century, a number of initiatives were bringing about a new decade of global change research. A key element was the Amsterdam Declaration of Earth Systems Sciences in 2001, which represented a major call for a new phase of global change research (IGBP 2001). Responding to the increasing complexity of global change and globalization processes, the declaration incited more integration of international programs, a focus on interdisciplinary projects, more international collaboration involving developing countries, and improved and shared data infrastructure. The Earth System Science Partnership (ESSP) emerged out of that process and aimed at providing bridges across the four existing international programs and paving the way to what became Future Earth (Ignaciuk et al. 2012).

Earth Systems Science developed as a consolidated field of research and Human Dimensions became an umbrella for a diverse social and environmental research community. These networks underpinned a number of international efforts including the IPCC, the Millennium Ecosystem Assessment (MA), and more recently the Intergovernmental Platform on Biodiversity and Ecosystem Services (IPBES).[10]

The launching of Future Earth has represented a major catalyst for the realignment of global change, climate change, and development agendas. Future Earth has been founded on a set of underlying principles aiming at challenging further the boundaries of international research programs and the modus operandi of scientific research on global change. The introduction of principles such as co-design and co-production, user-inspired and solution-oriented Earth system research created both anxiety and excitement in the broader research community. From the onset, Future Earth has aimed at integrating natural, economic, engineering, arts, humanities, and social sciences around questions that recognize the regional and global scale of the problems and that are oriented toward sustainable environmental solutions, climate change, and social inequality. These goals involve large-scale coordination of resources, data infrastructure, and forms of interaction with diverse stakeholder communities. The development of a basic science plan (Future Earth Initial Design Report) provided the foundations and initial framework for Future Earth (Future Earth 2013).

In 2011 the International Human dimensions program (IHDP) carried out on international online survey to assess the emerging priorities of the social sciences and humanities communities regarding global change research and global environmental assessments (Duraiappah and Rogers 2011).[11] Anthropologists, half of whom were working in environmental fields, represented around 13% of the participants. The first finding of the survey was that a significant segment of the participants were trained across social and natural sciences, and their activities tended to be primarily interdisciplinary research directly or indirectly linked to global environmental change.

Not surprisingly, the survey overwhelmingly showed that the social dimensions, while central, were still underrepresented in both global-change research and global assessments. Among the highest-priority research areas highlighted in the survey were: (1) equity/equality and wealth/resource distribution; (2) policy, political systems/governance, and political economy; (3) economic systems, economic costs, and incentives; and (4) globalization and social and cultural transitions. An overwhelming 80% of respondents called for funding programs that recognized the importance of these areas to advance understanding, models, and scenarios of environmental and climatic change. This community also recognized the need for scaling up social-sciences interpretations of global change, interpreted by the fact that 90% of the respondents supported (and were interested in participating in) global environmental change assessments that explicitly considered these issues.

However, an assessment of social sciences research capacity and publications from 1990 to 2011 by the International Social Sciences Council (ISSC) and UNESCO shows significant regional and international differences in the participation and publication of social science research related to climate change (see, for instance, Caillods 2013). While the analysis shows slow but progressive changes, the representation of the social sciences in global change research is heavily based on research published in the United States and Western Europe, where familiarity and support for interdisciplinary research is relatively consolidated. This analysis shows disciplinary culture, limited training capacity, and structural barriers (for example, funding) restricting broader participation of social sciences within and particularly outside North America and Western Europe.

In sum, in just two decades a diverse social sciences community has established a firm, albeit regionally uneven, position within global environmental and climate change research. This participation, however, has been far from inclusive across the diversity of social sciences and the humanities. The research agenda emerging as part of Future Earth intends to extend the scope of this research program and thus to open opportunities for expanding the contribution of anthropologists in the coming decade.

An Emerging Research Agenda for 2015–2025, the Future Earth Initiative

The research agenda proposed by Future Earth captures well the evolution of global environmental change research and the gradual role that the social sciences and humanities play and will play during the next decade.[12] As the range and scope of research questions have expanded, the opportunities for anthropology in particular have become significant.

The proposed research agenda provides stepping stones to an ambitious vision aiming at supporting the global sustainability challenges expressed in such international agreements as The Future We Want[13] and the post-2015 Sustainable Development Goals (SDG).[14] However, it is not my intention in this chapter to present this research agenda in detail; rather it is to outline the overall framework setting the scope of global change and sustainability research in the coming decade[15] and potential areas of engagement for anthropology.

Setting an Overarching Framework for Global Change and Sustainability

One can consider four overarching and interlinked components setting the scope for Future Earth's 10-year research agenda: research priorities, science-policy interface, science practice, and capacity mobilization.[16] The first component of this vision focuses on orienting international global change research that prioritizes global sustainability challenges, inclusive of addressing social inequalities. This vision includes the need to safeguard fundamental ecosystem functions and freshwater and marine systems and to provide fair access to water, energy, and food, the deep decarbonization of socio-economic systems, the transformation of urban environments to increase resilience of vulnerable areas and sectors of society to climate change, and the promotion of sustainable rural futures. The second component focuses on developing open access and inclusive observation and data platforms able to assess and inform trends and thresholds in social-environmental systems at different scales; supporting the development of new metrics for assessing development and economic growth; and developing a new generation of coupled earth system models and scenarios that can contribute to public and policy engagements and planning. The third element focuses on the practice of global change research and sustainability sciences. Future Earth puts forward an ambitious call for new ways of integrating fundamental and applied research and developing mechanisms for co-design and co-production that take into account different priorities and knowledge systems. The fourth element of this vision focuses on expanding and mobilizing research capacity and promoting new networks of scientists, scholars, policy makers, private sectors, and civil society leaders to cooperate on research and practice.

This vision rests on the ability of the research, policy, and practicing communities to align priorities and mobilize resources and training and to promote necessary institutional changes. It also rests on the support of national funding agencies for interdisciplinary, societal-driven research involving various types of partnerships and dependent on international coordination of research efforts and infrastructure. In principle, these challenges come from and align with the strategic vision of major national and international funding agencies (see for instance, NSF strategic plan 2014–2018; NSF RCN Social Observatories Coordination Network; EU Horizons 2020; UK NERC; and the Belmont Forum, among others[17]). The extent to which this research agenda will affect disciplinary funding remains to be seen, but it will certainly open opportunities for international collaborative projects and development of new research networks.

Following the initial design of Future Earth, the research agenda is organized around three main themes: Dynamic Planet (focusing on physical, ecological, and social processes underpinning regional and global environmental change), Global Sustainable Development (focusing on production-consumption-resource nexus, indicators of development, environmental change, and inequality), and Transformation toward Sustainability (focusing on barriers and opportunities for social change and sustainability transitions).

The first theme, Dynamic Planet, includes a series of questions related to observation and the understanding of change, risks, and thresholds and also projecting scenarios on various dimensions of social and environmental change. The research agenda proposes questions that bridge fundamental gaps in earth system research (from atmospheric to land to ocean research) to social processes contributing and affected by these changes. Three subthemes are proposed to capture issues of observing and attributing change; understanding processes, interactions, risks, and thresholds; and exploring and predicting futures.

The second theme, Global Sustainable Development, focuses on questions about the interrelationship and feedbacks between economic growth, social inequality, and environmental change. The emphasis on nexus and trade-offs of resources, energy, water, and food systems also opens the way for questions about consumption patterns, lifestyles, and narratives of development. Significant importance is given to the transformation of urban areas and the opportunities and historical contingencies of governance arrangements. Three subthemes are proposed to address issues of meeting basic needs and overcoming inequality, governing sustainable development, and managing growth, synergies, and trade-offs.

The third theme, Transition toward Sustainability, focuses on understanding and evaluating transformations in social systems, evaluating behavioral dimensions of transformation, and building development pathways to overcome historical and structural limitations for regional and global sustainability. This theme gives attention to the role of cultural values, beliefs, worldviews, and ethics associated with behavior change toward the environment. Explicit attention is given to political economy, development assumptions and narratives, trade-offs of technological changes, and opportunities created by network and communication technologies. Three subthemes are proposed to address issues of understanding and evaluating transformations, identifying and promoting sustainable behaviors, and transforming development pathways.

Potential for Anthropological Engagement

The framework described here provides structure for a rich set of questions and interfaces for anthropologists and social scientists in general to engage with and to chart research agendas of specific disciplinary interests.

As the goal is not to introduce these questions and agenda in detail, I provide a brief illustration of overarching themes and questions of potential interest to anthropologists that include social-environmental complexity; the well-being, inequality, and environmental change nexus; urban-rural systems and social interactions; perceptions, values, and narratives; and integrating knowledge systems, data and modeling.

The complex nature of interactions and feedbacks of global-change problems requires interdisciplinary collaborations involving the development of a new generation of complex systems theories, conceptual frameworks, data articulation, and modeling tools able to couple and integrate social data and environmental change processes. Complex systems perspectives are needed for the understanding of long-term trends in Earth-systems change resulting from interactions and feedbacks between social processes, land use, atmospheric emissions, biodiversity, and indicators of global change. There is explicit recognition for the role of archaeological, anthropological, and historical approaches that are sensitive to temporal and spatial scales and to historical dimensions of these interactions. Although interest in complexity in social-environmental systems dates to the early history of anthropology, this interest has been diffused in the discipline in spite of the nature of research themes commonly examined by anthropologists, from the intersection of globalization, migration, and environmental change to complex articulations between local identities and global agendas of conservation (Ogden et al. 2013). These issues open the opportunity for advancing conceptual frameworks and analytical tools able to bring complex system perspectives to the interpretation of social and political economic processes.

Significant attention is given to questions focusing on the nexus well-being, inequality, and environmental change. These questions call for new approaches to understand linkages between health, economic vulnerability, and changing ecosystems under various types of climate and environmental change scenarios. They explore how inequality affects the ability of households, communities, and states to mitigate and adapt to climate change. In addition, what types of opportunities and barriers exist for societies at different levels to chart pathways toward sustainable development? Another important area of research in global change relates to emergent patterns of disease distribution resulting from changes in ecosystems structure and function, land use, and circulation and migration patterns. Anthropologists working on these issues at local and regional level can bring different conceptual frames to link these processes to national and global political economies.

Few topics relate as closely to anthropological research on human-environment interaction as that of urban-rural interconnections and transformations. Increasingly, the focus is on the understanding the transformation of urban and rural landscapes as interconnected to migration, governance systems, and global political economy of resource circulation. Key areas of research include understanding emerging trends in urbanization and efforts to localizing rural-urban economies for sustainability of food systems. These processes vary significantly at a global scale, requiring understanding of how social change and patterns of migration and global chains of resource "teleconnect" urban and rural areas across the global with diverse implications for social and environmental vulnerability. Anthropologists are well positioned to contribute to understanding intersections and consequences of expanding frontiers of commodity production, extractive industries, resource concessions, and social-environmental change and the consequences for global change. These issues call for research on the evolution of governance and property regimes in complex rural-urban landscapes, including the articulation of protected areas, indigenous and local population territories, and agro-industrial operations. These issues overlap with anthropological interest on how local populations, networks of circulation and migration, contribute and interact with urbanization and rural transformation. Furthermore, questions of interest to anthropologists include those focusing on understanding how changes in intergenerational expectations and visions of "modernity" and well-being of urban and rural youth affect mobility, employment, and consumption patterns—and in turn how sustainability is understood and pursued.

The relationship of perceptions, values, narratives, and behavior has become central to different areas of research on global change and sustainability. The range of themes and questions should be attractive to a wide spectrum of anthropologists over questions exploring potential conceptual approaches to understand the coevolution of cultural values and beliefs, worldviews, and narratives, as well as ethics influencing people's attitudes, life choices, and behaviors toward sustainability in different contexts. This line of research includes understanding the role of the media, in its diverse manifestation, in framing behavioral changes related to sustainability and to the way narratives about urban life and identity are associated with social transformations in urban areas.

Another important area of research relates to understanding the implications of new ways of valuing nature and streamlining environmental accounting as a form of resource governance and development strategy. Anthropologists are uniquely positioned to contribute to the understanding of how different views of nature and ethics of conservation influence discussions regarding biodiversity and ecosystem services valuation. More generally, what are the implications of new ways of valuing nature for our relationship with the environment, and how is the valuation of nature affecting local communities and, in turn, broader narratives of development (Brondizio et al. 2010)? The latter part of the question relates to long-term interest in the role of social movements in shaping narratives and policies related to global environmental and climate change in different regions.

Underlying these different research frontiers is the crosscutting task of integrating knowledge systems, data, and modeling. Significant efforts are called for to develop coupled models and scenario-building tools that are able to capture complex and nonlinear social-environmental change at various scales. These efforts require new methods and technologies to access, integrate, organize, manipulate, and summarize, openly and transparently, the large, interdisciplinary datasets covering different forms of knowledge and data. Significant advances have been made in these regards, particularly at the local and regional levels, but novel approaches are needed to improve reliability and scaling of analytical approaches, models, and scenarios that can support and facilitate assessments and policy discussions at different levels.

In advancing on these fronts, the challenge is not only to break further barriers between the natural and social sciences but also between scientific and indigenous knowledge, qualitative and quantitative approaches. In fact, increasingly at the core of global change research agendas is recognition that advancing understanding and policies to mitigate environmental and climate change requires the inclusion of multiple knowledge systems (Tengö et al. 2014). Knowledge and tools to bring scientific and other knowledge systems to collaborate on climate change and environmental assessment are still insipient and limited. Anthropologists can find a particularly productive role in bridging the expectations of the scientific community and of representatives of local and indigenous networks to collaborate at regional and local levels on global change research.

Concluding Remarks

To return to my initial point: Complex global change problems pose challenges and opportunities that ultimately rest on the integration of social and physical sciences, the humanities, and other knowledge systems. Global change research is changing rapidly in scope and nature, demanding better epistemic fit between questions, data, theory, and analytical methods. Current and emerging themes in global change research require anthropologists to collaborate on advancing conceptual frameworks and analytical tools to address complexity and connectivity in social and ecological systems (Brondizio et al. 2009) and also to provide a critical voice in these debates. Global change research agendas and funding programs presuppose cross-disciplinary collaboration, coproduction of knowledge and stakeholder engagement, integrative cross-level methodologies, and familiarity with multiple disciplinary languages. Anthropology's traditional approaches, terminology, and strengthens face opportunities and challenges in the face of new questions, paradigms, technologies, and multidimensional datasets. These issues should be pushing the discipline to rethink how we train and prepare students to work in partnership and to contribute to the understanding of a rapidly changing world and ever more interdisciplinary funding landscape.

Interdisciplinary and increasingly transdisciplinary thinking are the modus operandi of global change research, requiring disciplines to collaborate on developing joint questions, conceptual frameworks and methodologies that contribute to a larger picture while motivating and advancing the equally important underlying specialized knowledge. Bridging the divides between and within natural and social sciences and humanities in achieving these goals also requires moving beyond the myriad of dichotomies that have structured paradigms and methodologies in the social sciences: culture-nature, structure-agency, materialism-idealism, rational-moral, and local-global, to cite just a few. Many of the internal debates in anthropology since the 1990s have, unfortunately, helped to reinforce such dichotomies often, gaining more visibility than have more relevant efforts to promote new ways of thinking about the interactions and interdependencies representative of social realities. Therein, in global change research, lies an opportunity and not a threat: to promote interdisciplinary and transdisciplinary collaborations and ways of unifying concepts and approaches that otherwise are catalogued in different compartments.

Anthropologists have been striving, with variable success, to reconcile fundamental differences in epistemology, analytical tools, and conceptual frameworks of human-environmental interaction and change. Since the 1990s, environmental anthropology has served as the broader interdisciplinary umbrella within which anthropologists working on various facets of human-environment interactions could find relative balance in reconciling perspectives within the discipline without resorting to deterministic approaches. It has brought together a plurality of specialized fields that have branched out of post-1950s cultural ecology, representing new paradigms and favoring particular variables: ecological anthropology, political, historical, and symbolic ecology, among several others. Ironically, as in Morin's reference at the beginning of this chapter, we have advanced in understanding the pieces of the puzzle—the social, the ecological, the political, the historical, the cognitive, the symbolic—but these pieces are still partially segregated in different compartments. As a consequence, environmental anthropology is now confronted with the interdisciplinary challenge of integrating the various intradisciplinary specialties into a new synthesis able to provide more integrative approaches to the growing and accelerated complexity of local and global societies and environments—in other words, to continue to be inclusive, reflexive, and committed to people and places while at the same time reuniting this rich array of specialties into more cohesive and holistic frameworks able to position anthropology within the broader discussions of global change (Brondizio, Fiorini, and Adams 2016).

Seeing the challenges posed by emerging global change research agendas as creative tensions offers opportunities to invigorate disciplinary scholarship and contributions to societal debates. The brief overview of emerging research areas is but a piece of a vast research landscape on global change and sustainability. As noted by Barnes and colleagues (2013), anthropology brings ethnographic insights, historical perspective, and a holistic approach to questions of climate and global change more broadly. It is time to take these contributions a step farther and to respond in earnest to the discipline's own call for engagement by opening disciplinary boundaries and taking the lead among the social sciences in the coming decade of global change and sustainability research.

Acknowledgments

For their support, I would like to thank the Department of Anthropology, the Anthropological Center for Training and Research on Global Environmental Change (ACT), the Vincent and Elinor Ostrom Workshop in Political Theory and Policy Analysis at Indiana University, and the Future Earth program. Some of the material presented here was produced as part of the United Nations University project on catalyzing action toward sustainability of deltaic systems with an integrated modeling framework for risk assessment (DELTAS), funded by the National Science Foundation (NSF award No. 1342898) under the Belmont Forum program. I appreciate the invitation to contribute to this volume, as well as the comments and editing suggestions provided by the editors.

Notes

1. I use the acronym GEC to refer to international Global Environmental Change programs such as IGBP, WCRP, IHDP, and DIVERSITAS.
2. The process of implementing Future Earth has involved a series of stages and consultation processes. A transition team was charged in 2011 with consulting a broad constituency and developing an implementation plan. A major step in this process was undertaken by the joint organization by the GEC programs of a major international conference in 2011 entitled Planet Under Pressure (Stafford-Smith et al. 2012). Future Earth was formally launched during the Rio+20 conference in June of 2012. This event was followed with the establishment of an interim secretariat in 2013 (housed at ICSU-Paris), the nomination of a Science Committee, and an interim Engagement Committee. From 2013 to its full implementation in 2015 Future Earth has been functioning in parallel coordination with the GEC programs (see Figure 5.1). By mid-2014, an international bid process led to the selection of a permanent secretariat involving five global hubs (Sweden, France, Japan, United States, Canada) and a series of regional hubs (under discussion as of October 2014). The permanent globally distributed secretariat is set to be in full function by the beginning of 2015. For more information, see www.futureearth.org.
3. The four Global Environmental Change international programs include the World Climate Research Program (WCRP), the International Geosphere Biosphere Program (IGBP), the International Program of Biodiversity Sciences (DIVERSITAS), and the International Human Dimensions Program (IHDP)—and, previously, the Earth Systems Sciences Partnership (ESSP, discontinued in 2012). WCRP continues to function independently but in collaboration with Future Earth.
4. The Science and Technology Alliance for Global Sustainability ["the alliance"] is comprises the Belmont Forum and International Group of Funding Agencies for Global Change Research (IGFA), the International Council for science (ICSU), the International Social Science Council (ISSC), the United Nations Environmental

Programme (UNEP), the United Nations Educational, Scientific and Cultural Organization (UNESCO), the United Nations University (UNU), and the World Meteorological Organization (WMO, observer).

5. It is beyond the scope of this chapter to discuss the history of GEC programs in detail; there are several sources (as cited here) that describe their histories in detail.
6. See Mooney and colleagues (2013) and IGBP History of global environmental change research at www.igbp.net/about/history.4.1b8ae20512db692f2a680001291.html for detailed histories of these programs.
7. Including, for instance, the World Meteorological Organization, the International Council for Sciences (ICSU), the International Social Sciences Council, Scientific Committee on Problems of the Environment (SCOPE), the Intergovernmental Oceanographic Commission of UNESCO (IOC), and, more broadly, UN organizations such as UNESCO, UNEP, and the United Nations University (UNU).
8. During these early stages GEC programs were also promoting interdisciplinary research but within broader physical sciences discipline umbrellas—for instance, in fields such as biogeochemistry, ocean-atmosphere interactions, land-ocean interface, among others.
9. IHDP Core Projects included Global Environmental Change and Human Security (GECHS), Urbanization and Global Environmental Change Project (UGEC), Industrial Transformation (IHDP-IT), Earth System Governance Project (ESG), Land-Ocean Interactions in the Coastal Zone (LOICZ), Global Land Project (GLP), Global Carbon Project (GCP), Global Water Systems Projects (GWSP), Global Environmental Change and Human Health (GECHH), and Global Environmental Change and Food Systems (GECAFS).
10. For different disciplinary and regional perspectives and interfaces between the social sciences and global environmental change, see parts 2 and 7 of ISSC and UNESCO 2013.
11. Using the research network of IHDP and ISSC (reaching an estimated 6,000 researchers), the survey received 1,276 responses from 103 countries. The most represented regions included Western and Central Europe (32.5%), sub-Saharan Africa (17.2%), and the United States and Canada (16.2%).
12. This research agenda evolved out of a series of international consultation and workshops between 2013 and 2014, including working meetings with all GECs programs and Core Projects.
13. www.un.org/en/sustainablefuture
14. http://sustainabledevelopment.un.org/index.php?menu=1565
15. See www.futureearth.org for the Strategic Research Agenda 2014.
16. This is my own categorization of the Future Earth research agenda and framework.
17. See, for instance, NSF: www.nsf.gov/about/performance/strategic_plan.jsp; NERC: www.nerc.ac.uk/latest/publications/strategycorporate/strategy/EU; Horizons 2020: http://ec.europa.eu/programmes/horizon2020/; Belmont Forum: https://igfagcr.org; NSF Social Observatories: www.socialobservatories.org.

References

Barnes, J., Dove, M., Lahsen, M., Mathews, A., McElwee, P., McIntosh, R., Moore, F., O'Reilly, J., Orlove, B., Puri, R., Weiss, H., and K. Yager. 2013. Contribution of anthropology to the study of climate change. *Nature Climate Change* 3: 541–44.

Brondizio, E. S., Fiorini, S., and Adams, R. 2016. History and scope of environmental anthropology, in H. Kopnina and E. Shoreman-Ouimet, *Routledge Handbook in Environmental Anthropology*. New York: Routledge.

Brondizio, E. S., Gatzweiler, F., Zagrafos, C., and Kumar, M. 2010. Socio-cultural context of ecosystem and biodiversity valuation, Chapter 4 in P. Kumar (Ed.), *The*

Economics of Ecosystems and Biodiversity (TEEB). United Nations Environmental Programme and the European Commission. London: Earthscan Press, pp. 150–81.

Brondizio, E. S., and Moran, E. F. (Eds.). 2012. *Human-Environment Interactions: Current and Future Directions*. Dordrecht: Springer Scientific Publishers, 17 chs., 434 pp.

Brondizio, E. S., Ostrom, E., and Young, O. 2009. Connectivity and the governance of multilevel socio-ecological systems: The role of social capital. *Annual Review of Environment and Resources* 34: 253–78.

Caillods, F. 2013. Regional divides in global environmental change research capacity. In *ISSC and UNESCO World Social Science Report 2013: Changing Global Environments*. Paris: UNESCO, pp. 125–32.

Crate, S. 2011. Climate and culture: Anthropology in the era of contemporary climate change. *Annual Review of Anthropology* 40: 175–94.

Crate, S., and Nuttall, M. (Eds.). 2009. *Anthropology and Climate change: From Encounters to Action*. Walnut Creek, CA: Left Coast Press, Inc. (epilogue).

Duraiappah, A. K., and Rogers, D. 2011. *Survey of Social Sciences and Humanities Scholars on Engagement in Global Environmental Change Research*. International Human Dimensions Programme on Global Environmental Change (IHDP), United Nations Education, Science and Culture Organization (UNESCO), International Social Sciences Council (ISSC).

Future Earth. 2013. *Future Earth Initial Design: Report of the Transition Team*. Paris: International Council for Science (ICSU).

IGBP. 2001. *The 2001 Amsterdam Declaration on Earth System Sciences*, www.igbp.net/about/history/2001amsterdamdeclarationonearthsystemscience.4.1b8ae20512db692f2a680001312.html.

Ignaciuk, A., Rice, M., Bogardi, J., Canadell, J. G., Dhakal, S., Ingram, J., Leemans, R., and Rosenberg, M. 2012. Responding to complex societal challenges: A decade of Earth System Science Partnership (ESSP) interdisciplinary research. *Current Opinion in Environmental Sustainability* 4(1): 147–58.

ISSC and UNESCO. 2013. *World Social Science Report 2013: Changing Global Environments*. Paris: UNESCO.

Mooney, H., Duraiappahb, A., and Larigauderiec, A. 2013. Evolution of natural and social science interactions in global change research programs. *PNAS* 110 (Supplement 1): 3653–56.

Moran, E. F., Skole, D. L., and Turner, B. L., II. 2004. The development of the international Land Use and Land Cover Change (LUCC) Research Program and its links to NASA's Land Cover and Land Use Change (LCLUC) Initiative, in G. Gutman et al., *Land Change Science: Observing, Monitoring, and Understanding Trajectories of Change on the Earth's Surface*. Kluwer Academic Publishers, pp. 1–16.

Morin, E. 2008. *On Complexity*. New York: Hampton Press, 127 pp.

Ogden, L., Heynen, N., Oslender, U., West, P., Kassam, K., and Robbins, P. 2013. Global assemblages, resilience, and Earth stewardship in the Anthropocene. *Frontiers in Ecology and the Environment* 7(11): 341–47.

Stafford-Smith, M., Gaffney, O., Brito, L., Ostrom, E., and Seitzinger, S. 2012. Interconnected risks and solutions for a planet under pressure: Overview and introduction. *Current Opinion in Environmental Sustainability* 4(1): 3–6.

Stern, P., et al. 1991 *Global Change: Understanding the Human Dimensions*. NRC, National Academy of Sciences.

Tengö, M, Brondizio, E. S., Malmer, P., Elmqvist, T., and Spierenburg, M. 2014. *A Multiple Evidence Base Approach to Connecting Diverse Knowledge Systems for Ecosystem Governance, AMBIO*. DOI: 10.1007/s13280-014-0501-3.

Part 2: Assessing Encounters Old and New

Chapter 6

Gone with Cows and Kin? Climate, Globalization, and Youth Alienation in Siberia

Susan A. Crate

> "The transition from a communist infrastructure to a market economy presents a great challenge to indigenous agropastoralists of the former Soviet Union" . . . the demise of the Soviet era, along with its imposed economic and subsistence infrastructures, presents a unique opportunity to understand what form households took in the pre-collective and collective era and what form they take today. (Susan Crate 2003: adapted from p. 499)

Such testimony supported research in 1999–2000 that investigated how rural Viliui Sakha—native horse and cattle breeders of northeastern Siberia, Russia—had adapted to the 1991 fall of the Soviet Union on a household food-production level. The main adaptation to the rapid change was through reviving forms of pre-Soviet household-level food production, based in cattle breeding and inter-depending with kin households (Crate 2003, 2006a). I termed the system *cows and kin* and found that it fit Robert Netting's smallholder-householder theory (1993), in which he argued that, in times of rapid change, the household system is the most resilient unit for subsistence production, having both integrity and longevity owing to its specific qualities of intimate ecological knowledge and implicit labor contracts.

Then, a decade later,[1] everything had changed once again, and inhabitants spoke about it freely:

> We used to get most of our household food from our animals, garden and nature. Now we get most of it from the store. Back then everything was deficit in the store—I remember getting in line the night before for some deficit item—a women's dress or other—now the stores are full and we are having a deficit of cow products! That's a big change, if you ask me. (Viliui Sakha householder, summer 2012)[2]

The change is visually stark. Cows are almost absent now from the village streets and yards, and from householders' daily lives and routines. At the same time, within this context of change, the critical role of the kin-based household and interhousehold system remains. In 1999–2000 these kin systems focused on labor reciprocity and revival of knowledge to realize home food production. Today they serve primarily to negotiate and secure economic needs, including job and market relationships, and to support the recent trend of youth leaving the rural areas to realize opportunity in the regional urban centers. Again, the 1999 narrative highlights the change:

> For the majority of contemporary Viliui Sakha households, kin, like cows, play a major role in subsistence. Almost all of the households surveyed said they have close kin (parents, grand-parents siblings, cousins, aunts, and uncles) in their immediate village. Over half of these are sibling or parental relationships. (Crate 2003: 513)

In 2012 the same paragraph would read:

> For the majority of contemporary Viliui Sakha, kin continues to play a major role in household economic success. Almost all of the households surveyed said they have close kin (parents, grand-parents siblings, cousins, aunts, and uncles) in their immediate village. Over half of these are sibling or parental relationships. Although no longer universally essential[3] to home food production, as was the case a decade ago, kin remains critical for employment, housing, market relationships, and other opportunity.

Our newer research protocol of investigating the local effects of and responses to climate change, initiated in 2006 and continued through 2013, showed how Viliui Sakha are adapting to a *complexity of change*, including the local effects not only of global climate change but also of the economic forces of globalization and of the demographic change affecting them as youth increasingly leave the rural areas. This chapter shows how ethnographers working in communities over a longer-term period of time can acquire important insights into change in its many forms, using methods that document attachment to place and local diversity, both cultural and biological. Furthermore, this powerful and unique contribution to how ethnographers understand change in contemporary contexts has a potentially critical role in interdisciplinary efforts to account for attributes of global change, both human and local.

Background

This chapter represents a "coming full circle" for me as an anthropologist engaged in research on climate change. Since 2006 I have focused my research on how Viliui Sakha communities perceive, understand, and respond to the local effects of global climate change. These enquiries within my own case research context led me to ask similar questions about peoples in other parts of the world and also about our role(s) as anthropologists in engaging with climate change-related issues (Crate and Nuttall 2009). In 2012, after several years of observing dwindling cow herds in my research villages, I began asking directed questions about household-level cow keeping. While working with our long-time village assistant, Katya, who was herself a cow keeper and producer of all Sakha foods, I remember clearly the second day, during our 13th interview. We were working with a 74-year-old household head and the first householder that year who said that his household still had cows. Katya and I instantly caught each other's eyes and exclaimed: "You still have cows?" at which point he added that he did but planned to slaughter them all in the fall. "I am the only one who cares for them, and every year I am more and more feeble and now my eyesight has gone."

Such a pattern of either having recently stopped, or about to stop, continued through the interviews, and, in the end, we found that only 9 out of 38 households in Elgeeii and 7 out of 16 households in Kutana (24% and 44%, respectively) continued cow keeping. By contrast, in 1999–2000, 124 out of 210 in Elgeeii and 59 out of 79 in Kutana (59% and 75%, respectively) were cow keepers. This means that in 12 years there has been a 35% and 31% drop in cow keeping, respectively.[4] If we factor in those about to slaughter their herd, there has been a 49% drop in numbers in Elgeeii and a 54% drop in Kutana since 1999–2000. Further, many without cows in 2012 had stopped in the previous last few years.[5]

Village size and access to hay lands explain some of the contrast between the villages. Overall, the smaller the village population, the more hay land per household and the more cow keeping, whereas the larger the village, the more paid jobs and the fewer cow-keeping households. This situation could explain why now there is less of a drop in cow-keeping households in Kutana (pop. 1,000) than in Elgeeii (pop. 3,000). But this logic fails for a third village (Table 6.1). Khoro, with a population of slightly over 200 and an overabundance of hay land per household, has—like Elgeeii—seen a precipitous drop in cow numbers since 2004. In part this drop can be explained by environmental factors, if we consider that most of Khoro hay land has

Table 6.1 Elgeeii, Kutana, and Khoro Cattle Numbers 2002–2011

	Elgeeii		Kutana		Khoro	
	All Cattle	*Milk Cows*	*All Cattle*	*Milk Cows*	*All Cattle*	*Milk Cows*
2002	1,886	660	——	——	——	——
2003	1,926	669	——	——		
2004	1,865	664	1,076	362	——	——
2005	1,742	682	——	——		
2006	1,743	643	754	275		
2007	——	——	823	292		
2008	1,560	584	773	282		
2009	——	——	868	290	337	107
2010	1,577	457	——	——	269	80
2011	1,511	470	978	330	247	74
Difference	-375	-190	-98	-32	-90	-33
% Decline	-20%	-29%	-9%	-9%	-28%	-31%

been flooded continually in this period. However, that Kutana has also lost hayland because of floods yet manages to retain cow keeping suggests that environmental factors are only part of the story. Statistics from village-level and regional records show a drop in overall numbers (Table 6.1).

The percentage of households continuing to keep cows and the percentage drop in number show that Kutana has managed to maintain higher numbers overall. Table 6.2 shows how demographics can explain part of the story. Statistics for the Suntar region show an overall greater drop in cow numbers compared to population (Table 6.3). Interviews and focus groups enabled us to tease out these factors.

Findings

Our analysis of 2012 field data discerned that the pivotal influences affecting cow keeping appear as environmental, demographic/intergenerational, and economic. I discuss each briefly in the following sections.

Environmental

Many households stopped keeping cows as a direct result of their hay land going underwater. For example, in Elgeeii 29 households out of 31, and in Kutana 10 out of 16 households, said their hay land was underwater. In all villages householders have the option to access village supplemental hay land if theirs is not productive. However, in Elgeeii, that land is across the river, and most people have no boat or other means to access it, whereas in Kutana it is too far away to travel to. In response, the government has made some efforts to drain the water from the fields, since the flooding affects overall agricultural production (Figure 6.1). However, draining these fields,

Table 6.2 Elgeeii, Kutana, and Khoro Population 2002–2011

	Elgeeii	**Kutana**	**Khoro**
	Population	*Population*	*Population*
2002	2,224	—	—
2004	2,234	976	—
2009	—	—	302
2011	2,022	1,011	272
Difference	-202	+35	-30

Table 6.3 Suntar region cattle numbers compared to population

	Suntar Region		
	All Cattle	*Milk Cows*	*Population*
1990	27,500	10,500	26.7
1999	17,260	6,720	26.0
2004	**21,956**	**8,146**	**25.7**
2005	20,571	8,056	25.6
2007	19,188	7,401	25.6
2008	17,873	6,973	25.4
2009	18,100	7,078	25.3
2010	**17,826**	**6,571**	**25.1**
% Change since 2004	-19%	-19%	-2%

Figure 6.1 Extensive government effort to drain hayfields

considering how sub-Arctic surface waters—including numerous lakes and several rivers—share water in the permafrost system and how there is little if any geographical relief, is a complex engineering problem. Most of these efforts have failed, and many residents report the draining has made matters worse. For example, one household with land in *Uulaakh* ("watery" or "full of water") reported how they now get less hay than when the land was flooded, because it is too dry to grow anything. But water on the land is just one of a set of environmental factors, mostly due to the local effects of climate change, challenging cow keeping (Figure 6.2).

Other factors include warming winters, cooling summers, lots of rain and, at the wrong times, lots of snow, more floods, sudden temperature changes, fewer birds and animals, and changing seasonality (Crate 2008, 2011). Of these factors, the last makes cow keeping particularly difficult, if we consider how Viliui Sakha have adapted to a confined seasonal regime in the Arctic. Because of such change, cow keepers must adapt to an elongated fall season characterized by a freezing/thawing pattern; an unusual winter season, with more snow than previously and *chiskhaan* ("freezing winds"); a late spring or spring arriving on time but staying cold, not gradually warming up as in the past and with less rain; and a cool summer with too much rain during hay season and cold nights. I next detail how each situation affects horse and cattle breeding.

An elongated fall delays the onset of winter, which means cows can go to pasture later, thereby prolonging hay stores, but it also delays the annual

Figure 6.2 Hayfields are increasingly flooded, owing to the interacting local effects of climate change

slaughter, performed once temperatures remain below freezing, requiring more hay to keep animals alive. Horse keepers are also negatively affected, since the elongated fall's freeze/thaw pattern creates an impervious ice cover that prevents horses from accessing fodder.

The winter season is increasingly characterized by almost continual snowfall and *chiskhaan*. Typically winters are snow and wind free, owing to the extreme continental climate. However, with it now snowing throughout winter and with the constant wind, cow care requires much more human labor, keeping paths cleared and attending to the barn, since cows cannot go out as much in these conditions. Horses, which normally roam freely and graze for fodder under the snow, are now prone to starvation, because of the issue of ice and accessibility, already mentioned, and the unmanageable snow depths. Then, come spring, the high snow levels melt into floods and standing water on the land, affecting hayfields.

Spring is either late or arrives "on time," according to inhabitants' lifelong adaptation to the annual seasonal changes of their environment, but then it alternates between cold and warm instead of gradually warming as before. There is also less spring rain. Many householders described spring 2012 as overly prolonged with an extended cold snap, which prevented grass from growing and cows from going to pasture, thereby depleting most householders' hay stores. In response, people either slaughtered their herds or bought hay from adjacent regions, which sold at a premium that many could not afford.[6] Finally, the extreme spring conditions of cold and lack of rain affect the early fundamental hay growth for the upcoming season.

Since 2004 summers have been unseasonably cool and rainy. The extreme continental climate allows for a very hot, albeit brief, summer that, in combination with very long hours of sunlight, provides the proper conditions for rapid plant growth and development. Now summer is not as hot, and nights are especially cold. This weather affects the growth of hay and fodder available for animals. Additionally, summer was a relatively dry season, especially in July and August during hay-cutting time; but the recent increase in summer rain impedes hay cutting and also results in poor hay quality, since people have no other choice than to cut hay in the rainy conditions.

Demographic/Intergenerational

> I am very surprised. I take my son hunting—he used to love going with me and staying out in nature. Now he will not stay—he complains that he is cold and uncomfortable and wants his computer and so he walks home. (Kutana inhabitant, 2011)

My earlier research in the region shows how intergenerational change exists in Viliui Sakha settlements (Crate 2002)—how I would emerge from an elder's house to see people doing exactly the manual labor that the elder had described, hauling water or herding their cows; then suddenly three teens would stroll by with pierced navels, orange and green hair, and high heels. Time seemed to stand still and to race forward. Today it continues to do so. In the context of more recent focus groups and interviews, we hear a common concern: where are the young people? Others comment about how all the young people now want to go only to the city. They decry the effects of television, commenting about how it is ruining youth's brains: "they watch and see the alcohol, the cigarettes, the easy way of life and want to mirror it." Furthermore, many householders consider this intergenerational change *the* central driver of change overall.

Some related to a difference in how young people grow up now, describing how they had had an active childhood spent running, jumping, and playing outside with groups of other children and in a natural world that was very clean and unspoiled. They recalled having a strict work regime, both inside and outside the home. In the summer there were hay camps, and during the hay season everyone had to hay—if people came to the village to help raise a house when the haying was not done, they were first sent to hay. Most young people today do not play outside but are occupied with computers, TVs, and other electronics. "I can't get my child away from the computer no matter how hard I try . . . and youth are stubborn and don't listen . . . not like before." Many claim they also do not read: "I don't see youth with books like we used to read."

Suntar's regional statistician showed how the numbers reveal youth out-migration as a very strong trend. He explained it as largely due to the politics of the post-Soviet period, which advocates for all young people to get a higher education in the urban areas. He predicts that in 20 years, assuming the present trend continues, the villages will cease to exist, and he supports his prediction with a demographic graph, explaining how the trend shows young people increasingly absent (Figure 6.3).

The out-migration of youth depletes the local work force for the more labor-intensive cow duties, including to hay in the summer, bring in the hay from the outlying fields, and do the daily work of cow care. The Suntar agricultural statistician described how in the past the work force was in place: "Cow care is too much work now—the current work force are all elderly, and the next generation is not interested. In the Soviet period we had a ready-made work force—700 kids would finish school and work for two years on the farms, and later 50% of them would go into farming." But the times have changed, and a majority of interviewees expressed their support for youth out-migration. Many parents are pushing their children to get a higher education and a salaried job.

Economics

Whereas in 1992, 10% of all surveyed households kept cows, and of those only one milk cow, in 2000, 55% of all households were keeping cows, and the average among them was three milk cows. When asked why they keep

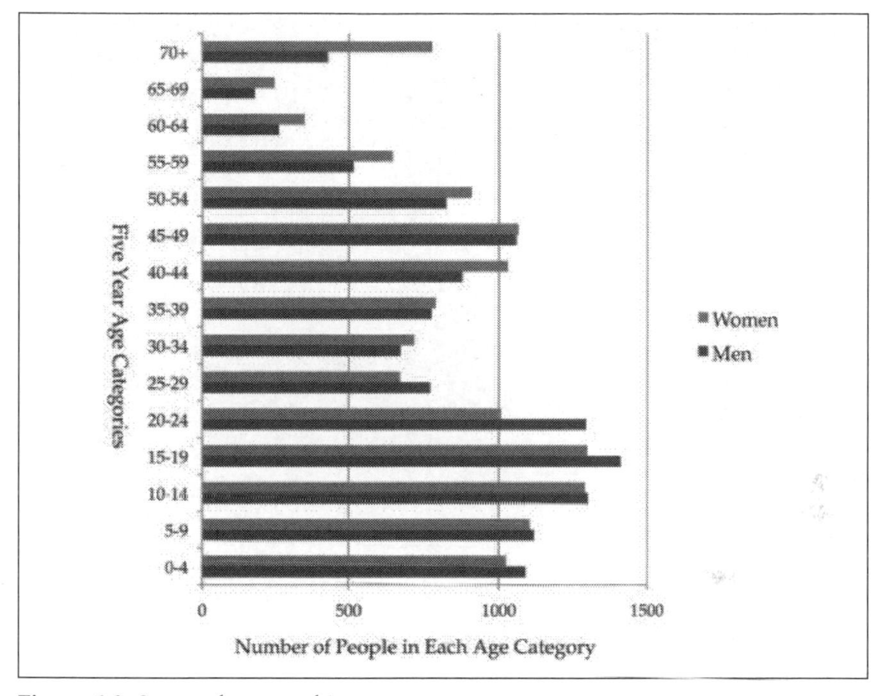

Figure 6.3 Suntar demographics

> cows most households said in order to have fresh meat and milk products, which they could not afford to buy otherwise and were no longer available for sale in village stores. (Crate 2003: 505–06)

Today households buy a large percentage of their food from the store for the simple reason that the stores are stocked with all they need. One householder commented: "Back then there was nothing in the stores—all was 'deficit'—we even made our own furniture. Now it is the complete opposite—we even have IKEA!" (Figure 6.4). Today the stores may be full, but there are now deficits in cow products. One interviewee recounted the recent past, when there was a bounty of cow products for sale in contrast to today, when people form lines at cow-keeping households for milk.

Householders are also able to buy more from the store owing to greater financial security.[7] Ten years ago salaries did not come on time. People received small advances and bought things on credit at stores. Householders said they now have money and enjoy spending it: "All you need now is money—if you have money, everything is here!" Here differences between the villages remain. In Kutana half of interviewed households said they produce and buy as before, whereas in Elgeeii only 20% do.

Figure 6.4 The inside of one of several stores in Elgeeii village, circa 1999. Today stores are overflowing with consumer products.

This consumer trend is visually evident. One interviewee lives in the middle house in a set of three houses, all kin. The first is her late father's house, built using local materials characteristic of construction during the Soviet period. It has a natural solid wood log base with wood siding and shudders, and handmade triple pane windows to maximize heat retention in the cold winters. Hers is in many ways similar but with a few modern features. The third house is her son's. It is newly constructed of modern materials, which, the mother commented, all arrived in a container from the city. I remember entering the third house for the interview and feeling as if something was wrong. After a few minutes I realized that the house lacked the massive central wood stove that was the customary heat source in all houses and essential to surviving the long Siberian winter. This modern house was heated with electric central heating.

Another change that reflects a changing local economy is the rise of horse numbers in the villages (see Table 6.4). This rise may seem contradictory, when one considers the environmental changes challenging horse keepers. Horses are not replacing cows, since they are not a source of milk and many dislike the taste of horse meat. However, horses *are* much more efficient in meat production and require less labor than cows do. Mares foal in January and February and can reach slaughter size as a yearling by the next December, whereas cows require a minimum of three to four years to grow

Table 6.4 Elgeeii, Kutana, Khoro, and Suntar region horse numbers 1999–2011

	Elgeeii	Kutana	Khoro	Suntar Region
	Horses	*Horses*	*Horses*	*Horses*
1999	—	—	—	8,254
2000	—	—	—	8,405
2001	—	—	—	8,774
2002	610	—	—	9,585
2003	669	—	—	9,725
2004	753	745	—	9,807
2005	746	—	—	9,807
2006	738	673	—	9,366
2007	—	712	—	10,433
2008	798	807	—	10,940
2009	—	851	209	10,924
2010	999	—	222	11,642
2011	1,191	1,017	229	—

to slaughter size. Moreover, horses do not require the daily care and most of the labor that cows do, which is traditionally performed by youth. Horse breeding is more challenging in the context of the difficulty horses have finding fodder and householders' need for supplemental hay, as described earlier.

In the midst of these economic changes, many interviewees make the case that cow keeping is now passé. Cow keeping is going out of fashion, and there is now a social status of being "freed from the *khoton* ('cow barn')." Consider Lida, a 50-year-old woman who has done cow keeping since her marriage 30 years ago and also before that helped her mother with daily care and her household with hay every summer. Lida is now "freed from the khoton." She spoke with downcast eyes of how she used to leave work with only one thing on her mind—the *khoton* and the cows. She detailed how she made a calendar each New Year to mark the days before spring and to determine how much hay her cows needed and how much she had left to be sure it would last. Her eyes suddenly looked up and came alive as she talked about how, since her family got rid of their cows the year before, she does not have to worry about those things, and life is good!

Implications

The interactions with and responses to the complexity of change have implications for community health, food security, and the future workforce. First, the change from local to store-bought food is and will continue to negatively affect health. Many who had recently or planned soon to quit cow keeping remarked about how they will miss having all the *urung as* ("white foods," referring to foods made from milk). These foods maintain health in the extreme climate and are also central to Sakha culture and lifeways—take, for example, *suorat* ("cultured milk"), made from the skim milk left after separating the crème fraîche from household cow milk. Sakha eat *suorat* to maintain healthy digestion. Even milk, considered a required ingredient for Sakha tea, and other milk foods are increasingly scarce. Futhermore, poorer households tend to sell all their milk to a central milk station to generate income and buy their family margarine, milk powder, and canned meat to save money.

Second, implications for food security follow directly. Today Viliui Sakha households continue to supplement with other resources from nature—breeding horses, growing gardens and plants in greenhouses, foraging for berries and mushrooms, and hunting and fishing. However, fewer households are engaged in these supplemental activities, and the main reason for doing so has changed. Whereas in 1999–2000 such households engaged largely to feed their families and kin because nothing was available in the stores, contemporary subsistence households justify doing it for food security. "We could buy everything from the store—we have enough to do that. We continue to keep cows and to garden, hunt, and fish in case there is a catastrophe!" If the trend of people stopping cow keeping and other subsistence activities continues, with most relying on the village stores to feed their families, what could the future look like?

And what about the very real possibility that the transportation of goods to the stores will be interrupted? Food security is underpinned by

> dynamic interactions between and within the biogeophysical and human environments [that] lead to the production, processing, distribution, preparation and consumption of food, resulting in food systems . . . food security is diminished when food systems are stressed. Such stresses may be induced by a range of factors in addition to climate change and/or other agents of environmental change (e.g., conflict, HIV/AIDS) and may be particularly severe when these factors act in combination. Urbanization and globalization are causing rapid changes to food systems. (Gregory, Ingram, and Brklacich 2005)

A third implication of the complexity of change is its effects on the future workforce. As much as inhabitants complain that their youth are constantly preoccupied with computer and other electronics and information systems, they also agree that these skills are essential to progress and securing a good job. However, there is not sufficient information technology employment for all youth in the villages or in the regional centers. What will be the future of the villages?

There Are Two Sides to Every Coin, or Could This Be a Natural Evolution to a Real Market?

> A milk cow goes for 50,000–60,000 rubles, the equivalent of US$2,000. If a person has a herd of 10, that is wealth. I could lose my job tomorrow. But

> today's cow keepers think of themselves as slaves and most want to stop and secure a salaried job—almost like cows are out of style and salaried jobs are in! The fact of the matter is we need both. I myself am busy all the time working with no time for cows. I would like to have local cow products and would gladly support someone to do that (Elgeeii householder 2012).

As in most research, findings are graced with outliers that lead to new questions. The decrease in household-level cow keeping is, in turn, perpetuating a local demand for cow products. Sakha, not unlike most world cultures, prefer the taste for locally produced and/or foraged foods. The quality and flavor of village-raised meat is markedly superior to store-bought. In terms of developing locally based economies, one of the four definitions of future sustainability that these communities generated in a 2003–2006 project (Crate 2006b), the disbanding from cow care of the majority and the development of the trade by a few could be a real market evolving in its own time. Furthermore, in each context there are stirrings of kin-based entrepreneurial efforts. "All are getting rid of their cows, which makes my market better, because people want local meat and milk—they are Sakha—the meat in the stores is tasteless" (Elgeeii entrepreneur cow keeper 2012). Those who are still keeping cows and who plan to continue keeping them are doing so by working with their kin to realize labor and products—in other words, by continuing the cows and kin adaptive strategy but on a larger scale and in a more efficient way. In this study's random sample there are three such entrepreneurial interdependent kin household operations in Elgeeii and two in Kutana. If kin is a constant and the cow-keeping kin households understand that there is an increasing market for their products as their neighbors slaughter their herds, could these cow-keeping operations be the future sources of local meat and milk in their communities?

Discussion: Similar Global Trends and the Call for Ethnography

> For the first time in history, more than half [the world's] human population, 3.3 billion people, were living in urban areas. By 2030, this is expected to swell to almost 5 billion. Many of the new urbanites will be poor. Their future, the future of cities in developing countries, the future of humanity itself, all depend very much on decisions made now in preparation for this growth. (United Nations Population Fund [UNPFA] 2007)

In this chapter I have shown with the case of Viliui Sakha how anthropology, specifically via the toolset of ethnography, can investigate and clarify the complexity of change affecting rural livelihoods. This complexity of environmental, demographic, and economic changes challenging contemporary Viliui Sakha livelihoods is stark yet consistent with the transition and adjustment processes in other world regions (Giddens 2002). Like most rural peoples in the contemporary world confronted by such changes, to a greater or lesser degree, other forces of change often command more of inhabitants' attention on a daily basis (Hovelsrud and Smit 2010). Although there are ecosystem, geopolitical, sociocultural, and other locale-specific nuances that shape the challenges Viliui Sakha face, the complexity of change is not unique but part of a global trend (see Andersson 2010). Starting in the mid-20th century, the world's rural areas have been in biocultural decline (Maffi and Woodley 2010), causing sociocultural (Kay, Shubin, and Thelen 2012), economic, and environmental crises (Bradshaw et al. 2007; Intergovernmental Panel on Climate Change [IPCC] 2012, 2013; Stedman, Patriquin, and Parkins 2012). Considering the critical contribution that rural areas make to planetary health (Tilman 1999; Tilman, Reich, and Knops 2006; Zavaleta et al. 2010) as nexuses of biological, cultural, and ethnic diversity, rural areas and the communities inhabiting them must be a focus of comprehensive research efforts to understand and address global sustainability.

Anthropologists recognize the need to consider climate change within the larger mix of change (Nuttall et al. 2005). Accounting for this larger mix is also not new in the field of interdisciplinary climate science. First, the local and regional effects of climate change result in a cascade of other ecological changes within the ecosystem. Using the frameworks of adaptation and resilience, interdisciplinary groups working on the Arctic Climate Impact Assessment argued that understanding the impact of climate change could be accomplished only by considering an ensemble of relevant, concurrent changes in the system (McCarthy and Martello 2005). Second, those ecosystem effects need to be considered within the context of other changes not climate related. For example, O'Brien and Leichenko (2000) demonstrated how the synergisms between two global processes, climate change and globalization, need to be reconsidered. They called this synergism "double exposure" and showed how considering their joint influence in analysis created different results than each taken alone. Other work to accommodate the host of impacts in the context of climate change refers to them as "multiple-stressors." For example, interactions among multiple stresses with a system and the system's resilience are important considerations in determining the vulnerability of Arctic peoples and ecosystems to potentially harmful effects (McCarthy and Martello 2005). The Arctic Resilience Interim Report states: "The need for 'integrative concepts and models' that can aid systemic understanding of the Arctic, including the cumulative impacts of a diverse suite of interconnected changes, is critical in the current period of rapid ecological, social, and economic change" (Arctic Council 2013: 4). Certainly, understanding the biophysical system changes is an equally important task, and this chapter's argument to bring attention to and expand our understanding of the sociocultural effects and responses is by no means intended to diminish that importance. Rather it is more an argument to level the playing field and bring our understanding of the sociocultural up to the same level of importance to which we have advanced our understanding of the biophysical.

Some attempts to account for the complexity of change involve modeling, which has caught the attention and curiosity of anthropologists (for example, Kirsten Hastrup's chapter this volume). Most climate modelers recognize the complexity in predicting human behavior. There have been advances in this area, especially when interdisciplinary teams combine their efforts and expertise to work toward a more comprehensive approach to understanding how regional and local particulars interact with and affect global processes such as climate change (Pielke et al. 2012). They take careful scrutiny of IPCC climate models and even scaled-down models that take into account regional weather patterns—to show how these attempts fall short at bringing in the human elements of change. As with the validation of global climate models (IPCC 2007), the most convincing validation of predictive models would be through simulation of past socioenvironmental interactions covering the last few decades and centuries: letting the model run from specified starting conditions in the past and making comparisons with long-term observations or reconstructions of reality. At present, attempts to do this are confined to bottom-up process-based models at local scales whereby human actions are essentially embodied as switches in land use (for example, Anderson et al. 2006; Welsh et al. 2009). The development and testing of interactive models that can simulate the evolving nature of interactions among social and environmental states are a major research priority (Dearing et al. 2010).

The question becomes just how to simulate evolving social interactions. Some attempts sound hopeful, such as developing such parameters as "contextual/social vulnerability" (Füssel 2009), but in the process of modeling them, the human component is reduced to a rigid nonhuman mechanical process. As anthropologists are increasingly called to collaborate with interdisciplinary teams, we need to develop more human-inclusive approaches to understanding change. Studies using a resiliency framework engage operationalizing constructs—largely borrowed from the discipline of ecology (Berkes, Folke, and Colding 2000). Anthropologists have taken the resiliency concepts to understand how a people's resilience and vulnerability in interaction with ecosystem feedbacks show how cultural factors and differences play a determining role in a group's adaptive success (see Nelson and Finan 2009; Nelson, West, and Finan 2009; Vásquez-León 2009). These studies highlight how adaptation to climate change is not a simple function of technical solutions but rather is determined by relationships based in a web of reciprocities, obligations, and assets (Roncoli et al. 2009; Crane et al. 2010).

Although most climate researchers recognize the need to understand the complexity of change in the context of interdisciplinary collaborations, exactly how to integrate social science effectively, especially ethnographic and narrative information, continues to elude. We need to make anthropological findings relevant in interdisciplinary and policy contexts. If resilience is a constitutive element of any working society that allows for systemic sociocultural adaptation to external factors (Hastrup 2009), then I argue that understanding societies and facilitating their adaptive resilience, most prominently through ethnographic methods, is essential (Crate 2011a). First, ethnography is a product of one of anthropology's founding skill sets, that of "being there" (Roncoli, Crane, and Orlove 2009). Second, as ethnography has matured, it has differentiated from a highly local and site-specific focus to a multisited one that accommodates both local specific elements and the larger forces engaging those elements (Marcus 1998). Climate researchers are increasingly using a "cross-scale multi-sited research design and an interdisciplinary mix of interactive and structured tools and techniques [where] the analytical focus can be expanded to encompass local communities and their multiple action spaces and also the higher spheres of decision making, where policy and science are shaped" (Roncoli 2006: 94). Multisited ethnography involves the analytical engagement of local to global connections, with ethnographers engaging in multisited work for a broader sense of the issues as they crosscut different locales and populations. The multisited approach reveals the ways that the climate science world and the social world are not separate but integral. Finally, ethnography has more recently gained the power to bridge understandings beyond the local to reach global audiences, for example, in Anna Tsing's (2005) "ethnography of global connection."

Considering the power and relevance of ethnography to the world's pressing issues, it behooves us to be diligent in finding ways to bring it to the interdisciplinary table. A brief overview of the literature shows how Viliui Sakha are not alone but rather part of a global phenomenon of rural transformation, increasingly exposed to environmental challenges, depopulation, and economic failings owing to contemporary climate and socioeconomic forces of change (see Crate and Nuttall 2009; Andersson 2010).

These challenges, in turn, threaten the future sustainability of our world to the extent that they diminish rural regions' former and future roles in the global diversity of human activities (Dietz, Ostrom, and Stern 2003). Rural communities represent *complex adaptive systems*, a dynamic interplay of "biophysical systems (for example, mechanisms of change and communication) and compatible key features of social systems: the holistic nature of culture, knowledge sharing through the senses, and the formative power of traditions, structures/materials, strategies and habits of mind" (Crumley 2012: 304). The foundations of these complex adaptive systems are diversity and place, with diversity being a generative feature of those complex adaptive systems (Gregory 2006). Key to that diversity, both biological and cultural, is that it is founded in a specific place. According to Crumley: "Our success as a species turns on our knowledge of and attention to place" (2012: 312). Increasingly scholars, from a diversity of disciplines, refer to the importance of place in having an important and unique function in positive identity development and psychological well-being (Gustafson 2001; Bott, Cantrill, and Myers 2003; Kyle et al. 2004). There is a reciprocal benefit between humans and the environment when nature is a part of an individual's identity. How these complex adaptive systems, founded on biological and cultural diversity and set in place, are challenged and compromised in these times characterized by unprecedented loss of diversity and place demands that we find effective ways to use qualitative approaches to understand fully how humans interact with a complexity of change in the contemporary world.

Ethnographers, carrying out fieldwork in one place, acquire a keen sensitivity to change and can identify the various mix of changes at work. Furthermore, ethnographic methods document attachment to place and local diversity, both cultural and biological. Exactly how ethnography can contribute to interdisciplinary change research is an ongoing challenge, but approaches are evolving. The Viliui Sakha case suggests that local studies highlight how a global phenomenon such as climate change is having ecosystem- and culture-specific effects. It also shows how climate is one factor in a complexity of change that is challenging livelihood but that, in local vernacular context, is often not the most urgent.

Conclusion

My earlier investigations into the cultural implications of climate change, some of which I discussed in my chapter, Gone the Bull of Winter, in this volume's first edition, were founded on my understandings of how Viliui Sakha had successfully adapted to the fall of the Soviet Union by developing the cows and kin system of household food production (Crate 2003) and establishing local institutions of long-term sustainability (Crate 2006b). In these investigations, unprecedented climate and seasonal change were inhabitants' unknowns, which made their present and their future increasingly uncertain. Through anthropological methods and analyses we[8] teased out the changes that most affected local livelihoods (Crate 2008), clarified the extent to which local perception and belief influenced inhabitants' understandings and responses, most notably via changes in water (Crate 2011b, 2013) and described the lack of locally relevant information on climate change that left communities without complete understanding of the implications of local change. My ongoing collaboration with regional permafrost scientist Alexander Fedorov, himself a native Sakha of an adjacent region, prompted our facilitation of eight knowledge exchanges to both enhance local understandings and to inform scientific findings with in-depth local observations that show the highly local ways that the global phenomenon of climate change was affecting communities (Crate and

Fedorov 2013a). These exchanges, in turn, initiated a booklet intended to replicate the knowledge-exchange process for inhabitants throughout the watershed region (Crate and Fedorov 2013b). In the process of this climate-focused work, while interviewing households (many with whom I had worked since 1991), I began to notice a sharp decline in cow keeping. This reality was especially stark for me in households who had kept cows for many generations, often predating the Soviet period. At first I associated the decline with the difficulty inhabitants reported they were having in haying, owing to one of the main changes of climate—the increasing water on the land. However, several of the newly cow-free households were not having difficulty harvesting enough hay. So I probed further about cow keeping and clarified at least two other factors at work: economic globalization and youth out-migration. In some cases the issues of cow keeping brought about by climate change were not even mentioned.

In tandem, I began a comparative project focused on how changing seasonality, another of the nine main changes Viliui Sakha were observing, was affecting two disparate Arctic contexts—in my existing research location with Viluiu Sakha and among coastal communities of Labrador/Nunatsiavut, Canada. Herein I found a similar ice-dependence and concern about the lessening of seasonal ice in both but in terms of ice in two different manifestations: permafrost and sea ice (Crate 2012). Additionally, in the Labrador communities I noticed that the similar drivers, of economic globalization and youth out-migration, were also challenging rural contexts. Similarly, in nascent projects on the eastern shore of Maryland, rural Wales, Kiribati, Peru, and Mongolia, I started seeing a similar pattern and termed it the *complexity of change*. In this chapter I shared the ethnographic account of how I teased out the complexity of change in the Viliui Sakha context.

The process of seeing how climate and seasonal changes are challenging Viliui Sakha, with many stopping cow keeping, opened my eyes to the other changes that are also playing a role, including (1) the intergenerational change as young people leave the rural regions and so are not in the villages to carry on the cow-care activities, and (2) the economic change giving households more stable financial resources and access to a bounty of consumer goods at the village stores, making household-level cow keeping less necessary and concomitantly passé. These latter changes led me to consider additional implications and also to perceive the process as perhaps some larger systemic evolution. Was the move from cows and kin to several households keeping larger herds in each community the result of households coming into real relationships with the market of supply and demand, after the artificial economy of the Soviet period? Was this phenomenon the slower process of the economy settling in after 20 years to find a right mode of functioning? Or perhaps the change from cows and kin was globalization uprooting the adaptive capacity that Viliui Sakha have developed and maintained to produce much of their food and fiber locally.

By comparing and contrasting with other rural people's plight, I saw that Viliui Sakha share a common reality of contemporary times. The complexity of change affecting rural areas has taken on a new threat to biocultural diversity and sense of place, which underpin complex adaptive systems. These qualities and conditions can be successfully documented through ethnographic methods. Although interdisciplinary efforts attempt to incorporate these human aspects, there is still a need to engage ethnography to fully account for them. The practice of ethnography by anthropologists is based on long-term rapport and familiarity with the locale, an appreciation and understanding of emic perspectives, and a variety of qualitative ways of knowing that privy the researcher in those human aspects.

For Viliui Sakha, a trend is clear, but there remain many unanswered questions that will be answered only in the process of time. The more we consider that the extent to which the trend we see in close-up through the ethnographic lens of Viliui Sakha life is full of nuances and surprises based in the local context of culture and ecosystem, the more the global trend becomes a diversity of on-the-ground realities. In other words, the picture is not black and white. With humans in the picture, there is an open stage for further developments to evolve and interact with a myriad of processes.

In my 2003 article, my research brought me to the following conclusion: "What can be said is that in the present context, cows and kin represents a unique adaptation which, in the wake of Soviet infrastructure collapse, offers a sound mode of household-level food production for contemporary rural Viliui Sakha (Crate 2003: 526). Continuing the ethnographic process to understand how the complexity of change is interacting with that unique post-Soviet adaptation suggests that although cows and kin may retain its status as a sound mode of household-level food production, other forces—including climate change, economic globalization, and the alienation of and out-migration of the next generation from village life—are bringing about new iterations of cows and kin to meet a new time.

Acknowledgments

First and foremost I acknowledge all Viliui Sakha collaborators and communities involved in the research represented in this chapter. Although the article is based on longitudinal research with these communities since 1991, I recognize a majority of funding support from the Office of Polar Programs of the National Science Foundation (NSF): one grant from the Arctic Social Science Program (0710935)—Assessing Knowledge, Resilience, and Adaptation and Policy Needs in Northern Russian Villages Experiencing Unprecedented Climate Change—and the other grant from the Arctic Science Program (0902146)—Understanding

Climate-Driven Phenological Change: Observations, Adaptations, and Cultural Implications in Northeastern Siberia and Labrador/Nunatsiavut (PHENARC). Any opinions, findings, conclusions, and recommendations expressed in this material are mine and do not necessarily reflect the views of the NSF. I gratefully acknowledge and thank the NSF for supporting our research. This chapter is based on an article published in *Sibirica 2014* and available online: http://journals.berghahnbooks.com/sib/. I am also grateful to Mark Nuttall for his comments and suggestions on the chapter.

Notes

1. The 2003 article was based on dissertation data collected between July 1999 and July 2000.
2. To protect research collaborators' anonymity, all names have been changed to pseudonyms or reduced to a generic descriptor.
3. I use the qualifier "universally" here to speak about the majority of households that no longer keep cows. Kin remains essential to the minority still involved in household-level cow keeping.
4. In Elgeeii, of the 9 households with cows, 2 plan to slaughter their herd this fall; 3 have them but keep them at another person's barn to grow them to slaughter size and after that will stop cow keeping; in Kutana, of the 7 households with cows, 4 plan to slaughter them in the fall of 2012—all of which makes the number of long-term cow-keeping households in each village 4 and 3, or 10% and 19%.
5. In Elgeeii, 10 households said they had them and stopped keeping them in the past 1–3 years; 13 said they had them but stopped keeping them in the past 4–10 years; and 8 said they never had cows. In Kutana, of the 9 without cows, 5 of them had them until 3–5 years ago.
6. For 2,000 rubles (60 USD) per 2-ton stack (usually 1,200 rubles, or 35 USD).
7. This is the case for most interviewees but not the case for some households whose financial situation has worsened in the last 10 years owing to the lack of employment, getting old, and having to stop cow keeping and therefore buy products.
8. I use "we" to refer to all research beyond my dissertation since I worked in collaborative community projects.

References

Anderson, N. J., Bugmann, H., Dearing, J. A., and Gaillard-Lemdahl, M.-J. 2006. Linking palaeoenvironmental data and models to understand the past and to predict the future. *Trends in Ecology and Evolution* 21(12): 696–704.

Andersson, M. 2010. Provincial globalization: The local struggle of place-making. *Culture Unbound: Journal of Current Cultural Research* 2: 193–214. DOI: 10.3384/cu.2000.1525.10212193.

Arctic Council. 2013. *Arctic Resilience Interim Report 2013*. Stockholm: Stockholm Environment Institute and Stockholm Resilience Centre.

Berkes, F., Folke, C., and Colding, J. 2000. *Linking Social and Ecological Systems: Management Practices and Social Mechanisms for Building Resilience*. New York: Cambridge University Press.

Bott, S., Cantrill, J. G., and Myers, O. E. 2003. Place and the promise of conservation biology. *Human Ecology Review* 10(2): 100–12.

Bradshaw, C. J. A., Sodhi, N. S., Peh, K. S. H., and Brook, B. W. 2007. Global evidence that deforestation amplifies flood risk and severity in the developing world. *Global Change Biology* 13(11): 2379–95.

Crane T. A., Roncoli, C., Paz, J., Breuer, N., Broad, K., et al. 2010. Forecast skill and farmers' skills: Seasonal climate forecasts and agricultural risk management in the southeastern United States. *Weather, Climate and Society* 2(1): 44–59.

Crate, S. A. 2002. Viliui Sakha oral history: The key to contemporary household survival. *Arctic Anthropology* 39(1): 134–54.

———. 2003. Viliui Sakha post-Soviet adaptation: A sub-Arctic test of Netting's Smallholder Theory. *Human Ecology* 31(4): 499–528.

———. 2006a. *Cows, Kin and Globalization: An Ethnography of Sustainability.* Walnut Creek, CA: AltaMira Press.

———. 2006b. Investigating local definitions of sustainability in the Arctic: Insights from post-Soviet Sakha villages. *Arctic* 59(3): 294–310.

———. 2008. Gone the bull of winter: Grappling with the cultural implications of and anthropology's role(s) in global climate change. *Current Anthropology* 49(4): 569–95.

———. 2011a. Climate and culture: Anthropology in the era of contemporary climate change. *Annual Review of Anthropology* 40(1): 175–94. DOI: 10.1146/annurev.anthro.012809.104925.

———. 2011b. A political ecology of *Water in Mind*: Attributing perceptions in the era of global climate change. *Weather, Climate, and Society* 3(3): 148–64.

———. 2012. Climate change and ice dependent communities: Perspectives from Siberia and Labrador. *Polar Journal* 2(1): 61–75.

———. 2013. From living water to the "Water of Death": Implicating social resilience in northeastern Siberia. *Worldviews* 17(2): 115–24.

Crate, S., and Fedorov, A. 2013a. Methodological model for exchanging local and scientific climate change knowledge in northeastern Siberia. *Arctic,* 66(3): 338–50.

———. 2013b. *Alamai Tiin: Buluu Ulustarigar Klimat Ularitigar uonna Atin Kihalghar (Alamai Tiin: Climate Change and Other Changes in the Viliui Regions).* Yakutsk: Bichik.

Crate, S., and Nuttall, M. (Eds.). 2009. *Anthropology and Climate Change: From Encounters to Actions.* Walnut Creek, CA: Left Coast Press, Inc.

Crumley, C. L. 2012. A heterarchy of knowledges: Tools for the study of landscape histories and futures, in T. Plieninger and C. Bieling (Eds.), *Resilience and the Cultural Landscape: Understanding and Managing Change in Human Shaped Environments.* Cambridge: Cambridge University Press, pp. 303–14.

Dearing, J. A., Braimoh, A. K., Reenberg, A., Turner, B. L., and van der Leeuw, S. 2010. Complex land systems: The need for long-time perspectives to assess their future. *Ecology and Society* 15(4), www.ecologyandsociety.org/vol15/iss4/art21/, accessed February, 4, 2015.

Dietz, T., Ostrom, E., and Stern, P. C. 2003. The struggle to govern the commons. *Science* 302: 1907–12.

Füssel, H.-M. 2009. Review and quantitative analysis of indices of climate change exposure, adaptive capacity, sensitivity, and impacts. *Development and Climate,* https://openknowledge.worldbank.org/bitstream/handle/10986/9193/WDR2010_0004.pdf?sequence=1, accessed February, 4, 2015.

Giddens, A. 2002. *Runaway World: How Globalisation Is Reshaping Our Lives.* London: Profile.

Gregory, P. J., Ingram, J. S., and Brklacich, M. 2005. Climate change and food security. *Philosophical Transactions of the Royal Society B* 360: 2139–48. DOI: 10.1098/rstb.2005.1745.

Gregory, T. A. 2006. An evolutionary theory of diversity: The contributions of grounded theory and grounded action reconceptualizing and reframing diversity as a complex phenomenon. *World Futures* 62(7): 542–50.

Gustafson, P. 2001. Meanings of place: Everyday experiences and theoretical conceptualizations. *Journal of Environmental Psychology* 21(1): 5–16.

Hastrup, K. (Ed.). 2009. *The Question of Resilience: Social Responses to Climate Change*. Copenhagen: Royal Danish Academy of Sciences and Letters.

Hovelsrud, G. K., and Smit, B. (Eds.). 2010. *Community Adaptation and Vulnerability in Arctic Regions*. London: Springer.

IPCC. 2007. *Climate Change 2007: Impacts, Adaptation, and Vulnerability*. Working Group II Summary for Policymakers. Geneva: IPCC Secretariat, www.ipcc.ch/pdf/assessment-report/ar4/wg2/ar4-wg2-spm.pdf, accessed February 4, 2015.

IPCC. 2012. *Managing the Risks of Extreme Events and Disasters to Advance Climate Change Adaptation*. A Special Report of Working Groups I and II of the Intergovernmental Panel on Climate Change. Cambridge: Cambridge University Press.

———. 2013. *Climate Change 2013: The Physical Science Basis*. The Working Group I Contribution to the Fifth Assessment Report, www.ipcc.ch/report/ar5/wg1/, accessed February 4, 2015.

Kay, R., Shubin, S., and Thelen, T. 2012. Rural realities in the post-socialist space. *Journal of Rural Studies* 28(2): 55–62.

Kyle, G., Graefe, A., Manning, R., and Bacon, J. 2004. Effects of place attachment on users' perception of social and environmental conditions in a natural setting. *Journal of Environmental Psychology* 24(2): 213–25.

Maffi, L., and Woodley, E. 2010. *Biocultural Diversity Conservation: A Global Sourcebook*. London: Earthscan.

Marcus, G. E. 1998. *Ethnography through Thick and Thin*. Princeton, NJ: Princeton University Press.

McCarthy, J. J., and Martello, M. L. (Eds.). 2005. Climate change in the context of multiple stressors and resilience, in *Arctic Climate Impact Assessment (ACIA)*. New York: Cambridge University Press, pp. 946–88.

Nelson D., and Finan, T. 2009. Praying for drought: Persistent vulnerability and the politics of patronage in Cearà, northeast Brazil. *American Anthropologist* 111(3): 302–16.

Nelson D., West, C., and Finan, T. 2009. Introduction to "In focus: Global change and adaptation in local places." *American Anthropologist* 111(3): 271–74.

Netting, R. Mc. 1993. *Smallholders, Householders: Farm Families and the Ecology of Intensive, Sustainable Agriculture*. Redwood City, CA: Stanford University Press.

Nuttall, M., Berkes, F., Forbes, B., Kofinas, G., Vlassova, T., and Wenzel, G. 2005. Hunting, herding, fishing, and gathering: Indigenous peoples and renewable resource use in the Arctic, in *Arctic Climate Impact Assessment Scientific Report*. Cambridge: Cambridge University Press, pp. 650–90.

O'Brien, K. L., and Leichenko, R. M. 2000. Double exposure: Assessing the impacts of climate change within the context of economic globalization. *Global Environmental Change* 10: 221–32.

Pielke, R. A., Wilby, R., Niyogi, D., Hossain, F., Dairuku, K., Adegoke, J., Kallos, G., Seastedt, T., and Suding, K. 2012. Dealing with complexity and extreme events using a bottom-up, resource-based vulnerability perspective, in A. Surjalal Sharma, A. Bunde, V. P. Dimri, and D. N. Baker (Eds.), *Extreme Events and Natural Hazards: The Complexity Perspective*, Geophysical Monograph Series, Vol. 196. American Geophysical Union.

Roncoli, C. 2006. Ethnographic and participatory approaches to research on farmers' responses to climate predictions. *Climate Research* 33: 81–99.

Roncoli C., Crane, T., and Orlove, B. 2009. Fielding climate change in cultural anthropology, in S. A. Crate and M. Nuttall (Eds.), *Anthropology and Climate Change: From Encounters to Actions*. Walnut Creek, CA: Left Coast Press, Inc., pp. 87–115.

Stedman, R. C., Patriquin, M. N., and Parkins, J. R. 2012. Dependence, diversity, and the well-being of rural community: Building on the Freudenburg legacy. *Journal of Environmental Studies and Sciences* 2(1): 28–38.
Tilman, D. 1999. Diversity by default. *Science* 283: 495–96.
Tilman, D., Reich, P. B., and Knops, J. M. H. 2006. Biodiversity and ecosystem stability in a decade-long grassland experiment. *Nature* 441: 629–32.
Tsing, A. L. 2005. *Friction: An Ethnography of Global Connection.* Princeton, NJ: Princeton University Press.
UNFPA (United Nations Population Fund). 2007. *State of World Population 2007: Unleashing the Potential of Urban Growth,* www.unfpa.org/swp/2007/english/introduction.html, accessed February, 4, 2015.
Vásquez-León, M. 2009. Hispanic farmers and farm workers: Social networks, institutional exclusion, and climate vulnerability in southeastern Arizona. *American Anthropologist* 111(3): 289–301.
Welsh, K. E., Dearing, J. A., Chiverrell, R. C., and Coulthard, T. J. 2009. Testing a cellular modeling approach to simulating Late Holocene sediment and water transfer from catchment to lake in the French Alps since 1826. *Holocene* 19: 785–98.
Zavaleta, E. S., Pasari, J. R., Hulvey, K. B., and Tilman, G. D. 2010. Sustaining multiple ecosystem functions in grassland communities requires higher biodiversity. *Proceedings of the National Academy of Sciences of the United States of America* 107: 1443–46.

Chapter 7

Climate Change in Leukerbad and Beyond: Re-Visioning Our Cultures of Energy and Environment

Sarah Strauss

Revisiting Leukerbad

As anthropologists, we seek to understand and translate, helping to make the experiences of one place/time/people intelligible to those who inhabit different lifeworlds. Climate change is a universal experience but one that unfolds in its own way for each individual and in each locale. In the first version of this volume (Strauss 2009), I presented the case of Leukerbad in Switzerland, where I have worked to understand the "social lives" (cf. Appadurai 1986) of water, weather, and climate since 1997.

Leukerbad is situated in the Dalatal, a side valley off of the Rhone river valley in the Swiss canton of Wallis (Valais in French; it is a bilingual canton). As with many other locations in the Alps, the retreat of glaciers following the 1850 highstand that marked the end of the Little Ice Age has been a dominant feature of the Leukerbad landscape. Like other Alpine communities, Leukerbad has few natural resources besides water, and, until the 1960s, most of its inhabitants made do with minimal wage labor and small subsistence farms. Despite the shift from a subsistence-oriented economy, weather and climate have continued to be important concerns for the economic health of the valley. The short-to-midrange consequences of altered precipitation patterns for the ski area, as well as the long-term effects of global environmental change on the glaciers that permit a stable flow of water through the Dalatal, depend on reliable precipitation. Current projections, however, suggest that the end of the 21st century will bring drastic changes to the snow levels between 1,500 and 2,000 m, where most of the Leukerbad community and tourist offerings are located (Beniston 2006; EEA 2009; Gobiet et al. 2014), and even within that range there will be reduced numbers of snow days as we move past mid-century. The Loetschental, one valley over from the Dalatal, is a UNESCO World Natural Heritage site, as home to the largest glacier in the western Alps; its microclimate, not to mention its local history and culture, is quite different from the Leukerbad context.

In the earlier version of this chapter I discussed the glacier stories that abound in this region of the Alps; these stories describe the movements of glaciers back and forth as climate cycles shifted and the risks inherent in living near such sometimes-slumbering giants (Jegerlehner 1989 [1906]). The people of Leukerbad seek information detailed enough to help them to make decisions about the wisest course of action for managing their water resources, given the fact that climate change is occurring. In the mid-2000s, people asked if they should buy new snow cannons and continue to make artificial snow for the lower-lying elevations. The ski area is located in terrain that ranges from 2,300 m at the top, down to village slopes at 1,500 m; it is certain that the top will be fine for decades to come but equally certain that the village runs are at the bottom margin of viability. In studying how climate change models are constructed, and what the limits of the modeling process are, I learned that we do not possess what most local mountain communities need right now, and the likelihood of climate science being able to provide such information at the decadal scale at the local level (say, for a 10 km grid) is going to be a long time in coming.

Even the best ocean-atmosphere coupled General Circulation Models, which take into account the greatest number of factors, can generate data that is relevant only for the global scale—because information is taken from averages at different levels from the upper atmosphere to the bottom of the ocean. Such data tell absolutely nothing about what will happen in any specific place on the planet. As computing power increases, more and more efforts to refine the scale of the models are taking place. The very best regional climate models can get to a 30 km grid size, which is great if you are in Kansas but not so useful in the Alps. If I were to position Leukerbad in the middle of a 30 km grid, it would become quite clear that the area considered to be "the same" for modeling purposes is enormously diverse from a hydrological perspective and therefore is likely to demonstrate very different effects from changes in climate. People living in these adjacent communities are also likely to have different ideas about viable adaptive or mitigating strategies in response to such changes. In Leukerbad, there has always been a plentiful supply of many different kinds of water, and indeed the entire economy of this place is built on water, in the forms of thermal baths, a ski area, a glacier-fed flow-through hydroelectric plant, and the touristic value of its surface water features—mountain lakes, glaciers, and the river that runs from them. In other parts of the Rhone valley (Valais/Wallis), 10–15 km away on the south-facing slopes, water scarcity at the very local level has been a problem solved in the past by community organizations to manage and especially redirect runoff into *bisses*, or small irrigation canals that moved water from the higher slopes down to specific agricultural zones, often at an extremely high cost to human life because of the dangerous cliffs that had to be traversed. The availability of water in this region is quite different from that in Leukerbad.

If it is the case that two adjacent regions have developed sufficiently different local cultures, as I have argued for the Dalatal and the Loetschental (Strauss 2009), we might expect that their perceptions of risks related to climate change, and their subsequent responses, might also be quite different. Indeed, for the Loestchental, the concern for climate change impacts on this resource of global significance[1] is different from the extremely local concerns that Leukerbad's inhabitants have for their own resources. While climate models do not distinguish difference at that local a level, and would offer the same degree of risk for each community—and thus might invoke a similar response from the policymaking community as to the most effective steps suggested for mitigating that risk—an anthropological analysis requires us to examine the evidence with a different eye and can help us to see why such broad-scale efforts may not succeed in "retraining" local communities to respond in a way that the scientific community finds appropriate.

Who Can Know?

So, in terms of using regional modeling data to help guide specific decisions in Leukerbad, it appears that we can do little. My conclusions (Strauss 2009) regarding the fatalistic attitude of many Leukerbadners about their future climate were borne out in recent observations of the complexity of climate-related decision-making processes in reference to a lower-altitude portion of their ski area. The Obere Maresse ski area was the only intermediate skiing left near the center of the village; it closed in 2008 because of aged equipment and lack of funding, although one can imagine that the constant requirement for making snow at that altitude, well below 2,000 meters, was also a consideration. Public outcry at the loss of these ski slopes led to a series of proposals, from a failed summer toboggan run initiative to the current double chairlift restoration effort;[2] nowhere have I seen any consideration of the projected climate change effects on the availability of snow for the next two to four decades, which would seem to be a significant element to consider in any new ski area investment. And when asked directly in 2014, the response from the team was essentially: "who can know what will happen?" If the snow turned out to be insufficient, they figured that summer hiking would still be available, and they would not make decisions based on climate forecasts. Yet the community must also consider the desires of their workers and guests in this tourism-driven economy, and the uncertainties of climate forecasts against the certainties of public outcry against a loss of recreational resource. This situation demonstrates how difficult it is, even in a "best-case" scenario, to take immediate action on forecast or even already experienced climate change effects.

The situation in Leukerbad is in fact one that I do often describe as "best case," and it highlights exactly how difficult it will be to engage local populations in temperate climates to change their behaviors in response to assertions of impending climate change-related impact. The population in Leukerbad is nearly 100% literate, the general sentiment among the Swiss is acceptance of climate change as a real threat to "business as usual," and the economy of the village and the valley where it is located is entirely dependent on water resources and weather conditions, whether in terms of snowpack and the ski area, general mountain tourism, hydroelectric energy production linked to glacier loss and variable precipitation patterns, or effects of surface water changes on the thermal springs that have made this village a destination for thousands of years.

The general outlook in the European Alps is for more winter precipitation but higher temperatures, so that low-lying areas are likely to have more rain and higher elevations, more snow (EEA 2009). But the summer seasons are predicted to be much drier, with less precipitation—and, because of the increased rain in winter, reduced snowpack lasting through the summer season. Without the snowpack to ensure more regular streamflow patterns, the lack of precipitation combines with the reduced streamflow to create extremely dry conditions conducive to widespread forest fires, as occurred in the extremely warm summer of 2003. In addition to having one of the worst forest fires in the history of the valley, the summer of 2003 also brought an unusual occurrence to Leukerbad—a severe storm took out a number of trees, which blocked the river and knocked out part of the water distribution system. For the first time in memory, Leukerbad suffered a water shortage—not because of the dry summer conditions and a lack of water, but because of a severe storm. While that storm was just one weather event, the climate models do suggest that there will be, overall, an increase in such extreme events in mountain regions in the climate of the future, and therefore a need to think about the many different ways that a changing climate could affect access to water resources (Beniston, Stoffel, and Hill 2011). My interviews in Leukerbad suggest that although people are pretty much in agreement that climate change is happening and will pose a problem in the future, for both their community and the world at large, they are still at a loss for what to do to address this problem in any meaningful way and are more likely to say that, for example, if their river dries up because the glacier melts, they will just move.

But to where? Leukerbadners live in the headwaters region of the major European river systems (Rhine and Rhone), and if they don't have water, no one else does, either. While all Swiss communities, including Leukerbad, have mandatory recycling and waste stream management, they also have utility pricing structures for water and electricity that tend to be considerably expensive, so the incentive for improving efficiency is greater. In recent decades, Leukerbad and other Swiss communities have started to look more like American communities, walking less and buying more, increasing home size and therefore power and water use along with the increase in carbon footprint. Switzerland also has stronger regulatory requirements for efficiency and much greater investment in public transportation and other infrastructure that makes a low-carbon lifestyle more viable than in most parts of the United States, for example. By reemphasizing more historically typical ways of living efficiently on the resources we have, it is still possible to buy time, even if we cannot predict the precise effects of climate change on our own local communities.

A Way Forward?

It seems that there is a lack of urgency regarding climate change for many people in temperate climates, particularly in the United States, but this situation has started to change, at least in terms of recognition that a problem does, in fact, exist. For a range of reasons that include sensationalism as well as the immediacy of the events, more attention has always been given to extreme weather and water issues; the effects are more direct and rapid, so communities and regional resource managers are more likely to place restrictions on individual consumers to address potential or actual water shortages, or make announcements for evacuation in the case of extreme events. Community resilience in the face of climate change starts with a general ability to respond to smaller/more rapid (shorter-term) environmental changes like these. Many of the changes that would be most useful for long-term climate crisis are the sorts of things that general public health and community service guidelines would support anyway—not special but significant ways to strengthen the ability of individuals and communities to weather any kind of storm: helping people to recognize what their resources are, how to access them, what the limits of expert knowledge are, generating templates for making informed choices and use of existing local communication networks to both disseminate information and provide assistance.

The kinds of community-based support systems that can help with addressing the impacts of climate change in locales like Leukerbad are the same kinds of support that would help us to build strong communities for any occasion—but for which funding is rarely available under circumstances that do not include "crisis" as their primary ingredient. Perhaps one outcome of the increased attention to climate change as a problem requiring attention will be that communities might be able to draw on additional, earmarked resources to develop the kinds of communication and information sharing services and networks that will serve them well, no matter what kinds of problems they face in the future.

Anthropologists have historically been good at documenting local contexts and demonstrating change over time, whether sociocultural, economic, or environmental. This kind of information—what Crate and Nuttall called "witnessing" in the first edition of this volume—is surely essential, but the next steps must still be taken: facilitating multidirectional communication about local impacts and concerns related to energy, water, and climate change, so that people who are making the forecasts and developing the models know what kinds of information is important to people in local communities regarding their meteorological "products" and local people have an idea where to look and how to make sense of the vast amounts of data on these topics that is available. In addition to communication and translation, another key element involves making it clear when the data do not actually exist for a particular place and a particular problem, so that false expectations about the ability of science to provide "answers" are not generated and those researchers engaged in the production of data have some idea where to apply their attentions whenever possible. Finally, we need to work to develop sustainable systems on all fronts, not only in terms of the material resources like energy and water but also in terms of the intangible aspects of our communities that make them places where we want to build our lives.

Wallis and Wyoming

In contrasting the places where I work and where I live, in Switzerland and in the American West, I can see many similarities and many differences. For example, although both Wyoming and Switzerland export energy, the mix is quite different—Valais produces 50% of the nation's hydropower, and hydro represents 60% of the electricity generated in Switzerland; the remainder is primarily nuclear, with some renewables thrown into the mix. The carbon footprint for producing electricity in Switzerland is almost nonexistent. Wyoming is one of the leading exporters of coal in the United States, producing nearly 40% of all U.S. coal, although it has the smallest population, so by measures of per capita carbon produced, the state looks awful, yet because of Wyoming's small size, we have recently been one of the most proactive states in developing carbon regulation legislation and trying to help set precedent for the kinds of considerations that need to be made as we transition from a fossil-fuel-based economy. Switzerland is among the most progressive countries in the world in terms of climate change policies; Wyoming is a western state whose legislature decided in 2014 that K–12 schools should not use the national Common Core K–12 education standards because of the inclusion of climate change curriculum. Although that decision is likely to be reversed in 2015, it generated international news attention.

There are many different models for behavioral and sociocultural change, and not all are built on the need for more information. We have at our disposal a number of strategies, some of which are default positions, while others require more effort for directed change: education through information, epiphany, history/stories, example, or experience. Other options include moral motivation, legal codification, environmental forcing, and acculturation. Climate scientists speak of anthropogenic forcing of the climate—things human societies do that push the climate system beyond what it would do on its own; we might also speak here of cultural forcing, exemplified by the popular "Nudge" theory of behavioral economists Thaler and Sunstein (2008), which could include all types of social-structural forcing, the shifting of institutional or political structures to take on new shapes as they are revised by new understandings of the problems we face and therefore the range of solutions that could be useful.

Let's look first at the delivery of technical information, since that is always such a popular effort. At this point, the fifth IPCC reports (for example, 2013, 2014), along with the United Kingdom's Stern report and numerous others, have given an enormous amount of evidence from a wide variety of disciplines that the climate of our planet is undergoing fairly rapid change. Those whose livings are made by application of the scientific method, of course, might find it difficult to imagine that still *more* concrete, scientifically obtained, information is not *more* useful in terms of convincing others about the imperative to act on this information to reduce the long-term effects of changes we have already clearly set in motion. But as with much scientific information about, for example, the health effects of our actions, whether at the individual level of smoking or the population level of structural and institutional changes that give children in urban and suburban areas little option but obesity, the American response (not the same as the European response, but globally dominant nonetheless) has been to shift immediately to a moralizing stance. Such an approach forces polarity to a greater degree than is either necessary or useful and pushes the debate to one that is easily dismissed because of the inherent uncertainties of local and regional impacts. There is very little scientific uncertainty regarding the fact that the global environmental change is happening, but whether specific outcomes will occur in specific communities is a lot harder to address through science, and spending extra energy proving one outcome or another is not, perhaps, the most useful way we can direct our time and energy.

Cultural change is the key to addressing the effects of climate change, and not merely increasing the environmental knowledge base of the world's citizens. But what kinds of change are we talking about? Light bulbs are one standard element, great if you had them to begin with and have an option of which ones to change to. Changing light bulbs is an individual choice, although one mediated by structural possibilities that allow distribution of new technologies at sufficient levels to create a true option. There are very few other examples of such options. Public transportation, which of course would dramatically reduce the carbon footprint of most nations, is not widely available outside a few scattered urban areas around the world, with the exception of much of Western Europe.

The primary key is a strategy of increasing efficiency on all fronts—in water utilization, energy consumption, and food production and distribution systems. Easy to say, and extremely effective where it has been implemented; in California and the Pacific Northwest region of the United States, building of several new power plants has been avoided because of clever pricing strategies for electricity that employ incentives for off-peak use and allow power companies to gain from the overall reduction in energy use (for example, Tidwell 2014). People living in the drought-prone conditions endemic to parts of the western United States have learned how to take three-minute showers and live with brown lawns (or better, recognize that lawns are native to places like Great Britain and have no place in a desert landscape). However, to make any kind of dent in the current problem of anthropogenic climate change, we will need every possible alternative to fossil fuel use put into practice.

Fueling Change by Building Community

We have given far too little thought to the extraordinary importance of fossil fuel energy for our global cultural systems. We grumble about high power bills and fret about the impact of emissions on the acceleration of global climate change, but we have done very little to even acknowledge the ways that nonrenewable energy sources, climate change, water scarcity, and chronic/endemic health problems have been tied together to form a synergistic beast that has been slouching inexorably toward systemic collapse. If instead of thinking that we become more "advanced"/ "modern"/"happy" through greater energy consumption, we turn the measure to energy efficiency, we might find some kind of sustainable solutions. Perhaps the major piece of generalizable local knowledge about the effects of climate change is that communities need to make decisions for themselves (Fiske et al. 2014). Scientists cannot say which decisions are "right" for everyone; instead people in local areas need to take back their communities and reclaim the right to share food, entertainment, and care of each other through what British permaculturist and advocate for the Transition Movement Rob Hopkins calls re-localization (2008).

Knowledge does NOT equal behavior change (Strauss 2009), so simply changing or translating across forms of knowing does not imply a cultural/ behavioral shift. Contingencies in everyday life make decisions to shift daily strategies in significant way a relatively rare event; the time lag can be considerable even if the "spirit is willing." From my personal experience, although I have been researching climate change for 20 years, and my husband works on developing new solar fuel-cell technologies, it was only a few years ago that we were able to muster the time and financial resources to overcome inertia and install a solar hot water system, among other lifestyle and resource-use changes. I have been in complete command of the technical and moral imperatives involved in major energy efficiency strategies, but that was not sufficient for me to overcome the other contingencies of everyday life with small children, two tenure-track jobs, and all the rest. So how can I expect action on the part of others whose lives are even more dependent on structural conditions and do not have the freedom that a tenured academic job permits?

I think that one of the most constructive strategies we can develop to turn things around is to recognize that there are many separately framed "problems" (climate change, water supply, energy resources, obesity, asthma) that in fact have rather singular solution (that is, the cost of addressing multiple issues simultaneously is dramatically less than it would be if you tried to solve these major concerns independently). By combining these efforts, we also might invoke fewer behavioral or cultural changes, and those would be integrated, which by definition will be easier to achieve. An example of this combination effect can be found in looking at chronic health problems alongside climate change and energy issues. In 1991 there was one mall in the entire country of Switzerland. Since then, malls have proliferated like mushrooms, and the average percentage of overweight/obese citizens has gone up by around 10% (Kyle et al. 2007), still somewhat lower than the rate of increase in the overweight/obese populations of the United States and other Western nations. At the same time, industrial as well as to some degree developing countries have reported significant increases of asthma, and these increases have been linked to industrial pollutants and especially the particulates associated with burning diesel fuels (Baer and Singer 2009). Declining during this same period has been the amount of physical exercise experienced by most people in developed countries, and the further decline of public transportation systems across the United States. Other notable trends include the post 1950s loss of community-based activity in favor of malls and consumer-oriented individualist pursuits. Hmmm. Less public transit, more overweight people. More asthma, more industrial pollutants. More malls, more energy consumption, more trash, loss of community. Warmer climate, changing water supplies. Which behaviors do we need to change? Which ones lead to the greatest synergistic effects on the others? Building community seems to be one framework for addressing a wide range of problems.

Anthropology can help us to see how all stakeholder groups—local communities, research teams, political entities—rely on their own histories, sociocultural webs of meaning, and institutional structures to propagate particular beliefs and values or respond to perceived constraints and opportunities. By shifting between a wide-angle perspective and a close-focus zoom, anthropology offers a way to understand the motivations and meanings that are important to local communities as they make decisions about management of natural resources—and also to see how those local views relate to broader regional, national, and even global interests.

Notes

1. www.leukerbad.ch/bergbahnen-sport/erlebnisparkoberemaressen.php
2. www.leukerbad.ch/bergbahnen-sport/erlebnisparkoberemaressen.php

References

Appadurai, A. 1986. *The Social Life of Things*. Cambridge: Cambridge University Press.

Baer, H., and Singer, M. 2009. *Global Warming and the Political Ecology of Health*. Walnut Creek, CA: Left Coast Press, Inc.

Beck, U. 1995. *Ecological Enlightenment*, M. Ritter (Trans.). Atlantic Highlands, NJ: Humanities Press International.

Beniston, M. 2006. Mountain weather and climate: A general overview and a focus on climate change in the Alps. *Hydrobiologia* 562: 3–16.

Beniston, M., Stoffel, M., and Hill, M. 2011. Impacts of climatic change on water and natural hazards in the Alps. *Environmental Science and Policy* 14: 734–43.

European Environment Agency. 2009. *Regional Climate Change and Adaptation: The Alps Facing the Challenge of Changing Water Resources*. EEA Report No. 8. Copenhagen: European Environment Agency.

Fiske, S. J., Crate, S. A., Crumley, C. L., Galvin, K., Lazrus, H., Lucero, L. Oliver-Smith, A., Orlove, B., Strauss, S., and Wilk, R. 2014. *Changing the Atmosphere: Anthropology and Climate Change*. Final report of the AAA Global Climate Change Task Force. Arlington, VA: American Anthropological Association.

Gobiet, A., Kotlarski, S., Beniston, M., Heinrich, G., Rajcak, J., and Stoffel, M. 2014. 21st-century climate change in the European Alps: A review. *Science of the Total Environment* 493: 1138–51.

Hastrup, K., and Rubow, C. (Eds.). 2014. *Living with Environmental Change: Waterworlds*. New York: Routledge.

Hopkins, R. 2008. *The Transition Handbook*. White River Junction, VT: Chelsea Green Publishing Company.

Intergovernmental Panel on Climate Change. 2013. Summary for Policymakers, in T. F. Stocker, D. Qin, G.-K. Plattner, M. Tignor, S. K. Allen, J. Boschung, A. Nauels, Y. Xia, V. Bex, and P. M. Midgley (Eds.), *Climate Change 2013: The Physical Science Basis. Contribution of Working Group I to the Fifth Assessment Report of the Intergovernmental Panel on Climate Change*. Cambridge: Cambridge University Press.

———. 2014. Summary for Policymakers, in contributions of John Agard, Lisa Schipper (Eds.), *Climate Change 2014: Impacts, Adaptation and Vulnerability*. Cambridge: Cambridge University Press.

Jegerlehner, J. 1989 [1906]. *Walliser Sagen*. Zurich: Edition Olms AG.

Kanton Wallis. 2000. Brochure. *Kandidat UNESCO Weltnaturerbe: Jungfrau/Aletsch/Bietschhorn region*.

Kyle, U. G., Kossovskya, M. P., et al. 2007. Overweight and obesity in a Swiss city: 10-year trends. *Public Health Nutrition* 10: 914–19.

Strauss, S. 2009. Global models, local risks: Responding to climate change in the Swiss Alps, in S. Crate and M. Nuttall (Eds.), *Anthropology and Climate Change: From Encounters to Actions*. Walnut Creek, CA: Left Coast Press, Inc., pp. 166–74.

Thaler, R., and Sunstein, C. 2008. *Nudge: The gentle power of choice architecture*. New Haven, Conn.: Yale.

Tidwell, M. 2014. BPA's "Case for Conservation": Helping public power utilities make the business case for energy efficiency. *ACEEE Summer Study on Energy Efficiency in Buildings*. Washington, D.C.: American Council for an Energy-Efficient Economy, pp. 5–337; 5–36.

Chapter 8

Storm Warnings: An Anthropological Focus on Community Resilience in the Face of Climate Change in Southern Bangladesh

Timothy J. Finan and Md. Ashiqur Rahman

Nature of the Problem

Anthropogenic activity since the mid-1800s has led to an unprecedented warming of the planet, and the effects on the planet's socioecological systems are increasingly manifest in the form of both extreme events and slow onset changes of the environment (Adger 1999; Adger et al. 2003). Moreover, the poorest segments of the global population are the most exposed and vulnerable to these events, and they are currently experiencing the direct and devastating impacts of global climate change (World Bank 2010; IPCC 2014; UNDP 2014). In fact, the increased occurrence of climate-related global disasters and the threat of reaching catastrophic "tipping points" have inspired the proposal of a new era, the Anthropocene, wherein man (*Anthropos*) makes an indelible imprint on the planet.

The international development community now considers climate change to be one of the highest priorities facing poor countries, and the major international donors and agencies have *en masse* adopted program strategies labeled *disaster risk reduction*, *climate change adaptation*, and *community resilience*. Large quantities of international resources are now allocated toward achieving these programmatic goals. In particular, the most recent "cutting edge" concept driving development theory and practice is that of resilience, and it has generated a global effort to understand how such resilience is built in households, communities, and livelihood systems (Adger et al. 2011; Tanner et al. 2015) and how it can be measured (Frankenberger and Nelson 2013; Mitchell 2013). Resilience and climate change are now inextricably linked, and the emergent discourse proposes that the outcomes of climate change will depend on the resilience of communities; thus, the urgent development task is to discover and integrate the intervention approaches that build such resilience.

We know that humans occupy center stage in the drama of climate change as both cause and victim. As such, an understanding of complex human systems is fundamental to formulating strategies of response to climate change impacts. This assumption creates ample space for the anthropological perspective and the application of anthropological tools. Since ideas drive actions, this chapter examines the conceptual frameworks that drive the global response to climate change adaptation and seeks to define how anthropology is prepared and poised to enhance the resilience-building mission that appears so urgent. Natural scientists have developed the tools to describe the natural system effects of climate change with some precision, as the series of IPCC reports (IPCC 2007; IPCC 2014) demonstrate. These reports, however, do not adequately address the impacts on local livelihoods cum socioecological systems. What we call climate change "with a human face," described and analyzed at subglobal levels, particularly the local level, is a preeminently anthropological enquiry. An assessment of household or community resilience requires knowledge of household decision making, of power relationships, of institutional contexts, of historical process—all of which call for anthropological concepts and methods. We contend that anthropology inserts this human face at the point where resilience is built (Barnes et al. 2013: 541).

This chapter intends to transmit a double message. First, it presents the analytical framework and methods that are used to identify the current and projected impacts of climate change on local communities and then argues that the anthropological perspective is particularly effective in understanding the social-environmental dynamics underlying climate change. Second, it offers a Bangladeshi case study of a social-ecological system that epitomizes the climate change threat and the governance options available to enhance resilience.

Vulnerability, Adaptation, and Resilience

Over several decades, the global development community has appropriated the terminology of vulnerability, adaptation, and resilience and imbued them with deep technical significance. This is the dominant narrative—the way that global actors and stakeholders—at all levels—think about climate change. We suggest that the substance of a climate change anthropology is woven from these three conceptual threads, but we acknowledge that each has its own historical lineage (see Oliver-Smith this volume).

Vulnerability traces its beginnings to the study of disasters, as the comprehensive Eakin and Luers (2006) review tells us. In disasters theory, risk is understood as the probability of an extreme event occurring, and vulnerability is seen as the level of susceptibility to the effects of the event. In the hands of development practitioners, vulnerability assumes a broader analytical mission. It begins with the definition of a social group (household, community, livelihood system) whose vulnerability is assessed in terms of its *exposure* to a stress (for example, flood), its level of *sensitivity* (or the immediate impact), and its *adaptive capacity* (or the ability to recover). In this formulation, the anthropological instinct is to focus on the determinants of sensitivity and adaptive capacity—household and community characteristics such as asset entitlements and resources; environmental, historical, and institutional contexts; and the logic of local level decision making.

The effective measurement of vulnerability—that is, to be able to say that one household or community is more or less vulnerable than another—is one of the challenges of development practice. Such an assessment requires intensive data collection and analysis using both qualitative and quantitative methods at the local level (for example, Hahn, Riederer, and Foster 2009; ARCC 2013). The anthropological tradition of local level fieldwork, household analysis, and holistic systems thinking well serves the design and implementation of a household or community vulnerability study. For example, the data on which the "capability of a household to respond" is assessed include household level survey data on natural, physical, economic, and social assets, the household decision-making strategies that allocate these assets toward different ends, the outcomes of such strategies (in terms of well-being), and the institutional/historical context that defines power relationships and external linkages. At the same time, we must recognize that vulnerability is a "capability" not a process of change; thus the vulnerability assessment is a static analysis of that capability to respond in the face of pressure.

Adaptation, our second conceptual thread, is a process of change, and it has a rich history in anthropological theory. Adaptation has long been used in anthropology to describe successful or "functional" interactions of human cultures in localized environments as part of a long-term evolutionary process (for example, Cohen 1974: 46; Finan and Nelson 2009). In a shorter time span, Nelson and colleagues see adaptation as "the decision-making process and the set of actions undertaken to maintain the capacity to deal with future change or perturbations to a social-ecological system without undergoing significant changes in function, structural identity, or feedbacks of that system while maintaining the option to develop" (Nelson, Adger, and Brown 2007: 113). Through this lens, adaptation is often measured as the "survival value" of a decision-making process—or, in climate change anthropology, a system outcome that might involve the adoption of a new technology or reorganization toward greater diversification. Although the concept has roots in ecological and evolutionary science, it is widely used by social scientists currently to denote the past or future outcomes of adjustment to climatic change and variability. Adaptation, in terms of climate change, implies a desired pathway of change as a social-ecological system adjusts to the adverse effects of climate change without sacrificing system integrity (Lei et al. 2014: 614). Again, the anthropologist brings the requisite focus on change and a respect for historical processes that govern adaptation.

If vulnerability is a capability of response to stress and adaptation is the longer-term outcome of those responses, resilience has become a measure of strength of that adaptive process. Also borrowed from ecology, resilience is adaptive success—the capacity of a household or community to transform itself in response to stress without a change in structure and function and in such a way that improves its capability to absorb future stresses. Adger (2000) distinguishes between ecological and social resilience and highlights their synergy by suggesting that a resilient ecosystem may reinforce the resilience of the social system (and vice versa). In effect, the resilience concept becomes more attractive to climate change anthropologists owing to its implicit recognition of human agency and power. Resilience theory puts people, their assets, freedoms, and capabilities fundamental to human development and successful livelihoods at the center of analysis (Tanner et al. 2015). The effects of on-the-ground climate change forces are thus measured in terms of community resilience—the ability to protect its livelihood systems (in general form and function) while adjusting itself in such a way that it is *better* prepared to meet future shocks and stresses. The development community, for its part, has fully embraced the challenge of building community resilience (for instance, Mitchel 2013) and identifying the measures of resilience so that effective evaluation can be conducted (FSIN 2014).

Because resilience is a result of highly complex household and community-level decision making and resource allocation, it is a prime target of anthropological interest. We would offer that anthropology uniquely provides an effective lens and the methodological tools to understand the conceptual relationships between vulnerability, adaptation, and resilience. Human beings in institutional and environmental contexts make decisions regarding the mobilization and allocation of resources aimed at achieving welfare goals of one sort or another. These decisions are sometimes constrained by such factors as powerlessness, inequity, and suppression (similar to what Blaikie and colleagues call "root causes" [1974]), but they also demonstrate an ingenious ability to cope and survive. As a fundamental human process, then, adaptation bolstered by enhanced resilience is a function not only of natural system adjustment but also of power, culture, race, class, gender, ethnicity, and the other building blocks of anthropological theory, method, and practice.

An anthropology of climate change has both the conceptual credentials and toolkit needed to understand how changes in the natural system will revise current terms of engagement at the level of communities and households. In some cases, global climate change will manifest itself in abrupt events, such as floods and storms; in other cases, change is experienced as trend lines, such as changing rainy seasons, more frequent drought, higher temperatures, and encroaching desertification. In either case, human populations will sense the pressure and will begin to respond, be it proactively or reactively. Anthropology and its holistic tools and ways of knowing are critical to grasping the direction this change dynamic will take. We illustrate with a case study.

Freshwater Beel Aquaculture in Coastal Bangladesh

Bangladesh has emerged as a global poster child for global climate change. The Fifth IPCC Assessment (WG II) supports the previous (2007) assessment with further evidence that the low-lying coastal areas such as Bangladesh face the greatest and most immediate challenges from climate change through sea-level rise (IPCC 2014).

The projected rise in sea levels, particularly in "contained" water bodies such as the Bay of Bengal, is one aspect of global change that now appears inevitable, and residents in coastal Bangladesh already talk of ongoing land loss to the sea. The most conservative projection (low GHG emissions) reported in the IPCC Fifth Assessment is estimated at 44 cm by the end of the 21st century increasing to 98 cm under the most dire emission assumptions (IPCC 2014: 369). More significant, the rise in sea level in the northern Bay of Bengal combined with warmer sea surface temperatures will likely result in more severe storm and cyclone activity with accompanying high water surges and saltwater intrusion.

In terms of the vulnerability-adaptation-resilience framework, let us first examine *exposure*. In Bangladesh, the magnitude of global warming impacts, especially for coastal populations, is catastrophic. The coastal region covers 32% of Bangladesh, where 35 million people live. A recent study by the Bangladesh Center for Advanced Study (BCAS) found that 81% of households already experience high levels of salinity in rice paddies compared to 2% household a decade ago. As a result, about 80% of the households faced a severe food crisis 10 times or more during the last 10 years (Rabbani et al. 2013: 6). In addition to agriculture, fishing livelihoods are threatened owing to salinity changes in freshwater breeding grounds, and the fragile Sundarbans mangrove forests with their unique biodiversity are in danger of disappearing altogether (Ahmed, Alam, and Rehman 1999; Ali 1999).

Bangladesh is one of the most densely populated and impoverished countries in the world. More than 40% of its estimated population of 170 million lives on less than US $1.25 a day, and its total annual per capita income (GDP/per capita) for 2013 was estimated at US $829 (World Bank 2014). One of the world's mightiest riverine systems, the Ganges—Brahmaputra-Meghna (GBM)—carves its way through Bangladesh, forming a massive delta that covers around one-third of the country's area and is home to 70% of the population. Annually, strong southwestern monsoons move across the Bay of Bengal dumping an average of 2 m of rainfall on most of the country. The GBM system, with a relatively low volume in the dry winter months, fills with water accumulated within the vast Himalayan watershed and floods the deltaic lowlands as it moves to the bay. From a livelihood perspective, moderate annual flooding restores the fabled fertility of the country's paddy land and replenishes the freshwater fish stocks so critical to the food security of the population, especially the poor. However, excessive flooding leads to extreme levels of damage, including high levels of mortality and livelihood disruption. Thus the natural and human systems have negotiated an uneasy balance, where livelihoods are dependent on the annual renewal of the resource base, yet anomalies and extremes can have devastating consequences. The livelihood stress introduced to this deltaic system by climate change is manifest, at least initially, in the extreme flooding, cyclonic events, storm surges, excessive flooding, and backwater effects (saltwater intrusion).[1]

The coastal regions of Bangladesh comprise 36,000 km^2 in area, and because of this area's fertile agriculture and fishing, the population density is particularly high. From Khulna in the west to Chittagong in the east, the delta is a highly complex and dynamic hydrological system driven by a delicate interplay of natural and human influences. Such factors as seasonal flooding, water salinity, and tidal movements affect the diversity and distribution of valuable species and the quality and quantity of agricultural land, which in turn determine the organization and outcomes of local livelihood systems. The *sensitivity* of livelihoods here is high owing to the population density, the widespread poverty of farmers and fishers, and the complex and fragile nature of the ecological systems on which their livelihoods depend.

The *adaptive capacity* of coastal livelihoods is limited by historical, social, and political factors such as the concentrated ownership of resources, pervasive social inequality, lack of political voice and representation, embedded corruption, high rates of illiteracy, alarmingly high levels of child malnutrition, widespread exposure to arsenic toxicity in the drinking water, and the social exclusion of women from public life.[2] Half of all agricultural workers are landless; most are extremely cash-poor and burdened by crushing debt; employment is seasonal and low paid; and the very poor reside in areas prone to flooding and storms. Thus the process of adaptation to climate change is not simply a mechanical adjustment to natural perturbation but a profound sociocultural confrontation with root causes of structural vulnerability.

Let us shift to a local reality and focus on the threat of climate change and the options for building resilience. The southwestern coastal area has hundreds of water bodies called *beels*. These are small geological depressions located at various distances from a coastline and connected to the complex hydrology of a region through canals and minor rivers. They vary in surface area from several acres to several hundred and in quality from brackish and saline to fresh water. Our focus here is on the freshwater beels. During the dry winter months (November to March), beel water levels recede into small isolated pools, and drier areas are cultivated. During the monsoon season (roughly May to September), the beels expand into small lakes. Although beel land is privately owned and managed—mostly by the villagers who reside on the edges—the water in the beel is considered *khas* land, that is, public lands open to all (Barkat, Zaman, and Raithan 2001).

Historically during the dry months, farmers cultivated *boro* rice (the high-yielding irrigated crop) in the beels or pastured their livestock on native vegetation. During the monsoon months, the beels fill with rain and flood waters from nearby rivers and canals of the delta, providing a public access fishing source for the surrounding communities, particularly the landless farmers and laborers. Thus two distinct groups—with different livelihood characteristics and vulnerabilities—coexisted under this institutional arrangement and managed the beel resources.

In the case of the freshwater beels, the production of the giant prawn (*macrobrachium rosenbergii,* in Bangla, *golda*) transformed the local economy over the last 30 years.[3] Farmers who, after seeing the success of shrimp production in the saline beels, experimented with prawn fry in their household ponds and introduced Golda aquaculture in the 1970s. (Rowshanara et al. 2004; Ahmed, Demaine, and Muir 2008), Later in the 1980s, a now legendary innovative farmer, Keramat Ali of Fakirhat (in Bagerhat District), introduced the first prawn enclosure on beel land. As this technology developed and spread, farmers with land rights to the dry beel areas built rectangular enclosures on the beel floor by raising earthen walls called *ghers,* trapezoidal in shape and about 1 m high and .5 m across the top. These enclosures keep rainfall in and flood waters out. During the dry season, boro paddy land is cultivated as was always done; after the monsoon season starts, farmers put postlarvae in a small flooded trench along the inside of the gher walls. As the rains fill up the trench, the postlarvae spill out into the open field, where they grow to adulthood. A monsoon rice crop (*aman*) is cultivated inside the gher, and most farmers practice fish aquaculture, adding various carp species to the mix. Additionally, vegetables and other crops for sale and home consumption can be cultivated along the gher walls. There are now an estimated 130,000 gher producers along the southwest coast occupying an estimated 50,000 ha of land (Khondaker 2007; Ahmed et al. 2008), and many more households derive benefit from provisioning inputs to this system, particularly from the supply of postlarvae, feed, and labor.

From a livelihoods perspective, the gher technology diversifies income, intensifies the productivity of scarce resources (land), and employs greater quantities of abundant labor resources. Unlike *bagda* shrimp production (in saline beels),[4] gher production is the economic mainstay of mainly small-scale farmers, since the intensive management requirements of gher production precludes the construction of large ghers, thus limiting large capital investment interests. Whereas the average saline aquaculture farm is 4.0 ha in size, the prawn fields range from 0.06 to 1.0 ha in size with an average of 0.23 ha (Ahmed et al. 2008: 809). The "vertical" intensification and diversification of production and the integration of resource-scarce farmers into an international market have greatly increased income-earning opportunities for extremely poor farm households.[5] This transformation has particularly benefited women, who now grow vegetables on gher walls and participate in the ancillary income-generating activities that provide the postlarvae fry and feed to the system (Finan and Biswas 2005). In effect, the introduction and widespread adoption of gher technology increased the adaptive capacity of many families in the coastal beel villages by increasing and diversifying household incomes and assets. Input markets also generate seasonal employment for some 400,000 fry collectors as well as for a large number of feed suppliers and marketing agents.

Despite this seemingly successful development story, the effects of climate change loom as dark clouds on the horizon. The governance and management of the gher structures, the biology of prawn production, and the uncertainties of global markets combine with climate change impacts to challenge the sustainability of this livelihood. On the land management side, the gher technology has dramatically altered the landscape of coastal freshwater beels where farmers have built irregular checkerboard patterns of ghers across the beel floor, crossing natural (and public) water canals that feed and flush the hydrologic system. The disruption of the natural hydrological cycle has resulted in severe waterlogging and cases of conflict among adjacent gher owners. In addition, the ghers interfere with the spawning patterns of fish species that traditionally entered the beels.[6]

The risks to gher farming from climate change are potentially catastrophic and include flood, drought, sea-level rise, and sea-surface temperature changes (Ahmed et al. 2014: 27). Ahmed and colleagues (2012) have documented that changes in salinity and in water temperature have negatively affected postlarvae broodstock in rivers and coastal tidewaters, reducing the overall supply of wild PL used to stock the ghers. Ahmed and associates (2014) cite multiple climate change threats to the gher livelihood focusing primarily on sea- level rise, which has altered local ecosystems and increased saltwater intrusion through both surface and groundwater sources. Neither the prawn nor the current rice varieties can tolerate theses level of salinity. These authors also point to the increased frequency of cyclones due to warming sea temperatures. Specifically, Cyclone Sidr (2007) and Cyclone Aila (2009) incurred great damage to the gher social-ecological system by destroying trees and dikes. Seawater storm surges during Aila reached distances 50 km away from the southwestern coasts.

This case study vividly demonstrates the complexity of environment and societal interactions in the context of climate change. We contend that the climate change scenario of increased storms and seawater surges, saltwater intrusion into groundwater aquifers, and the impacts of altered ecosystems on the supply of wild postlarvae stock, exacerbates the negative impacts on the social-ecological system caused by current land use and management activities. There is wide agreement in the literature cited here that the intensive integrated gher technology has brought benefits to scarce-resourced farmers (and their communities) through increased incomes, expanded labor opportunities, and integration into global economies. Nonetheless, the current system lacks the resilience that will allow it to respond to the imminent ecosystem changes provoked by climate change.

Climate Change Anthropology

This case study portends a sobering future for local, gher-based livelihoods along the southwestern coasts of Bangladesh. At this point, nature has already set her course and society must find the adaptive strategy—the means to build the resilience that will protect and strengthen these livelihoods. Climate change anthropology is thus challenged to help define the pathways of adaptation. A previous study of adaptation to climate extremes in southern Arizona (Finan et al. 2002), suggests that building adaptive and transformative capacity requires two major responses: technological adjustments and social reorganization involving greater participation by the state. So, in the case of southwestern Bangladesh, community and livelihood resilience in beel villages will depend on some combination of technological and complex organizational response including a reworking of public policy. In particular, we contend that the urgent action will focus on changes in governance—or what is called adaptive comanagement of this shared resource.

This perspective on adaptive comanagement is echoed in recent studies of climate change impacts in Bangladesh that frame the problem facing shrimp and prawn integrated livelihoods in terms of *social-ecological systems* and call for multilevel governance of complexity (Pouliotte et al. 2009; Ahmed 2013; Ahmed et al. 2014). Ostrom provides a theoretical framework that emphasizes the complexities of social-economic systems and their multiple informational needs: "Understanding a complex whole requires knowledge about specific variables and how their component parts are related. Thus, we must learn how to dissect and harness complexity, rather than eliminate it from such systems" (Ostrom 2009: 420).

Indeed, beels are complex sociohydrological systems influenced by the larger riverine delta dynamics. Thus decisions made by one gher owner might adversely impact the entire beel system and the other gher farmers. Beel dwellers and environmentalists already recognize and are concerned about waterlogging (because of the blocking of canals) and its impact on both human and prawn health. To build resilience against the stressors of sea-level rise, storm surge, and saltwater intrusion, organizational adjustment will be necessary. In addition, the residents around the beels live in equally complex social systems where access to resources, social and political status, and power interact in highly dynamic ways. Following Ostrom, this social system complexity must also be "harnessed" in order to build resilient structures.

Adaptive comanagement requires an equal exchange of relevant information (Ostrom 2009). In the case of the beel socioecological system, that information includes the distribution of livelihood profiles within the villages that surround beels. Livelihood analysis (for example Scoones 1998; Bebbington 1999) provides an appropriate framework for describing the complexity across a social landscape made up of farmers, laborers, fisher people, and shopkeepers, each engaged in livelihood activities. The output of the livelihood assessment identifies vulnerabilities across these livelihood systems and, from a community perspective, describes how these livelihood systems are articulated, interdependent, and synergistic. In effect, the livelihood approach defines the "response space" (Thomas et al. 2005) of the beel communities—that is, their differing capabilities to adjust to changes in the natural system.

Adaptive comanagement requires further anthropological understanding of the social and political networks that support local elites and their respective informal institutions of power. Bangladeshi society is highly stratified, and local elites play a dominant role over the control of local resources. Their power is often corrupt and coercive, using the infamous *mastan*, or local musclemen (goons) to enforce their will (Pandey 2004). These unequal relationships of power affect access to beel land, capital, and output markets. In addition, these local beel systems are linked to larger scale institutions, such as global markets, policy-setting government agencies, and the plethora of NGOs that work in the region. All these stakeholders exert their influence over the activities and the well-being of beel-based livelihoods.

Therefore, within the context of the social-ecological complexity of the beels, adaptive comanagement has become an issue of governance. Recent literature (Pouliotte et al. 2009; Ahmed et al. 2010; Ahmed et al. 2014) argues strongly for the creation of public, private, and civil society (NGO) partnerships, community-based action agendas, and the dissemination of new technology (such as salt-tolerant cultigens) to prepare freshwater bodies for the climate changes stressors. We would add to this list by proposing new models of beel management based on common property approaches that negotiate complexity (Ostrom 2009).

The gher livelihood practiced within the beel social-ecological system cannot be sustained without negotiation with fellow producers in the same system. Without abandoning the principle of private property ownership, one must establish new collaborative institutional rules to deal with climate stressors. From the perspective of an anthropology of climate change, community-based coadaptive management implies three necessary conditions: access to multidisciplinary knowledge, understanding of underlying power relationships, and the development of multiscalar partnerships. For each of these, the anthropological perspective and toolkit provide critical inputs. The foundation of successful negotiation is shared information on beel hydrology, livelihoods vulnerability, and stakeholder networks. Collective decision making that involves beel users, value chain actors, local universities, government agencies, and NGOs is critical to building resilience within an adaptive co-management approach.

Conclusion

Climate change anthropology provides its greatest value when it can provide the human face of climate change—to translate its effects into the realities of real-time households and communities managing their complex resource systems under stress. Here we have proposed that the resilience of a social-ecological system is the relevant target for the anthropological perspective and toolkit. The case of beel freshwater systems using the gher technology provides insights into the contribution that anthropology can make to local level problem solving. In these coastal systems, the urgency of response to climate change cannot be understated, and we agree with current analyses that adaptive comanagement provides the most space for maneuverability in climate change response. Governance as a form of multilevel negotiation based on shared multidisciplinary information is one adaptive pathway to consider.

Notes

1. In northwestern Bangladesh, climate change projections actually portend great incidence of drought during the low-water winter season.
2. In traditional rural society in Bangladesh, Muslim women cannot participate in public affairs without the permission or presence of the husband or male relative. Even going to the market is frowned upon in conservative parts of the region.
3. The more common peneid varieties of shrimp, called *bagda*, are grown in large saline beels in livelihood systems dominated by large capital investment.
4. There is a major livelihood difference between the capitalist production of "shrimp" in the large saline ponds and the production of freshwater "prawns" in the beels. This distinction is not, however, the focus of the present chapter.
5. The Bangladeshi fishery sector is also critically important at the national level where it contributes 4 percent to total GDP and 23 percent to agricultural GDP. In 2011, total fish production was more than 3.3 million metric tons and 2.7 million metric tons came from the freshwater fisheries (Sumi et al. 2015: 46). Prawn is the main variety of fresh water fishery generates foreign exchange. The value of crustacean exports in Bangladesh is estimated at $350 million, behind only garments and remittances as the major earner of foreign exchange.
6. Other environmental problems involve the by-catch loss in postlarvae collection and the diminishing numbers of the apple snail, whose flesh has been a major source of feed for the prawn (see Finan and Biswas 2005, Ahmed and Troell 2010).

References

Adger, W. N. 1999. Social vulnerability to climate change and extremes in coastal Vietnam. *World Development* 27: 249–69.

———. 2000. Social and ecological resilience: Are they related? *Progress in Human Geography* 24(3): 347–64.

Adger, W. N., Brown, K., Nelson, D. R., Berkes, F., Eakin, H., Folke, C., Galvin, K., Gunderson, L., Goulden, M., O'Brien, K., Ruitenbeek, J., and Tompkin, E. 2011. Resilience implications of policy responses to climate change. *WIREs Climate Change* 2: 757–66.

Adger, W. N., Huq, S., Brown, K., Conway, D., and Hulme, M . 2003. Adaptation to climate change in the developing world. *Progress in Development Studies* 3(3): 179–95.

ARCC (African and Latin American Resilience to Climate Change). 2013. *Uganda Vulnerability Assessment Report*. Prepared by TetraTech ARD for USAID. Washington, D.C.

Ahmed, A. U., Alam, M., and Rahman, A. A. 1999. Adaptation to climate change in Bangladesh: Future outlook. In S. Huq, Z. Karim, M. Asaduzzaman, and F. Mahtab (Eds.), *Vulnerability and Adaptation to Climate Change for Bangladesh*. Dordrecht: Springer Netherlands. DOI: 10.1007/978-94-015-9325 0_9.

Ahmed, N. 2013. Linking prawn and shrimp farming towards a green economy in Bangladesh: Confronting climate change. *Ocean & Coastal Management* 75: 33–42.

Ahmed, N., Buntings, S. W., Rahman, S., and Garforth, C. J. 2014. Community-based climate change adaptation strategies for integrated prawn–fish–rice farming in Bangladesh to promote social-ecological resilience. *Reviews in Aquaculture* 6: 20–35.

Ahmed, N., Demaine, H., and Muir, J. F. 2008. Freshwater prawn farming in Bangladesh: History, present status, and future prospects. *Aquaculture Research* 39: 806–19. DOI:10.1111/j.1365-2109.2008.01931.x.

Ahmed, N., Occhipinti-Ambrogi, A., and Muir, J. F. 2012. The impact of climate change on prawn postlarvae fishing in coastal Bangladesh: Socioeconomic and ecological perspectives. *Marine Policy* 39: 224–33.

Ahmed, N., and Troell, M. 2010. Fishing for prawn larvae in Bangladesh: An important coastal livelihood causing negative effects on the environment. *Ambio* 39: 20–29. DOI: 10.1007/s13280-009-0002-y.

Ahmed, N., Troell, M., Allison, E. H., and Muir, J. F. 2010. Prawn postlarvae fishing in coastal Bangladesh: Challenges for sustainable livelihoods. *Marine Policy* 34: 218–27.

Ali, A. 1999. Climate change impacts and adaptation assessment in Bangladesh. *Climate Research* 12: 109–16.

Barkat, A., Zaman, S., and Raihan, S. 2001. *The Political Economy of Khas Land in Bangladesh*. Dhaka: Association for Land Reform and Development.

Barnes, J., et al. 2013. Contribution of anthropology to the study of climate change. *Nature Climate Change* 3: 541–44.

Bebbington, A. 1999. Capitals and capabilities: A framework for analyzing peasant viability, rural livelihoods, and poverty. *World Development* 27(12): 2021–44.

Blaikie, P., Cannon, T., Davis, I., and Cohen, Y. A. 1974. Culture as adaptation, in Y. A. Cohen (Ed.), *Man in Adaptation: The Cultural Present*. Chicago: Aldine Press, pp. 45–70.

Cohen, Y. A. (Ed.). 1974. *Man in Adaptation: The Cultural Present* (2nd ed). New York: Aldine de Gruyter.

Eakin, H., and Luers, A L. 2006. Assessing the vulnerability of social-environmental systems. *Annual Review of Environment and Resources* 31: 365–94.

Finan, T. J., and Biswas, P. 2005. *Challenges to adaptive management: The GOLDA project in southwest Bangladesh*. Report presented to the SANREM CRSP, University of Georgia, Athens.

Finan, T. J., and Nelson, D. R. 2009. Decentralized planning and adaptation to drought in rural northeast Brazil: An application of GIS and participatory appraisal toward transparent governance, in W. N. Adger, I. Lorenzoni, and K. O'Brien (Eds.), *Adapting to Climate Change: Thresholds, Values, Governance*. Cambridge: Cambridge University Press.

Finan, T. J., West, C. T., McGuire, T., and Austin, D. 2002. Processes of adaptation to climate variability: A case study from the U.S. Southwest. *Climate Research* 21: 299–310.

Frankenberger, T. R., and Nelson, S. 2013. *Background Paper for the Expert Consultation on Resilience Measurement for Food Security*. FSIN (Food Security Information Network), http://www.fsincop.net/home/en/

FSIN (Food Security Information Network). 2014. *Resilience Measurement Principles: Toward an Agenda for Measurement Design*. Technical Series 1. Rome: World Food Programme.

Füssel, H.-M. 2007. Vulnerability: A generally applicable conceptual framework for climate change research. *Global Environmental Change* 17: 155–67.

Hahn, M. B., Riederer, A. M., and Foster, S. O. 2009. The livelihood Vulnerability Index: A pragmatic approach to assessing risks from climate variability and change—A case study in Mozambique. *Global Environmental Change* 19(1): 74–88.

IPCC. 2007. Assessment of adaptation practices, options, constraints, and capacity, in M. L. Parry, O. F. Canziani, J. P. Palutikof, P. J. van der Linden, and C. E. Hanson (Eds.), *Contribution of Working Group II to the Fourth Assessment Report of the Intergovernmental Panel on Climate Change*. Cambridge: Cambridge University Press, p. 976.

———. 2014. *Climate Change 2014: Impacts, Adaptation, and Vulnerability. Part B: Regional Aspects. Contribution of Working Group II to the Fifth Assessment Report of the Intergovernmental Panel on Climate Change*, V. R. Barros, C. B. Field, D. J. Dokken, M. D. Mastrandrea, K. J. Mach, T. E. Bilir, M. Chatterjee, K. L. Ebi, Y. O. Estrada, R. C. Genova, B. Girma, E. S. Kissel, A. N. Levy, S. MacCracken, P. R. Mastrandrea, and L. L. White (Eds.). Cambridge: Cambridge University Press, 688 pp.

Khondaker H. R. 2007. Freshwater prawn resources of Bangladesh and role of department of Fisheries for prawn farming development, in M. A. Wahab and M. A. R. Faruk (Eds.), *Abstracts-National Workshop on Freshwater Prawn Farming: Search for New Technologies*. Mymensingh: Bangladesh Agricultural University and Bangladesh Fisheries Research Forum, 9 pp.

Lei, Y., et al. 2014, Rethinking the relationships of vulnerability, resilience, and adaptation from a disaster risk perspective. *Natural Hazards* 70: 609–27.

Mitchell, A. 2013. *Risk and Resilience: From Good Idea to Good Practice*. OECD Development Cooperation Working Paper.

Nelson, D. R., Adger, W. N., and Brown, K. 2007. Adaptation to environmental change: Contributions of a resilience framework. *Annual Review of Environment and Resources* 32(11): 10.

Ostrom, E. 2009. A general framework for analysing sustainability of social-ecological systems. *Science* 325: 419–22.

Pandy, K. P. 2004. Political culture in Bangladesh: Does leadership matter? *Social Change* 34(4): 24–31.

Pouliotte J., Smit, B., and Westerhoff, L. 2009. Adaptation and development: Livelihoods and climate change in Subarnabad, Bangladesh. *Climate and Development* 1: 31–46.

Rabbani, G., Rahman, A., Khandaker, M., and Shoef, I. J. 2013. Loss and damage from salinity intrusion in Sathkira District, coastal Bangladesh. *Loss and Damage in Vulnerable Countries Initiative* (case study report). Bonn: United Nations University Institute for Environment and Human Security.

Rowshanara, M., Islam, K., Ghos, S., Khan J., Pavel, H. T., and Jobon, A. H. 2004. *Paddy Centered Livelihood: An Anthropological Study in Six Villages of Southwest Bangladesh*. Dhaka: IRRI.

Scoones, I. 1998. *Sustainable Rural Livelihoods: A Framework for Analysis*. Working Paper 72. Brighton: Institute for Development Studies.

Sumi, K., Sharker, M. R., Ali, M. I., Pattader, S. N., Ferdous, A., and Ali, M. M. 2015. Livelihood status of Gher farmers of Beel Dakatia in Khulna district, Bangladesh. *International Journal of Aquatic Science* 6(1): 45–53.

Tanner, T., Lewis, D., Wrathall, D., Bronen, R., Cradock-Henry, N., Huq, S., Lawless, C., Nawrotzki, R., Prasad, V., Rahman, A., Alaniz, R., King, K., McNamara, K., Nadiruzzaman, M., Henly-Shepard, S., and Thomalla, F. 2015. Livelihood resilience in the face of climate change, *Nature Climate Change* 1: 23–26.

Thomas, D., Osbahr, H., Twyman, C., Adger, N., and Hewitson, B. 2005. *Adaptive: Adaptations to Climate Change among Natural Resource-Dependent Societies in the Developing World: Across the Southern African Climate Gradient*. Technical Report 35. Norwich: Tyndall Centre for Climate Change Research, East Anglia University.

United Nations Development Program (UNDP). 2014. *Human Development Report 2014, Sustaining Human Progress: Reducing Vulnerabilities and Building Resilience*. New York: United Nations.

World Bank. 2010. *World Development Report, 2010: Development and Climate Change*. Washington D.C.: World Bank.

———. 2014. *World Development Indicators 2014*. Washington, DC: World Bank.

Chapter 9

Correlating Local Knowledge with Climatic Data: Porgeran Experiences of Climate Change in Papua New Guinea[1]

Jerry K. Jacka

In a 2008 article in *Current Anthropology* Susan Crate outlined a research agenda for anthropological climate change research that incorporates two dimensions: an emic approach that highlights the cultural models that indigenous peoples use to perceive the effects of climate change and an etic approach that takes findings from Western scientific knowledge to "complement local understandings and facilitate positive action" (2008: 575). In my initial contribution to the first edition of *Anthropology and Climate Change* (Jacka 2009) I pursued the first component of Crate's agenda by exploring the cosmological implications of climate change among the people of Porgera in the highlands of Papua New Guinea. In this chapter I undertake the second part of the research agenda by correlating local knowledge with climatic data to examine the convergences and divergences of local knowledge and scientific knowledge. I argue that understanding the varying responses that local knowledge takes to climatic anomalies is essential for future global adaptation strategies designed to mitigate the effects of a changing climate.

In the highlands of Papua New Guinea extreme El Niño Southern Oscillation events create droughts and frosts that detrimentally affect the local tropical subsistence food base. In response highlanders pursue a suite of ecological and social practices to mitigate extreme climatic events. They are keen observers of climatic norms and anomalies. In this chapter I document these practices and compare oral testimonials of climate change with local climatic data. Overall, local perceptions of climate change closely mirror temperature and rainfall records from official sources. However, while strong anomalies in both wet and dry climate events affect subsistence, only extreme drought events enter into local discourse and knowledge that prompt mitigation measures. I argue that understanding these kinds of varying local responses to climatic anomalies is essential for future global adaptation strategies designed to mitigate the impacts of a changing climate.

As Sarah Strauss and Ben Orlove (2003) indicate in their introduction to the edited volume *Weather, Climate, Culture*, it is crucial to distinguish between climate, season, and weather or, in other words, between long-term, medium-term, and short-term climatological forces. While climate, a long-term trend, has received the most attention in the press, it is the extreme weather and seasonal events that are apt to cause much greater human suffering in the coming years. Pielke and Bravo de Guenni (2004) argue that the dominant model in climate science, like that used by the Intergovernmental Panel on Climate Change (IPCC), is overly concerned with modeling future scenarios of climate change without modeling the more immediate human vulnerabilities to climate change (see also Feddema et al. 2005, but cf. with the IPCC's Fifth Assessment Report). Thus there is a danger in focusing on the average at the expense of the extremes, particularly for people whose livelihoods are dependent on subsistence agricultural practices.

In this context, I situate my analysis via medium-term trends, in particular, effects of seasonal weather fluctuations related to the El Niño Southern Oscillation (ENSO) and the implications for more frequent El Niños for Porgeran people of Papua New Guinea. Over the years the IPCC has indicated that ENSO events have become more persistent in recent decades. The IPCC's Third Assessment Report noted that El Niño events "became more frequent, persistent, and intense during the last 20 to 30 years compared to the previous 100 years" (IPCC 2001: 6). The AR4 (Fourth Assessment Report) supports this statement, claiming that "more intense and longer droughts have been observed over wider areas since the 1970s, particularly in the tropics and subtropics" (IPCC 2007: 6). The latest report (AR5) argues that "ENSO will remain the dominant mode of interannual variability with global influences in the 21st century and due to changes in moisture availability ENSO-induced rainfall variability on regional scales will intensify" (Christensen et al. 2013: 1243). In the PNG highlands, El Niños are frequently accompanied by severe and prolonged droughts that are in turn associated with frosts (above 1,500 m) that destroy the staple of highlands populations, the fragile vines of the sweet potato (Brown and Powell 1974; Waddell 1975; Wohlt et al. 1982; Sillitoe 1996). The 1997 El Niño, for instance, affected crop production throughout the country's population of 5 million people, resulting in over half a million people receiving direct food aid from government and NGO aid groups (Ellis 2003).

Understanding the dynamics of climate change in PNG is of far greater urgency than just the impact on the populations living in and around the Porgera Valley. With the largest extant block of tropical rain forest in the Western Pacific and Southeast Asia (Primack and Corlett 2005: 8), the forests of the island of New Guinea deserve special attention for their role in providing carbon sinks for the increase of anthropogenic carbon dioxide in the atmosphere (Detwiler and Hall 1988; Achard et al. 2002)—one of the key greenhouse gases linked to global climate change (IPCC 2007). An increase in El Niños and droughts could result in more massive fires like the ones that raged through the tropical rain forests of Indonesia in 1997, 1998, and 2005 (Power et al. 2013). These changes in forest cover perpetuate a positive feedback cycle, such that the loss of forest cover contributes to warming, which contributes to more droughts and fires leading to more loss of forest cover (Feddema et al. 2005).

Changing Ecologies, Changing Landscapes

When I arrived in Porgera (Figure 9.1) in late 1998, the 1997 El Niño had been over for a year, but the effects, consequences, and memories of the event were still prevalent. People talked of how all the local springs dried up and the nearest water was over 3 km away. The smoke from the forest fires in Indonesia had caused the sun and moon to turn red, foreshadowing apocalyptic events that the recently Christianized Porgerans had heard of from the Book of Revelation. A solar eclipse and a strong earthquake that also affected the region at this time upset fundamental cosmological principles that people linked to larger moral and social problems in the valley. Large numbers of migrants, who had come from higher-altitude areas, were still living in the Porgera Valley, waiting for their gardens back home to begin producing again. Many of the migrants reported leaving behind the sick and elderly who could not travel to Porgera, many of whom were now dead.

Elevations in the Porgera Valley range from 900 to over 3,700 m above sea level. In the region where I worked (at about 1,800 m), frosts had not affected the gardens, but drought had dried up all the sweet potato mounds, creating severe famine conditions. Above 2,200 m, frosts and droughts had combined to completely destroy higher-altitude gardens. The migrants were all from the high-altitude grasslands to the south of the Porgera Valley, where the lowest elevations are around 2,300 m. The severity of the famine was mitigated by the presence of a multinational mining venture in the valley undertaken by Porgera Joint Venture (PJV), operator of the Porgera Gold Mine, which annually has produced about 900,000 oz. of gold since opening in 1990. PJV, in concert with the PNG government and AusAID (Bourke and Allen 2001), distributed bags of rice throughout the valley, while people's gardens withered in the sun and frost.

I eventually came to understand that the events of the 1997 El Niño were merely an exceptional climatic anomaly in an environment that Porgerans believed had been under change for some time. The most discussed transformation was the increase in temperature that everyone felt had been underway for the last few decades (cf. Salick and Byg 2007: 13). Further evidence for this was seen in the movement of species from one ecological zone to another. In an indigenous ecological perspective, Porgerans characterize their landscape in three ecological zones: *wapi*, which stretches

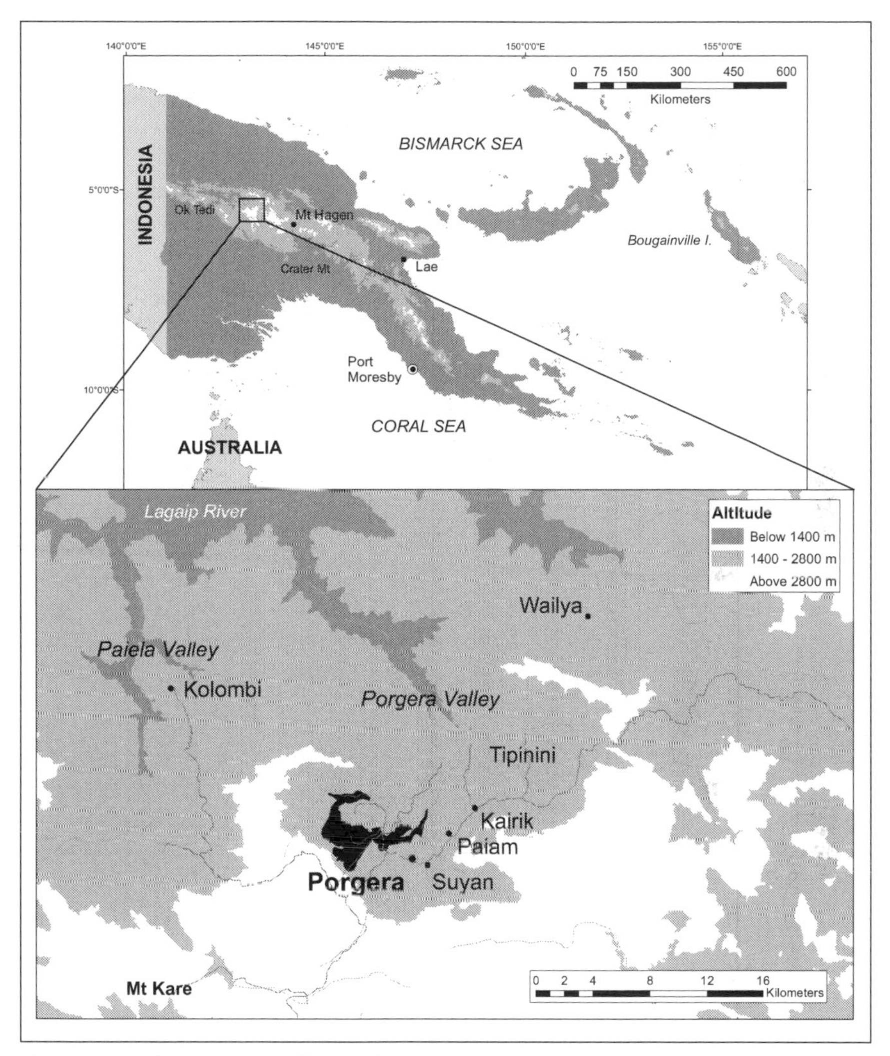

Figure 9.1 The Porgera Valley and Porgera Mine in western Enga Province, PNG

from the lowest reaches of the valley up to around 1,600 m; *andakama*, starting at the upper end of the wapi and reaching to about 2,200 m; and *aiyandaka*, the highest-altitude ecological zone. A number of features characterize these zones. Wapi is hot, humid, has a lot of insects, and is marked by species such as fruit pandanus (*Pandanus conoideus*) and wild bananas (*Musa* sp.). *Andakama* literally means the place of houses and clearings and is where people traditionally lived and gardened. The aiyandka is a cold, rainy zone with waterlogged soils and marked by a particular variety of wild nut pandanus (*Pandanus brosimos*), *wapena*. People also mentioned that several species of birds that used to be found only in the wapi could now be found in the andakama. Domesticated banana species that used to grow only in the lower section of the andakama could now be planted and would grow well in the aiyandaka. In essence, people perceived that species had shifted upward in response to the general warming trend.

Porgerans also reported that foreign species invasions accompanied the warming temperatures. A grassy weed with detachable seeds—locally called "poor man's grass"—that gets stuck in people's clothing grows rampant in fallowing sweet potato gardens. Elders report having seen this grass for the first time when they were in their teens. A beetle that attacks and kills casuarina (*Casuarina oligodon*) trees, the most important tree species for domestic use as firewood, was reported to have arrived in the valley only in the last decade. Many people pointed out to me the large stands of dying casuarina around the valley caused by beetle infestation.

When I was in Porgera between 1998 and 2000, I didn't collect or have access to climatic data, but what I did gather were numerous oral stories about climatic changes and severe weather events (see also Jacka 2009). On analyzing my climate oral histories, I became interested in how to reconcile local perceptions of climate change with data on rainfall and temperature. I also wondered how environmental anthropologists could use climate data to conceptualize severe disturbances and explore human responses to these disruptions. In the following section, I address these relations between the observed and the perceived weather patterns in Porgera.

Correlating Local Knowledge with Climatic Data

In addition to drought and severe weather, extreme seasonal trends are another area of concern locally. On average, Porgera shows little seasonality. Monthly rainfall records I obtained from PJV (1974 to 2011) show a low average of about 250 mm of rain per month from June to September, which increases to a high of about 350 mm per month during the rest of the year—a pattern that one group of climatologists has termed "wet and very wet" (McAlpine, Keig, and Falls 1983: 61). Recent El Niños in 1982 and 1997 have fit the pattern of drought and frost described here. The 1987, 1991, and 1993 El Niños, however, while also logging some of the lowest rainfall amounts on record for Porgera, have paradoxically also resulted in some of the highest monthly rainfall totals. This fact could be accounted for by a pattern that some highlands researchers have noted of extremely heavy rainfalls preceding major El Niño events (Bourke 1989; Brookfield and Allen 1989).

To compare the relationship between ENSO events and rainfall in Porgera, I overlaid the ENSO index as calculated by the National Oceanic and Atmospheric Administration's Climate Prediction Center[2] on a rainfall index I created[3] with monthly precipitation data (Figure 9.2). In Figure 9.2, positive standard deviations are periods where more rain fell than normal, while negative standard deviations indicate less rainfall. As can be

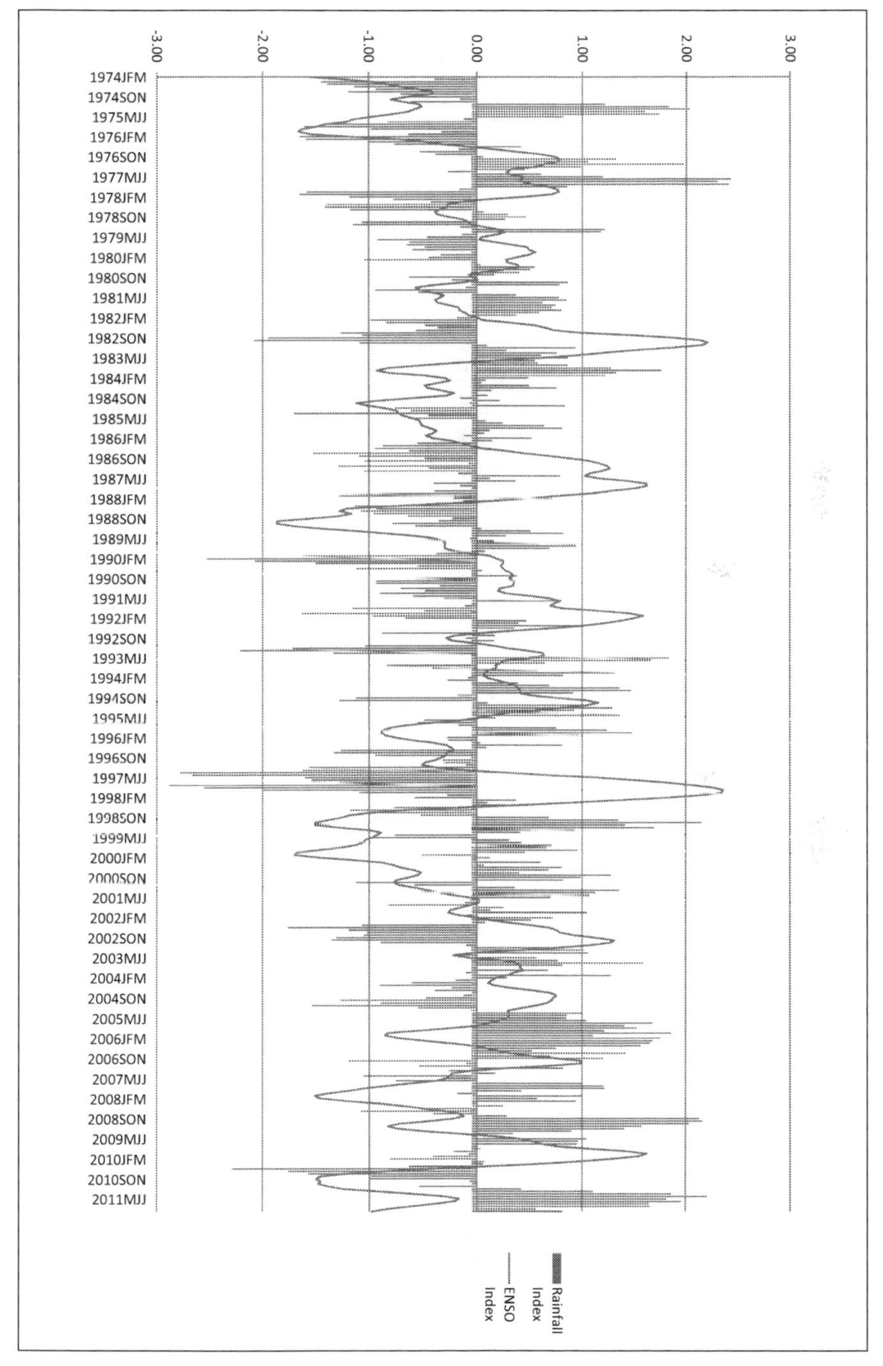

Figure 9.2 Rainfall index and ENSO index

seen, the strong El Niños of 1982 and 1997 (>2 SD) are associated with severely reduced rainfall in Porgera. In general, very strong El Niños mean less rainfall; however, there are periods (1990, 1993, and 2010) when this relationship is not as strong.

While ENSOs and droughts in the western Pacific typically generate the most attention among the international community and subsistence horticulturalists, droughts and extreme wet periods in particular sequences result in the lowest yields of sweet potatoes. As Robert Bourke (1989) has shown, the most severe crop losses occur after periods of extreme rainfall that are followed by droughts, even minor ones. Wet periods that occur in the three to ten weeks after initial planting reduce tuberous root development and limit it to the top 15 to 25 cm of the soil profile of a mound. A drought after this period of time, when the tubers are developing, results in a much reduced output. And, of course, a drought accompanied by a freeze completely kills the plants at this critical stage of tuber development. Bourke argues that droughts alone fail to cause famines and that extreme waterlogging of the soil is a better predictor of crop failures. In general, oral testimony by Porgerans highlights that extreme climatic events of too much or too little rain affect their livelihoods, but nevertheless nearly all of their oral history focuses on droughts.

In an attempt to quantify both the duration and intensity of wet and dry periods, I modified Allen's (1989) dry season index for PNG and derived three climate types (normal, wet, and dry) from the rainfall index to determine what constitutes an extreme climate event[4] (Table 9.1). The wet periods are indicated by the gray bars and the dry by the white bars; both show the duration of the climate anomaly, while the number indicates the summed standard deviation index score and hence the intensity of the climatic anomaly (the ellipses in the gray bars indicate continuation from the previous year). As Figure 9.3 shows, between the beginning of 1974 and the end of 2011, there have been seven extremely wet periods and eight extremely dry periods. Of the eight dry periods, only three—1982, 1997, and 2002—overlap (although not perfectly) with official El Niño periods according to NOAA's Climate Prediction Center.

Porgerans note that annually a few months into the wetter season (around January) is a time of famine in the valley that is alleviated only by the fact that nut pandanus season begins about this time. Famine in this context must be understood from its cultural perspective in that there are often many kinds of food available, but just not the beloved sweet potato (cf. Sillitoe 1996: 79–86). Occasionally, however, there are times when an overabundance of rainfall affects Porgeran subsistence. The wet period data correlate closely with Porgeran perceptions: five of the seven wettest periods occur around the beginning of the year. I should note that wet periods in Porgera are not linked to La Niña events in any significant way. There is some overlap, but the data are not as consistent as strong positive ENSO values and severe El Niños that remain drivers for dry periods in Porgera.

Table 9.1 Extreme wet and dry periods in Porgera

	DJF	*JFM*	*FMA*	*MAM*	*AMJ*	*MJJ*	*JJA*	*JAS*	*ASO*	*SON*	*OND*	*NDJ*
1974							-5.5					
1975				8								
1976			-5									
1977								8				
1978												
1979												
1980												
1981												
1982										-7		
1983								7.5				
1984	...											
1985												
1986												
1987												
1988												
1989												
1990				-9								
1991												
1992												
1993			-5.5									
1994												
1995												
1996												
1997							-19.5					
1998											6.5	
1999		...										
2000												
2001												
2002									-8			
2003												
2004												
2005								20				
2006					...							
2007												
2008											9	
2009		...										
2010								-7				
2011							11.5					

Temperature data from 1988 through 2003 from Porgera that I received from PJV show less variability than precipitation does. As would be expected in the aseasonal tropics there is little month-to-month variation and very small monthly standard deviations. Average monthly highs range between 19.7° and 20.6°C, and average monthly lows range from 11.3°

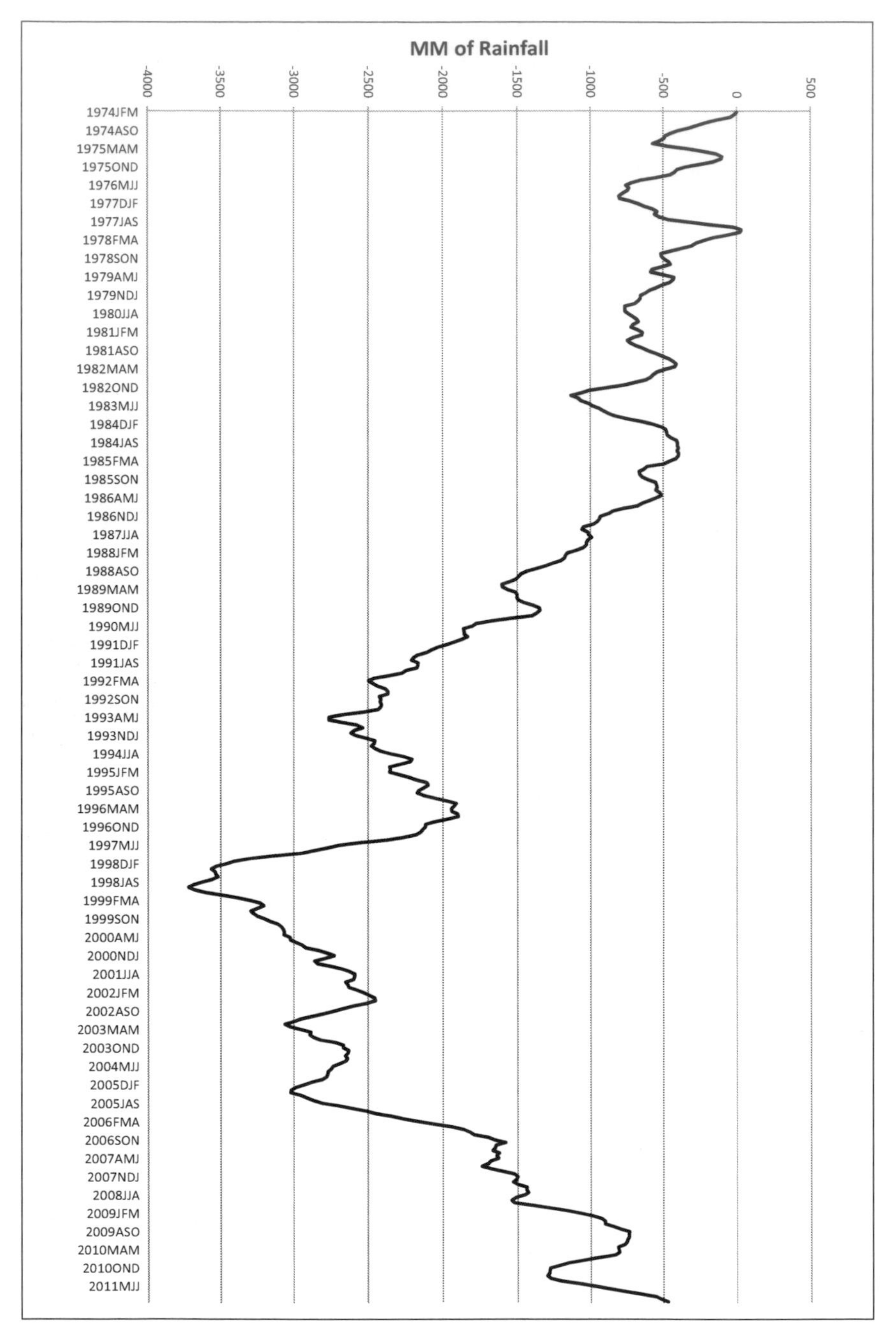

Figure 9.3 Cumulative deviations from mean rainfall

to 12.3°C. The general pattern of variability between lower and higher temperatures is that the wet season in Porgera typically has higher average temperatures than the drier season. For example, the lowest mean monthly temperature (9.8°C) occurred in the middle of the 1997 El Niño in September. To examine whether the increasing temperatures and drier climate recorded in oral histories were represented in the climatic data, cumulative differences from the mean were calculated for both rainfall and temperature. In Figure 9.3, downward slopes indicate negative deviations from the mean of the three-month running average rainfall, whereas upward slopes indicate positive deviations from the mean. The data in this example show that between the 1977 OND[5] average until the 1998 SON average, there was a 21-year general decline in mean rainfall, with the exception of a 3-year period between 1993 AMJ and 1996 AMJ. Since the end of the 1998 El Niño, the climate has been wetter than average over the long term. The climatic data demonstrate the importance of paying close attention to local knowledge about climatic patterns. In early 2000, at the end of the time when I was collecting data on perceptions of climate change, there had been a general decrease in rainfall over a two-decade period. Statements that the climate was "becoming drier" correlate closely with the rainfall data. However, in 2006, people were complaining about waterlogged soils and decreased crop production in their gardens. Given that they had just come out of the longest, most intense wet periods (19 consecutive months when rainfall surpassed 0.5 standard deviations each month), this is not surprising.

Figure 9.4 demonstrates a more complex picture of temperature deviations. From January 1988 to September 1998, the maximum temperature shows a warming trend that is especially prevalent from August 1994 to September 1998. The minimum temperatures, however, show a reverse trend, generally decreasing from January 1988 to November 1997. In essence, local statements are partially correct. It was indeed getting "hotter" in Porgera from the perspective of the late 1990s, if one was talking about the daytime temperatures. Nighttime temperatures were becoming "colder," though. Since people spend little time outside in the early dawn hours because of fear of nighttime spirits (Jacka 2010), I can assume only that their perceptions of a warming trend were based on their experiences and observations of afternoon temperatures. Additionally, climate interviews coincided with a surge in millenarian and apocalyptic Christian narratives of the world ending in the year 2000, and many interviewees mentioned "the fire in the sky" that God was bringing to usher in the Second Coming and the impending time of tribulations (see also Stewart and Strathern 1997). Asked about the fire in the sky that was coming, Epe Des, the Seventh Day Adventist pastor, remarked: "The Bible says that 2,000 years after Jesus died, the ground will end. At this time

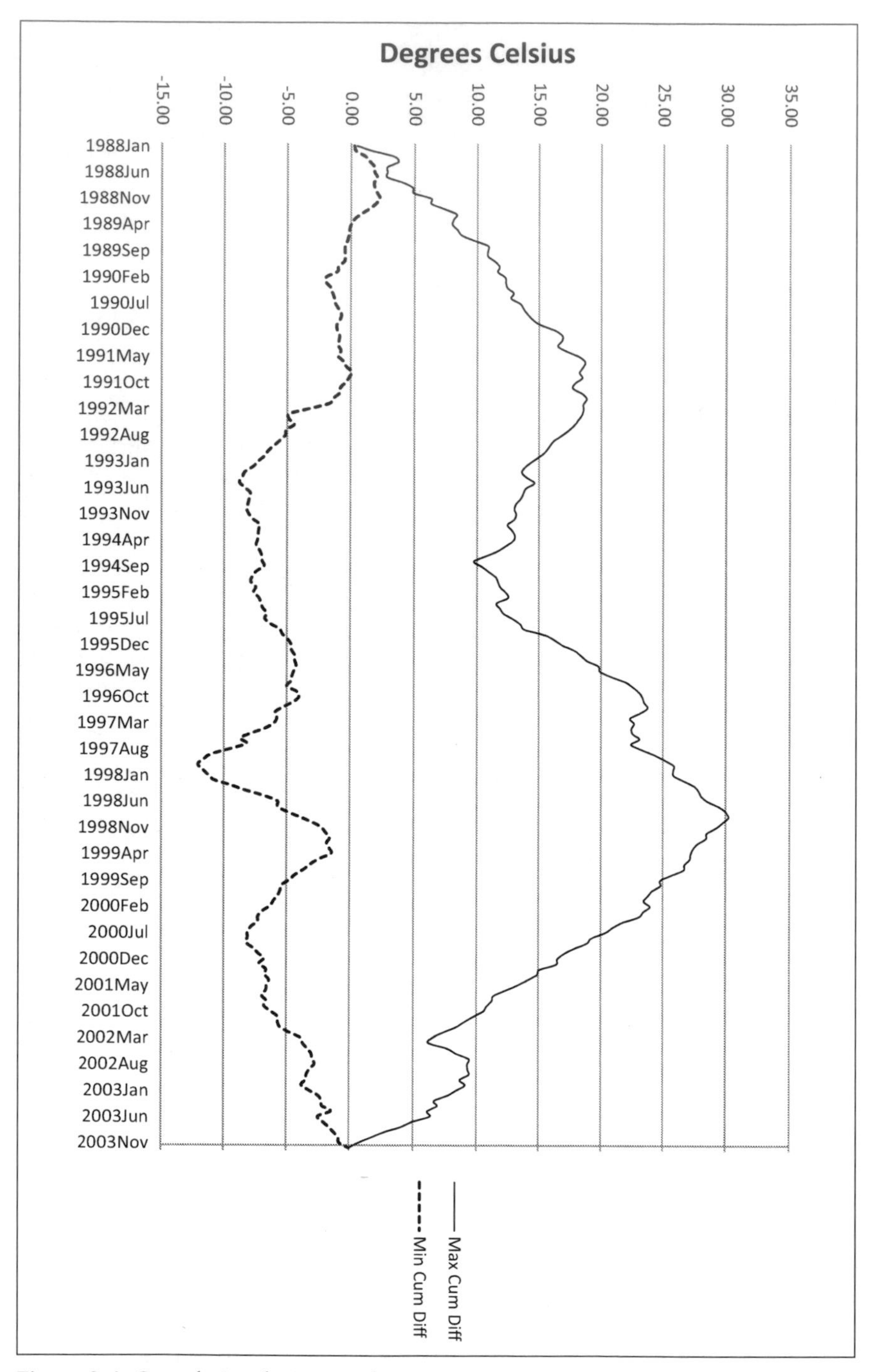

Figure 9.4 Cumulative deviations from temperature

there will be a great power that comes from heaven, this power will be a big fire. Jesus will come riding on a chariot with horses, there will be angels blowing horns, and then God will send a huge fire down and burn up the ground, the trees, the gardens, everything." Epe's comments had been stated to me just after the drought and the fires of the 1997–1998 El Niño. The validation of the Second Coming was read through a number of events that occurred leading up to this: a tidal wave at Aitape, an eclipse, the drying up of streams and rivers, the rising of a blood-red moon caused by the fires in Indonesia, and finally, an earthquake.

Conclusion

In this chapter I linked local perceptions of climatic anomalies and climate change with local precipitation and temperature data, and local discussions of drought and its accompanying responses with global data on the timing and intensity of ENSO occurrences. Overall, local perceptions of climate change corroborate the climatological data. Climate interviews conducted in 1999 and 2000 stressed an overall warming and drying trend, which was reflected in the monthly maximum temperatures (daytime temperatures) and in monthly precipitation data. While I conducted less formal interviews on climate change in 2006, numerous people were complaining about the reduction in crop yields due to a wetter climate. Cumulative deviations from the mean rainfall in the early 2000s do, indeed, indicate greater than average rainfall over this period. In summary, in the absence of climatological data, oral histories and local knowledge are strong proxies for climate change research. Policy makers, funding agencies, and researchers should explore in greater depth these linkages between local knowledge and climate data in order to help mitigate the effects of climate change on indigenous knowledge producers.

Notes

1. Research in Porgera was funded by the Wenner-Gren Foundation for Anthropological Research and North Carolina State University. The National Research Institute of Papua New Guinea provided research permission. I thank all these institutions for their support. My greatest thanks goes to the people of Tipinini in the eastern Porgera valley, especially Epe Des, Peter Muyu, and Ben Penale.
2. www.cpc.ncep.noaa.gov/products/analysis_monitoring/ensostuff/ensoyears.shtml
3. To create the index, I standardized seasonal rainfall with *Z* scores—the observed rainfall from the three-month mean was subtracted from the mean for that period, and the result was divided by the standard deviation of that period. For example, January, February, and March (JFM; February, March, and April [FMA]; March, April, and May [MAM], and so on) rainfall amounts were averaged to provide an overall mean and standard deviation for 12 three-month periods each year over the years 1974 to 2011.
4. I defined normal climate months as periods when the three-month running mean rainfall averages were between 0.5 and -0.5 standard deviations from the overall mean for those three months. Normal climate months were scored 0. I scored wet months as

follows: 0.5 for standard deviations between 0.5 and 0.99, 1.0 for standard deviations between 1.0 and 1.49, and so forth. I scored dry months the same way, albeit they received a negative score. To classify as an extreme wet or dry period, four consecutive months had to have an absolute score greater than 0.5 for each month, with at least two months greater than one standard deviation. I then summed the modified index scores for each month to indicate the intensity of the wet or dry period.

5. OND is the three-month mean for October, November, and December. Running three-month means are standard in climatological analysis.

References

Achard, F., et al. 2002. Determination of deforestation rates of the world's humid tropical forests. *Science* 297: 999–1002.

Allen, B. 1989. Frost and drought through time and space, Part I: The climatological record. *Mountain Research and Development* 9(3): 252–78.

Bourke, R. M. 1989. The Influence of soil moisture on sweet potato yield in the Papua New Guinea highlands. *Mountain Research and Development* 9(3): 322–28.

Bourke, R. M., and Allen, M.G. (Eds.). 2001. *Food Security for Papua New Guinea: Proceedings of the Papua New Guinea Food and Nutrition 2000 Conference, PNG University of Technology, Lae, 26–30 June 2000.* Canberra: Australian Centre for International Agricultural Research.

Brookfield, H., and Allen, B. 1989. High-altitude occupation and environment. *Mountain Research and Development*: 201–09.

Brown, M., and Powell, J. M. 1974. Frost and drought in the highlands of Papua New Guinea. *Journal of Tropical Geography* 38: 1–6.

Christensen, J. H., et al. 2013. Climate phenomena and their relevance for future regional climate change, in T. F. Stocker et al. (Eds.), *Climate Change 2013: The Physical Science Basis. Contribution of Working Group I to the Fifth Assessment Report of the Intergovernmental Panel on Climate Change.* Cambridge: Cambridge University Press, pp. 1217–1308.

Crate, S. 2008. Gone the bull of winter? Grappling with the cultural implications of and anthropology's role(s) in global climate change. *Current Anthropology* 49(4): 569–95.

Detwiler, R. P., and Hall, C. 1988. Tropical forests and the flobal carbon cycle. *Science* 239(4835): 42–47.

Ellis, D. M. 2003. Changing Earth and sky: Movement, environmental variability, and responses to El Niño in the Pio-Tura region of Papua New Guinea, in S. Strauss and B. Orlove (Eds.), *Weather, Climate, Culture.* Oxford: Berg, pp. 161–80.

Feddema, J. et al. 2005. The importance of land-cover change in simulating future climates. *Science* 310(5754): 1674–78.

IPCC (Intergovernmental Panel on Climate Change). 2001. *Climate Change 2001: Synthesis Report. Summary for Policymakers.* Geneva: IPCC Secretariat.

———. 2007. *Climate Change 2007: The Synthesis Report. Summary for Policymakers.* Geneva: IPCC Secretariat.

Jacka, J. K. 2009. Global averages, local extremes: The subtleties and complexities of climate change in Papua New Guinea, in S. Crate and M. Nutall (Eds.), *Anthropology and Climate Change: From Encounters to Actions.* Walnut Creek, CA: Left Coast Press, Inc., pp. 197–208.

———. 2010. The spirits of conservation: Ecology, Christianity, and resource management in Highlands Papua New Guinea. *Journal for the Study of Religion, Nature and Culture* 4(1): 24–47.

McAlpine, J., Keig, G., and Falls, R. 1983. *Climate of Papua New Guinea*. Canberra: CSIRO, Australian National University.

Pielke, R. A., Sr., and de Guenni, L. B. 2004. How to evaluate vulnerability in changing environmental conditions. In P. Kabat et al. (Eds.), *Vegetation, Water, Humans, and the Climate: A New Perspective on an Interactive System*. New York: Springer, pp. 537–38.

Power, S., et al. 2013. Robust twenty-first-century projections of El Niño and related precipitation variability. *Nature* 502: 541–45.

Primack, R., and Corlett, R. 2005. *Tropical Rain Forests: An Ecological and Biogeographical Comparison*. Oxford: Blackwell Publishers.

Salick, J., and Byg, A. (Eds.). 2007. *Indigenous Peoples and Climate Change*. Oxford: Tyndall Centre for Climate Change Research.

Sillitoe, P. 1996. *A Place against Time: Land and Environment in the Papua New Guinea Highlands*. Amsterdam: Harwood Academic Publishers.

Stewart, P., and Strathern, A. (Eds.). 1997. *Millennial Markers*. Townsville: Centre for Pacific Studies, James Cook University of North Queensland.

Strauss, S., and Orlove, B. (Eds.), 2003. *Weather, Climate, Culture*. New York: Berg.

Waddell, E. 1975. How the Enga cope with frost: Responses to climatic perturbations in the central highlands of New Guinea. *Human Ecology* 3(4): 249–73.

Wohlt, P. B., et al. 1982. *An Investigation of Food Shortages in Papua New Guinea: 24 March to 3 April, 1981*. Boroko: Institute of Applied Social and Economic Research.

Chapter 10

Speaking Again of Climate Change: An Analysis of Climate Change Discourses in Northwestern Alaska

Elizabeth Marino and Peter Schweitzer

Since 2009, when the first edition of this book was released, climate change outcomes in Northwestern Alaska and throughout the Arctic continue to be significant and to create multiple challenges for Arctic residents owing to increases in overall temperatures, storminess, and erosion rates (Huntington 2000; ACIA 2005: 997; Hinzman et al. 2005; McNeeley 2012; Cochran et al. 2013). In some cases these ecological outcomes in combination with changing patterns of development and human occupancy have created critical disaster scenarios (Marino 2012, 2015). Particularly threatening is the increase in habitual flooding to rural coastal and river settlements, which compromises the integrity and long-term viability of a number of Alaskan Native villages (USGAO 2003, 2009). In Alaska, climate change outcomes translate into flooding disasters that happen with greater and greater frequency. When flooding becomes a recurrent or habitual ecological state, it is not only people's homes but also their livelihoods, cultural practices, ceremonies, and methods of meaning-making that are compromised.

Questioning whether or not climate change is occurring is no longer relevant considering the strong scientific consensus that global climate change is a reality and that change is in the direction of warming on a global scale (ACIA 2005: 994; Anderegg et al. 2010). We define global climate change here as changes in overall climate patterns across a given space and time. In other words, global climate change is a distinct phenomenon of a global scale but one that has diverse local effects.

Despite general scientific consensus about the drivers of climate change, there remain many uncertainties about both the ecological outcomes of climate change and human responses to it. Important and unresolved research questions include: what will the diversity of climate change outcomes be on local scales; how will local, national, and international institutions respond and adapt to those outcomes (Agrawal 2010); and will institutional interventions improve at-risk communities' adaptive-capacity to respond to diverse outcomes (Marino and Ribot 2012). Critical to understanding outcomes of intervention is an analysis of how discourses of climate change evolve and who participates in creating and legitimizing those discourses (Foucault 1972, 1977; Woolgar 1986; Marino and Ribot 2012; Rebotier 2012). As consciousness of "climate change" emerged on a popular scale, "climate change has woven its way into the general consciousness worldwide" (Lorenzoni and Pidgeon 2006: 75). "Global climate change" did not necessarily come into worldwide consciousness through local experience but rather through global public discourses (ibid.).

This chapter explores how global discourses of climate change can both open up new opportunities for political engagement of historically marginalized groups and also limit the kinds of participation seen as legitimate from these same groups. We also discuss the limitations of restricting discourses about the Arctic and Arctic communities—even communities at great risk of flooding—to a discussion of climate change. This research was initially part of an interdisciplinary project investigating the relationship between fresh water, climate, and humans in the Arctic and has been continued through a research project on migration and relocation in response to flooding in rural Alaska. Our work involved spending time in five communities in northwestern Alaska documenting, among other things, changes observed in the landscape and how those changes were perceived and understood at the local scale. We built on this initial research with a project that investigated habitual flooding and climate change adaptation response in Shishmaref, Alaska, a community on the Seward Peninsula that is seeking to relocate because of flooding and increasing erosion linked, in part, to climate change outcomes.

Climate Change in the Arctic

Scientists have shown that the climate in the Arctic is changing more rapidly than other areas of the world as a result of global climate change, an effect known as polar amplification. From 1954 to 2003, the mean annual atmospheric surface temperature in Alaska and Siberia has risen between 2° and 3°C, with warming particularly salient during winter and spring (ACIA 2005: 992). In response, snow and ice features have diminished, and permafrost is melting as its boundaries move north, causing erosion and foundation problems for structures in Alaska (ACIA 2005: 997).

Confounding these amplified effects of climate change in the Arctic is a distinct lack of long-term data. As the Arctic Climate Impact Assessment (ACIA) makes clear: "The observational database for the Arctic is quite limited, with few long-term stations and a paucity of observations in general, making it difficult to distinguish with confidence between the signals of climate variability and change" (2005: 22). Although in some sense the Arctic lacks scientific longitudinal data, indigenous inhabitants have observed this environment for thousands of years. Nonnative scientists in the Arctic, therefore, have become interested in indigenous observations about climate and environmental change (Alexander et al. 2011). Indeed "scientific examinations of nontraditional data sets collected by naturalists, hobbyists, or indigenous peoples have been instrumental in linking changes in biological communities to recent climatic changes" (Sagarin and Micheli 2001: 811).

Iñupiat Perspectives

Iñupiat and other indigenous groups have been highly successful at living in northern latitudes for millennia. Close observation, categorization and understanding of the local environment, and interactions among its ecological features are central features of indigenous knowledges in the Arctic (Ray 1975; Nelson 1899; Burch 2006). Iñupiat communities in Northwestern Alaska, with whom we worked, continuously detailed knowledge of the local environment. Conversations frequently revolved around weather patterns, harvesting of species—both plant and animal—and areas traveled.

Extensive knowledge concerning the behavioral influences and feedback effects of plants, animals, and weather on one another permeates Arctic discourses. How chum salmon arrive after king salmon and how more snow and a later melt mean a more prolific berry season are examples of common and yet highly complex knowledge among Northern people. Toponyms are also rich in local knowledge. Iñupiaq place names on the Seward Peninsula—such as *Chikuchuilaqpiaq* ("place where the river never freezes") and *Iŋmilaq* ("place without water")—demonstrate an intimate awareness of conditions on the land and exemplify how long-time Arctic residents' local observations are encoded in the oral record (Marino 2005).

Many Arctic residents, including those living in Iñupiat communities, continue to witness a changing landscape that is altering subsistence practices, threatening livelihoods and modern infrastructural development, and increasing the risk of life-threatening conditions. For example, because of thinning sheet ice in the winter some Iñupiat community members find hunting more dangerous and more difficult (Adger 2004; Tenenbaum 2005; Wisniewski 2010). In our research we found that traditional water sources are increasingly threatened by northward moving and growing beaver populations. Local experts report an emerging, and hitherto unknown, problem with beetles and other insects infesting and killing trees on the Seward Peninsula.

To date, the most significant and widely publicized problem in Alaska, in part, due to global climate change, is the erosion of coastal areas, which threatens the viability of many Alaska Native villages (USGAO 2003, 2009; Bronen 2009, 2011). In total, 184 out of 213 Alaska Native Villages have experienced problems with erosion and flooding (USGAO 2003). In Shishmaref, dramatic erosion (due in part to later freezing of the ocean ice and increased erosion) has triggered debate among local residents and federal and state government agencies about relocating the entire community. Failure to do so could lead to a major flooding disaster, inundating the village and causing loss of life and property. In 2002 the community voted to relocate from the island. Local community advocates in Shishmaref have worked to develop and lobby for funding and policy mechanisms that would allow for redeveloping a community site in a safer location. So far, that funding has not materialized.

Invasion of the Photo Snatchers: An Opportunity for Political Participation of Northern Residents?

Arguably, arctic indigenous peoples are some of the most severely affected by global climate change worldwide, leading Neil Adger to use Inuit populations' "right to be cold" (a phrase originally coined by Sheila Watt-Cloutier; see Watt-Cloutier 2015 for a history of climate change as a human rights issue) as the quintessential example of how human rights are compromised by global anthropogenic climate change (Adger 2004). With the rise in public discourses about climate change, the overwhelming desire to document the phenomenon, and the identification of Arctic residents as some of the first "victims" of climate change, rural Alaska has been besieged with unprecedented numbers of journalists, photographers, scientists, and politicians over the last 25 years. All seem eager to engage in a discussion or, even better, to get a photo of people who have first-hand experience with climate change.

During our fieldwork in rural, primarily Iñupiat villages on the Seward Peninsula, the vast majority of summer visitors were either scientists or journalists studying or reporting on global climate change. While we were making dinner with a Shishmaref resident, who had already been featured in a Canadian documentary about climate change and been quoted and photographed for *People* and *Time* magazines, two television crews, one from Japan and one from Colorado, simultaneously filmed a story on climate change in his kitchen. Another resident from the village of White Mountain has made the front page of *USA Today* discussing picking berries. A bed-and-breakfast owner in White Mountain has nearly exclusively relied on climate change scientists to fill the beds in the summer and keep her business going. Clearly the "discourse of global climate change" has also affected indigenous communities in Alaska—but to what ends?

Historically, politicians have ignored many of the needs and challenges in rural Alaska because of the sheer smallness of rural Arctic populations and the expense of solving infrastructural and development needs in such a remote region (IAWG 2009). For example, rural Alaskan residents are some of the last American citizens to lack modern sanitation systems, and many homes lack running water, which can lead to poor health outcomes for residents (Eichelberger 2010).

Because of this chronic lack of political attention, climate change has provided opportunities for Northern residents to make their voices heard in political discussions and political arenas that may have previously been difficult to engage. Community members from Shishmaref, for example, have repeatedly testified to the U.S. Congress on the effects of climate change on their village and have lobbied for federal funding for relocation. Community members likewise participated in the United Nations Climate Change Conference in Copenhagen in 2009. Northern residents are also often invited to climate change initiatives hosted by scientists and scientific institutions. In 2014 the National Center for Atmospheric Research held a symposium on climate change outcomes for indigenous communities; four Alaska Native village representatives and two Alaska relocation scholars (including the lead author of this chapter) were invited, and funded, to participate.

This media and political attention has been an important tool for vocalizing the adaptation requirements of communities threatened with habitual flooding or other climate-driven challenges. When rural Alaska community members approach Congress, Alaska state legislators, or national and international scientific bodies, the opportunity arises to advocate for community needs. However, it is still unknown whether this attention and these global discourses will create either of the needed outcomes in terms of opportunities for political change and/or adaptation opportunities for communities in crisis. Perhaps more germane is whether these global discourses are inclusive enough to accommodate cultural differences in environmental perceptions. Our experience and the experiences of indigenous scholars and scientists who work on climate change issues (Cochran et al. 2013) suggest they are not—yet. Therefore the next question is this: *can* the global climate change discourse be flexible enough to allow for the multiplicity of perspectives and then solutions for climate change challenges within diverse vernaculars (Callison 2014)?

Climate Knowledge in Context

Anthropologists working among Inuit cultures throughout the circumpolar north note that epistemologies of environmental knowledge are frequently built around understanding a complex system of environmental influences occurring within a specific geographical space—for example, understanding the way shorefast ice responds to a multitude of influences off of the coast of Sarichef Island. Epistemologies of environmental knowledge are also especially dependent on personal experience—knowledge that is firsthand (Morrow 1990; Briggs 1991; Omura 2005)—which is an experience markedly different from climate modeling, for example, and which often entails removed, generalizable, and impersonal analysis. Although there is evidence from the Arctic that scientific analysis and traditional ecological knowledge—a term problematic but ubiquitous in the Arctic climate change literature (Nadasdy 1999; Wenzel 2004)—are largely comparable data sets (Huntington et al. 2004), cross-pollination of these data sets has been slow in coming (Cochran et al. 2013; Callison 2014). We see instead Arctic communities being used as examples of crisis—which we fear diminishes the profound contribution indigenous knowledge holders in the North could make to climate change discourses more broadly.

Another discrepancy we found between media discourses of climate change and Iñupiat discourses came with framing catastrophe. Comparing media interpretations of change and local discourses of change reveals that, in Shishmaref, residents are hesitant to hyperbolize conditions of flooding and environmental shift. Journalists, however, are quick to use catastrophic language. In Shishmaref there is a high priority on the accuracy of information, which includes resisting exaggeration. Quotes from Shishmaref residents in the following media excerpts stand in contrast to the more calamitous language from the journalists' perspective.

> Thousands of years ago, hungry nomads chased caribou here across a now-lost land bridge from Siberia, just 100 miles away. Many scientists believe those nomads became the first Americans. Now their descendants are about to become global-warming refugees. Their village is about to be swallowed up by the sea.
>
> "We have no room left here," said 43-year-old Tony Weyiouanna. "I have to think about my grandchildren. We need to move." (Verrengia 2002)

Another example:

> When the arctic winds howl and angry waves pummel the shore of this Iñupiat Eskimo village, Shelton and Clara Kokeok fear that their house, already at the edge of the Earth, finally may plunge into the gray sea below.
>
> "The land is going away," said Shelton Kokeok, 65, whose home is on the tip of a bluff that's been melting in part because of climate change. "I think it's going to vanish one of these days." (Sutter 2009)

In the next example, Wallach comments explicitly that Weyiouanna is unemotional about his statements and that Weyiouanna speaks with "the indifference of an engineer."

> "I don't think we have much choice now," he tells me on the eve of the new ballot. "Some might vote no—people so tied to the island they don't want to leave. We'll just have to make adjustments." Like a wholesale migration to the mainland, an adjustment he discusses with the indifference of an engineer, not someone who's lived here all his life. (N.d.)

In these examples, residents say "I don't think we have much of a choice now," "The land is going away," "I think it's going to vanish one of these days," "There's no room left" and "We need to move." These are all experiential statements of erosion that are grounded in physical realities. Surrounding these quotations is the more catastrophic language of "wholesale migration," "angry waves," "edge of the earth," "and swallowed by the sea" used by journalists.

Heather Lazrus and Carol Farbotko have written about the tendency for outsider media coverage to frame indigenous and politically marginalized populations as helpless "victims" of a catastrophic climate that fails to reflect insider interpretations. Working in Tuvalu, Farbotko and Lazrus (2012) claim that outsiders act as "ventriloquists for climate change narratives," framing islanders' stories to create a cautionary tale that ultimately sells newspapers and promotes outsider agendas. Similar to discourses by Shishmaref residents, Tuvaluan narratives, for example, are often less catastrophic and are framed by ideas of global citizenry and human rights (Farbotko and Lazrus 2012) instead of being framed by helplessness and victimization.

Discourses of Climate Change

Iñupiat participation in global climate discourses is critical to explaining and understanding climate change in the Arctic. Iñupiat participation in these discourses may also open opportunities for adaptation that would otherwise be unavailable to rural Arctic residents. However, these discourses may limit the extent and typology of participation. It is exceedingly problematic if Iñupiat voices are legitimate in global discourses only when they represent "victims" of climate change. Framing populations as victims compromises the agency of individuals and communities, and it diminishes the important roles of participation that Arctic residents should be allotted because of their particular expertise—and not only because of some communities' situated perspectives on climate change risk. As Michael Bravo (2009: 277) has noted for the Inuit of Nunavut, the local reception of climate change narratives is intricately tied to the complex histories of dealing with outsiders. This fact suggests that Inuit individuals and communities make strategic decisions about how to speak of climate change using both complex knowledge sets and complex, situated histories. To us this reality means that we must listen closely.

There are places where and occasions when Iñupiat discourses about change and scientific discourses about global climate change converge. Watt-Coultier, as head of the Inuit Circumpolar Council, was largely responsible for framing climate change as a human rights issue in the first place (Callison 2014). Similarly, when residents in Shishmaref lobby for their community to the U.S. President or Congress they engage scientific discourses about the climate. A danger exists, however, that in many cases Iñupiat narratives of climate change will be reduced to a series of anecdotes about disenfranchised groups of people in the world. While politically moving, these anecdotes ignore complex, sophisticated, and at times possibly un-"scientific" ways of organizing the world (Cochran et al. 2013).

It has been an enduring legacy of anthropology to document and accept a multiplicity of worldviews and divergent explanations for the same phenomena. But the anthropology of climate change must resist providing colorful anecdotes to scientific investigations and insist on documenting the complexities inherent in ecological knowledge, even if the outcomes surprise us.

In the end, taking Iñupiat discourses about (climate) change seriously includes pushing beyond the conventional dichotomies of "nature" and "society/culture" (Hastrup 2014) in order to engage fully with local ways of making sense of shifting environments. At the same time, Mike Hulme's (2011) warning not to "reduce the future to climate change" seems to be a particularly good fit for Shishmaref and its residents. Climate change is one aspect of this vibrant community; it is not the most important one.

Note

We would like to thank the many people in White Mountain, Shishmaref, Golovin, Elim, Nome, and Kotzebue who participated in this research. We feel honored to have had the opportunity to spend time on the west coast of Alaska, in Iñupiat territory. It is, in part, the goal of this research to create space for the experts who have worked with us to have other forums in which to share their knowledge about climate change—to be writing and speaking, not just being written and spoken about.

References

ACIA. 2005. *Arctic Climate Impact Assessment.* Cambridge: Cambridge University Press.

Adger, W. N. 2004. The right to keep cold. *Environment and Planning A* 36(10): 1711–15.

Agrawal, A. 2010. Local institutions and adaptation to climate change, in *Social Dimensions of Climate Change: Equity and Vulnerability in a Warming World.* Washington, D.C.: World Bank, pp. 173–98.

Alexander, C., Bynum, N., Johnson, E., King, U., Mustonen, T., Neofotis, P., Oettlé, N., et al. 2011. Linking indigenous and scientific knowledge of climate change. *BioScience* 61(6): 477–84.

Anderegg, W. R. L., Prall, J. W., Harold, J., and Schneider, S. H. 2010. Expert credibility in climate change. *Proceedings of the National Academy of Sciences* 107(27): 12107–109.

Bravo, M. T. 2009. Voices from the sea ice: The reception of climate impact narratives. *Journal of Historical Geography* 35: 256–78.

Briggs, J. L. 1991. Expecting the unexpected: Canadian Inuit training for an experimental lifestyle. *Ethos* 19(3): 259–87.

Bronen, R. 2009. Forced migration of Alaskan Indigenous communities due to climate change: Creating a human rights response, in A. Oliver-Smith and X. Shen (Eds.), *Linking Environmental Change, Migration,and Social Vulnerability*, SOURCE No 12/2009. Bonn: United Nations, pp. 68–74.

———. 2011. *Climate-Induced Community Relocations: Creating an Adaptive Governance Framework Based in Human Rights Doctrine.* New York University Review of Law and Social Change 35: 356–406.

Burch, E. S., Jr. 2006. *Social Life in Northwest Alaska: The Structure of Iñupiaq Eskimo Nations.* Fairbanks: University of Alaska Press.

Callison, C. 2014. *How Climate Change Comes to Matter: The Communal Life of Facts*. Durham, NC: Duke University Press.

Cochran, P., Huntington, O. H., Pungowiyi, C., Tom, S., Chapin III, F. S., Huntington, H. P., Maynard, N. G., and Trainor, S. F. 2013. Indigenous frameworks for observing and responding to climate change in Alaska. *Climatic Change* 120(3): 557–67.

Cruikshank, J. 2005. *Do Glaciers Listen? Local Knowledge, Colonial Encounters, and Social Imagination*. Vancouver: UBC Press.

Eichelberger, L. P. 2010. Living in utility scarcity: Energy and water insecurity in Northwest Alaska. *American journal of public health* 100(6): 1010.

Farbotko, C., and Lazrus, H. 2012. The first climate refugees? Contesting global narratives of climate change in Tuvalu. *Global Environmental Change* Special Issue: *Adding Insult to Injury: Climate Change, Social Stratification, and the Inequities of Intervention* 22: 382–90.

Foucault, M. 1972. *The Archaeology of Knowledge and the Discourse on Language*. New York: Pantheon Books.

———. 1977. History of systems of thought, in D. F. Bouchard (Ed.), *Language, Counter-Memory, Practice: Selected Essays and Interviews by Michel Foucault*. Ithaca, NY: Cornell University, pp. 199–204.

Hastrup, K. (Ed.). 2014. *Anthropology and Nature*. New York: Routledge.

Hinzman, L., Bettez, N., Bolton, W. R., Chapin, F. S., et al. 2005. Evidence and implications of recent climate change in northern Alaska and other Arctic regions. *Climatic Change* 72: 251–98.

Hulme, M. 2011. Reducing the future to climate change: A study of climate determinism and reductionism. *Osiris* 26: 245–66.

Huntington, H. 2000. Native observations capture impacts of sea-ice changes. *Witness the Arctic* 8(1): 1–2.

Huntington, H., Callaghan, T., Fox, S., and Krupnik, I. 2004. Matching traditional and scientific observations to detect environmental change: A discussion on Arctic terrestrial ecosystems. *Ambio*: 18–23.

Immediate Action Working Group (IAWG). 2009. *Recommendations Report to the Governor's Subcabinet on Climate Change*, www.climatechange.alaska.gov/docs/iaw_finalrpt_12mar09.pdf, accessed September 12, 2011.

Lorenzoni, I., and Pidgeon, N. 2006. Public views on climate change: European and USA perspectives. *Climate Change* 77: 73–95.

Marino, E. 2005. *Negotiating the Languages of Landscape: Place Naming and Language Shift in an Iñupiaq Community*. Unpublished Master's thesis. University of Alaska Fairbanks.

———. 2012. The long history of environmental migration: Assessing vulnerability construction and obstacles to successful relocation in Shishmaref, Alaska. *Global Environmental Change* 22(2): 374–81.

———. 2015. *Fierce Climate, Sacred Ground: An Ethnography of Climate Change in Shishmaref, Alaska*. Fairbanks: The University of Alaska Press.

Marino, E., and Ribot, J. 2012. Special issue introduction: Adding insult to injury—Climate change and the inequities of climate intervention. *Global Environmental Change* 22(2): 323–28.

McBean, G., Alekseev, G., Chen, D., Førland, E., Fyfe, J., Groisman, P. Y., King, R., Melling, H., VosePaul, R., and Whitfield, H. 2005. Arctic climate: Past and present, in *ACIA, Arctic Climate Impact Assessment: Scientific Report*. Cambridge: Cambridge University Press, pp. 21–60.

McNeeley, S. M. 2012. Examining barriers and opportunities for sustainable adaptation to climate change in Interior Alaska. *Climatic Change* 111(3-4): 835–57.

Morrow, P. 1990. Symbolic actions, indirect expressions: Limits to interpretations of Yupik society. *Etudes/Inuit/Studies* 14(1-2): 141–58.

Nadasdy, P. 1999. The politics of TEK: Power and the "integration" of knowledge. *Arctic Anthropology* 36(1/2): 1–18.

Nelson, E. W. 1899. *The Eskimo about the Bering Strait.* Washington, D.C.: Smithsonian Institution Press.

Omura, K. 2005. Science against modern science: The socio-political construction of otherness in Inuit TEK (traditional ecological knowledge), In N. Kishigami and J. Savelle (Eds.), *Indigenous Use and Management of Marine Resources.* Osaka: National Museum of Ethnology, pp. 323–44.

Ray, D. Jean. 1975. *The Eskimos of the Bering Strait.* Seattle: University of Washington Press.

Rebotier, J. 2012. Vulnerability conditions and risk representations in Latin-America: Framing the territorializing urban risk. *Global Environmental Change* 22(2): 391–98.

Sagarin, R., and Micheli, F. 2001. Climate change in nontraditional data sets. *Science, New Series* 294(5543): 811.

Sutter, J. D. 2009. Climate change threatens life in Shishmaref, Alaska. *CNN Tech: Climate Change,* http://articles.cnn.com/2009-123/tech/shishmaref.alaska.climate.change_1_climate-change-shishmaref-coastal-erosion?_s=PM:TECH, accessed February 6, 2012.

Tenenbaum, D. J. 2005. Global warming: Arctic climate—The heat is on. *Environmental Health Perspectives* 113(2): A91.

United States General Accounting Office (USGAO). 2003. *Alaska Native Villages: Most Are Affected by Flooding and Erosion, but Few Qualify for Federal Assistance.* GAO-04-142 report to Congressional committees. Washington D.C.

———. 2009. *Alaska Native Villages: Limited Progress Has Been Made on Relocating Villages Threatened by Flooding and Erosion.* Report to Congressional Committees, http://www.gao.gov/new.-items/d09551.pdf, accessed August 25, 2011.

Verrengia, J. B. 2002. *In Alaska, an Ancestral Island Home Falls Victim to Global Warming.* Associated Press, September 10, http://forests.org/shared/reader/welcome.aspx?linkid=15600&keybold=climate%20AND%20%20seal%20AND%20%20level%20AND%20%20rise.

Watt-Cloutier, S. 2015. *Right to Be Cold: One Woman's Story of Protecting Her Culture, the Arctic, and the Whole Planet.* Toronto: Penguin Random House Canada.

Wenzel, G. W. 2004. From TEK to IQ: Inuit Quajimajatuqangit and Inuit culturalecology. *Arctic Anthropology* 41(2): 238–50.

Wisniewski, J. 2010. *Come On Ugzruk, Let Me Win. Experience, Relationality, and Knowing in Kigiqtaamiut Hunting and Ethnography.* Unpublished Ph.D. dissertation. Department of Anthropology. Fairbanks, Alaska: University of Alaska Fairbanks.

Wisniewski, J., and Marino, E. 2007. *Being as Knowing: Knowledge as Experience in a Sentient World.* Paper presented at the annual meeting of the Alaska Anthropological Association, March 14–17, Fairbanks, Alaska.

Woolgar, S. 1986. On the alleged distinction between discourse and Praxis. *Social Studies of Science* 16(2): 309–17.

Chapter 11

Too Little and Too Late: What to Do about Climate Change in the Torres Strait?

Donna Green

This chapter updates my previous contribution to this book on the ability of Indigenous Australians, living in rural and remote Australia, to adapt to climate change. The earlier work outlined how the issue had first been identified just under a decade ago, in about 2006, in joint discussions with Indigenous leaders and climate scientists, at a workshop held in Darwin, Northern Australia. The outcome of this workshop was clearly identified needs to involve Traditional Owners in research on their country and to work with them to understand what adaptation strategies would be most useful for them. (Traditional owners are people who have social, economic, and spiritual affiliations with, and responsibilities for, a specific region of land and or sea.) The workshop also identified the need to focus on the Torres Strait as quickly as possible because of the very alarming effects of sea-level rise already occurring there. This chapter provides an update about these issues by summarizing my previous work and discussing the research developments that have occurred in the last decade.

Climate Impacts on Indigenous Land

Australia is a land of extremes. It is the driest inhabited continent, but it also suffers massive flooding caused by monsoonal rain that regularly cuts off transport and communication infrastructure to remote communities for over half the year (Green et al. 2009). Climate change will make these extreme weather events more common (Government of Australia 2015), with increasing temperatures being of primary concern to human health. Most of Australia's population lives in major cities located on or near the coast in the south and east of the continent, but for a brief period in 2013 national attention was focused elsewhere, on the extreme temperatures that broke records over large areas of the remote central, western, and northern parts of the country. The extremes forced the Australian Bureau of

Meteorology to add a new series of colors to their forecast maps, in order to increase their temperature scale to include a new high temperature of 129°F.

Many Indigenous Australians live in areas projected to have significant increases not only in absolute temperatures but also in regions where an increase in a run of hot days and hot nights, a heatwave, is likely. Coastal areas cooled by sea breezes are, however, not exempt from climate impacts—as rising sea levels combine with storm surges to affect coastal zones. Owing to a series of successful land claims in the Northern Territory in recent years, Aboriginal people have gained freehold title to almost 90% of the coastline. As these coastlines become increasingly eroded by larger storm surges, coastal communities will need to find ways to protect their shoreline infrastructure.

Indigenous Australia's Perception of Climate Change

Aboriginal people and Torres Strait Islanders have noticed environmental changes occurring over the last few decades (Green, Billy, and Tapim 2010). These observations have now been documented in a manner similar to those of other Indigenous peoples around the world (Green and Raygorodetsky 2010). Through raising awareness of the similarity in these Indigenous observations of unexplained changes in animal, plant, and weather patterns, and the effects suggested by climate scientists, interest has been increasing in many of these communities in having access to information relevant to building resilience to future climate impacts. Work with Torres Strait Islanders is a case in point.

In 2004 I initially focused on how much climate change would affect remote Indigenous communities and what they wanted to do about it. Successful Native Title land claims in the north of Australia suggested that communities maintaining the strongest links to land would have the most intact Traditional Environmental Knowledge (TEK). This suggestion was important, because I wanted to explore whether TEK passed down through oral histories could guide region-specific and culturally appropriate adaptation strategies. Additionally, I thought that it might be able to provide useful historical environmental observations for climate scientists—if the holders of the knowledge were willing to share it with them.

The first step was to understand how Traditional Owners felt about the likely effects of climate change and if they considered those impacts a problem worth dealing with. In collaboration with colleagues at the Commonwealth Scientific and Industrial Research Organisation, I organized a workshop that, for the first time, brought Traditional Owners from across northern Australia together with researchers and scientists. The workshop aimed to identify and to value both Western scientific and TEK forms of expert knowledge about environmental change and to acknowledge that both would be vital to design climate adaptation strategies that would take Traditional Owners' priorities into account.

As the scientists explained the latest temperature and sea-level-rise projections, it became clear that climate change was likely to exacerbate the Traditional Owners' existing social and cultural problems, such as the potential for further dispossession from their land. Their initial apprehension and doubts about the importance of climate change were superseded by a unanimous demand for information to enable them to adapt better. Specifically, they requested localized "on-country" workshops to allow further community discussions about likely climate change impacts and the recording of TEK about weather and climate.

The assembled group even agreed to prioritize urgent action in the Torres Strait, a decision influenced by a photo presentation made by one of the Islanders. The photos—taken just a month before the workshop, during record king tides—showed Islanders wading down streets knee-deep in water, damaged coastal infrastructure, and flooded rubbish tips (trash dumps) (Green 2007). Although not suggesting a direct causal link between the record tides and climate change, the photos were a powerful reminder of how exposed and vulnerable these communities were and what damage extreme weather events (such as those likely to be more frequent owing to climate change) could cause to the islands.

Climate Change Impacts on the Torres Strait Islands

The Islanders are traditional seafaring people who pride themselves on their intimate understanding of the seasonal shifts in the ocean and weather. *Ailan Kastom* (Islander culture) refers to a distinctive Torres Strait Islander way of life, incorporating traditional elements of Islander belief systems combined with Christianity (Sharp 1993). *Ailan Kastom* permeates all aspects of island life. It governs how Islanders take responsibility for and manage their land and sea country, how and by whom natural resources are harvested, and how they allocate seasonal and age-specific restrictions on catching particular species (Johannes 1984; Mulrennan and Scott 1999). But at the heart of *Ailan Kastom* is the connection between the people and their land.

Events such as the timing of the king tides are predictable for the Islanders. However, they had noticed that in recent years the waves occurring in these king tides seemed higher and more powerful than they had been before (Green and Steffensen 2010). On several of the islands, coastal tracks were being washed away and houses and long-established graveyards inundated. In addition to the psychological distress and cultural disruption caused by the flooding, the Islanders' remoteness makes repairing this damage extremely expensive, and Islanders lack access to the necessary resources to engage consultants to conduct assessments or to carry out maintenance work.

The Islanders understand that the problem extends beyond the initial flooding. They are concerned about indirect effects of climate change, seeing how, for example, inundations could jeopardize public health through contamination of freshwater supplies or the flooding of their land-fill rubbish tips (Figures 11.1 and 11.2). A full suite of indirect effects is hard to assess and quantify; these effects are, however, crucial to consider in designing comprehensive adaptation strategies.

Figure 11.1 Child playing in storm surge water on Warraber (photo by author)

Figure 11.2 Damage to the sea wall on Saibai caused by storm surge in 2006 (photo by author)

During the workshop Islanders expressed their concerns that their TEK was no longer as reliable as it had once been. The Islanders reported shifts in animal and plant behavior that did not accord with their experience (Green, Billy, and Tapim 2010; Green and Raygorodetsky 2010). These TEK observations showed that Islanders were acutely aware of changing temperatures and rainfall patterns, of shifting bird migrations and breeding seasons, and of changing abundance and distribution of particular species. For instance, a new species of mosquito had appeared on some islands, and changes in marine fauna habitats were perceived to be disrupting Islanders' traditional subsistence hunting patterns. Given the profound cultural importance of totemic sea animals—such as turtles and dugong—for many Islanders this issue takes on particular significance (Sutherland 1996; Department of Environment and Heritage 2005; Marine and Coastal Committee 2005).

Some of the key threats to the islands from climate change effects have initially been identified in a natural hazards report commissioned by the state government's Island Coordinating Council (ARUP 2006). This report shows how some important cultural heritage sites—such as graveyards, monuments, and sacred sites—are vulnerable to storm surges on several of the islands. Important infrastructure, such as airstrips and communal eating areas are also at risk of erosion (Figures 11.2 and 11.3). Although climate change effects may increase the total impact of a storm surge only incrementally—through such factors as sea-level rise and increased cyclone intensity—the extreme vulnerability of these sites means that any additional factor affecting the area inundated could have very serious consequences (SEA 2011).

Figure 11.3 Airstrip erosion on Warraber (photo by author)

Public Recognition and Local Adaptation

Public awareness about climate change in Australia has grown enormously in the past few years, with most people now recognizing the threat it poses to iconic natural ecosystems such as the Great Barrier Reef and Kakadu. Many of these threats are detailed in a large scoping study that outlines the risks to Indigenous communities in the tropical north of Australia (Green, Jackson, and Morrison 2009). Yet relatively few Australians realize that the very ability of Indigenous people (who have lived in these regions for thousands of years) to continue to exist there is also threatened. This lack of public awareness is due to a complex combination of factors, one of which is the lack of information about the status of these remote communities (O'Neill, Green, and Lui 2012). However, isolation appears not to be a limiting factor by itself, the knowledge of Islander relocation and adaptation available for elsewhere in the region (Dean, Green, and Nunn 2015).

In mid-2006, the major metropolitan newspapers began to report on the plight of several low-lying Torres Strait Islands (Minchin 2006; Michael 2007). Shortly after, the threat was officially acknowledged by the Australian government, which signed off on the Intergovernmental Panel on Climate Change's *Fourth Assessment Report.* This report acknowledged, for the first time, the likelihood that around half the people living on the Islands would have to be relocated in the long term (Hennessy et al. 2007). The Fifth Assessment Report (IPCC 2014) detailed how little adaptation action had been undertaken in the intervening period, although adaptation research agendas had been drawn up that related to Indigenous concerns (NCCARF 2012).

During this time, I brought together a group of climate scientists to write a paper that assessed the likely impacts and adaptation options for the Torres Strait region (Green et al. 2010a). I also worked with a public interest lawyer to publish two papers exploring the legal dimensions of climate effects in the Torres Strait (Green and Ruddock 2009; Ruddock and Green 2011). This work was presented internationally at the Threatened Island Nations: Legal Implications of Rising Seas and a Changing Climate conference at the Columbia Center for Climate Change Law in 2011.

With momentum slowly building, a website dedicated to providing content about the plight of Torres Strait Islanders was set up. This website was used to raise awareness, link into international campaigns, and gain political attention. Over several years of activity, including several broken commitments from State and Federal politicians, funding was finally committed to begin the most pressing short-term activities, such as rebuilding sea walls on the most affected islands (TorresStraitClimate 2014). However, Islanders know that, in the longer term, some people will probably have to move to other areas of their land or off their island completely.

Understandably, the Islanders see relocation as an action of last resort and are already working on adaptation strategies to delay, and ideally avoid, having to leave their ancestral homelands. For example, on low-lying islands, such as Boigu and Saibai, all new houses are being built on stilts, and nearly all the others have been raised so that they will not be affected by flooding. The Mer Island Council is negotiating to subdivide land away from the shoreline and higher up to provide areas safer for new housing. Emergency management plans are being drawn up in consultation with government agencies, while on some islands, where the resources are available, sea walls are being reinforced and extended.

This work is carried out through state and Federal programs that most recently have provided funds to develop adaptation tools and to test multistakeholder planning approaches (Bohensky and Butler 2014). These activities, however, do not mean that the necessary funds will be released to carry out the activities identified as "of high priority" as a result of these workshops, which is a source of frustration to Islanders, who know they do not have much time to prepare.

In 2014 the Torres Strait Climate Change Strategy 2014–2018 was released (TSRA 2014). This document includes an action plan in a number of areas, including culture and traditional knowledge; environment and ecosystems; settlements and infrastructure; people and communities; and communication and capacity building. It also provides an assessment of progress since 2010, identifying that, for all bar one, the activities had either not been initiated or were less than 75% complete (TSRA 2014: 21–27). Again, funding limitations significantly restrict many crucial activities.

My experience in the Torres Strait, and increasingly in other parts of Australia, has shown that when Indigenous people have access to information about how climate change is affecting their lives, often in tandem with seeing changes to their lands, they act. For this reason, I decided to continue to research and publish in the area of climate impacts on Indigenous Australians but this time to focus specifically on effects on human health and how factors such as air pollution could serve to exacerbate existing Indigenous disadvantage.

Future Research Directions: Heat /Health Impacts on Indigenous People

To provide tailored information for Indigenous Australians, over the last four years I have focused my research on how climate change is likely to affect disproportionately Indigenous people's physical health and well-being. This new work underscores the significance of connecting physical and mental health of Indigenous communities (Green and Minchin 2014). The arguments in this chapter lead on from earlier work where we argued that there were disproportionate burdens for Indigenous Australians owing to the fact that the multidimensional impacts of climate change on the health of country is inextricably linked with human health (Green, King, and Morrison 2010).

This research also stemmed from collaborative work with Aboriginal communities who were returning to their country. In one case, we looked at how climate mitigation funding could be used as a source of climate adaptation funding (Green and Minchin 2012). In this commentary, we discussed how new Indigenous-led carbon offset projects carried out in their own country fulfill this dual role. These projects are able to do more than just mitigate emissions, since they also achieve economic, social, and cultural co-benefits. This newer interdisciplinary work continued alongside "on-country" workshops and research that created videos of Traditional Owners talking about their concerns about environmental change in their country (Sharing Knowledge 2015). The complexities of multicausal environmental change—where climate effects are just one of a suite of factors including feral animals and a lack of proper cultural practice on country—reinforce the necessity of a suite of approaches to maintaining Indigenous health and well-being (Green and Webb 2015).

Work was also carried out with epidemiologists to assess whether there were physiological concerns specific to Indigenous people as a result of climate extremes. We reported the results of this research in Webb and colleagues (2014), which showed that there was a significant and disproportionate signal for heart disease in Indigenous people during heat extremes in the Northern Territory. Follow-up work considering the impact of climate extremes on respiratory diseases and the role of humidity is ongoing.

The most recent direction of this research has been to consider how these climate impacts are likely to be exacerbated by other environmental health concerns relating to the air pollution occurring from industrial development. Many of these remote and rural areas are located near mine sites, smelters, and other industrial activity, and research overseas suggests that environmental injustice in toxic exposures often occurs. Our study identified similar outcomes in Australia, with the first national level quantitative environmental justice assessment of industrial air pollution revealing a clear national pattern of environmental injustice based on the locations of industrial pollution sources (Chakraborty and Green 2014).

References

ARUP. 2006. *Natural Disaster Risk Management Study*. ARUP: Brisbane.

Bohensky, E., and Butler, J. 2014. *Torres Strait Futures: Results and Recommendations*. NERP Tropical Ecosystems Hub Project 11.1 Final Factsheet.

Chakraborty, J., and Green, D. 2014. Australia's first national level quantitative environmental justice assessment of industrial air pollution. *Environmental Research Letters* 9(4). DOI: 1088/1748-9326/9/4/044010.

Dean, A., Green, D., and Nunn, P. 2015. Drowning in red tape: Climate finance in an atoll nation, in E. Stratford (Ed.), *Islands: Perspectives from Human Geography*. New York: Routledge.

Department of Environment and Heritage. 2005. *Sustainable Harvest of Marine Turtles and Dugongs in Australia: A National Partnership Approach*. Commonwealth Government: Canberra.

Government of Australia. 2015. *Climate Change in Australia: Technical Report*. Canberra: Government of Australia.

Green, D. 2007. *Sharing Knowledge*, http://sharingknowledge.net.au/resources/videos-%20 Copy.html

Green, D., Alexander, L., McInnes, K., Church, J., Nicholls, N., and White, N. 2010a. An assessment of climate change impacts and adaptation for the Torres Strait Islands, Australia. *Climatic Change* 102(3): 405–35.

Green, D., Billy, J., and Tapim, A. 2010. Indigenous Australians' knowledge of weather and climate. *Climatic Change* 100(2): 337–54.

Green, D., King, U., and Morrison, J. 2010. Disproportionate burdens: The multidimensional impacts of climate change on the health of Indigenous Australians. *Medical Journal of Australia* 190(1): 4–5.

Green, D., and Minchin, L. 2012. The co-benefits of carbon management on country. *Nature Climate Change* 2(9): 641–43.

———. 2014. Living on climate-changed country: Indigenous health, well-being, and climate change in remote Australian communities. *Ecohealth* 10.

Green., D., Jackson, S., and Morrison, J. 2009. *Risks from Climate Change to Indigenous Communities in the Tropical North of Australia*. Canberra: Report for the Department of Climate Change.

Green, D., and Raygorodetsky, G. 2010. Indigenous knowledge of a changing climate. *Climatic Change* 100(2): 239–42.

Green, D., and Ruddock, K. 2009. Could litigation help Torres Strait islanders deal with climate impacts? *Sustainable Development Law & Policy* 9(2): 9.

Green, D., and Steffensen, V. 2010. *Climate Change in the Torres Strait*, torresstraitclimate.org/the-video/

Green, D., and Webb, L. 2015. Climate change, health, and well-being in Indigenous Australia, in C. Butler, J. Dixon, and A. Capon (Eds.), *Healthy People, Places, and Planet: Reflections Based on Tony McMichael's Four Decades' Contribution to Epidemiological Understanding*. Canberrra: ANU Press.

Hennessy, K., Fitzharris, B., Bates, B., Harvey, N., Howden, M., Hughes, L., Salinger, J., and Warrick, R. 2007. Australia and New Zealand: Climate change 2007—Impacts, adaptation, and vulnerability, in M. Parry, O. Canziani, J. Palutikof, C. Hanson, and P. van der Linden, (Eds.), *Contribution of Working Group II to the Fourth Assessment: Report of the Intergovernmental Panel on Climate Change*. Cambridge: Cambridge University Press.

IPCC. 2014. *Climate Change 2014: Impacts, Adaptation, and Vulnerability. Part B: Regional Aspects—Contribution of Working Group II to the Fifth Assessment Report of the Intergovernmental Panel on Climate Change*, V. R. Barros, C. B. Field, D. J. Dokken, M. D. Mastrandrea, K. J. Mach, T. E. Bilir, M. Chatterjee, K. L. Ebi, Y. O. Estrada, R. C. Genova, B. Girma, E. S. Kissel, A. N. Levy, S. MacCracken, P. R. Mastrandrea, and L. L. White (Eds.). Cambridge: Cambridge University Press, 688 pp.

Johannes, R. E. 1984. Marine conservation in relation to traditional lifestyles of tropical artisanal fishermen. *The Environmentalist* 4 (supplement 7): 30–35.

Marine and Coastal Committee. 2005. *Sustainable Harvest of Marine Turtles and Dugongs in Australia—A National Partnership Approach*, Natural Resource Management Ministerial Council, Marine and Coastal Committee. Canberra: Commonwealth Government.

Michael, P. 2007. Rising seas threat to Torres Strait islands. *Courier Mail* 2, August, Brisbane.

Minchin, L. 2006. Going under. *Sydney Morning Herald* 12, August, Sydney.
Mulrennan, M., and Scott, C. 1999. Land and sea tenure at Erub, Torres Strait: Property, sovereignty, and the adjudication of cultural continuity. *Oceania* 70: 46–76.
NCCARF. 2012. *National Climate Change Adaptation Research Plan for Indigenous Communities*. Brisbane: NCCARF.
O'Neill C., Green, D., and Lui, W. 2012. How to make climate change research relevant for Indigenous communities in Torres Strait, Australia. *Local Environment* 7(10): 1104–20.
Ruddock, K., and Green, D. 2011. What legal recourse do non-state islands have to obtain resources to adapt to climate change? *Macquarie Journal of International and Comparative Environmental Law* 7(2): 69.
SEA. 2011. *Torres Strait Extreme Water Level Study.* TSRA and System Engineers Australia final report.
Sharing Knowledge. 2015. *The Sharing Knowledge Project*, www.sharingknowledge.net.au
Sharp, N. 1993. *Stars of Tagai: The Torres Strait Islanders*. Canberra: Aboriginal Studies Press.
Sutherland, J. 1996. *Fisheries, Aquaculture, and Aboriginal and Torres Strait Islander Peoples: Studies, Policies, and Legislation*. Canberra: Environment Australia.
TorresStraitClimate. 2014. http://torresstraitclimate.org/the-video
TSRA. 2014. *Torres Strait Climate Change Strategy 2014–2018*. TSRA: Thursday Island.
Webb, L., Bambrick, H., Tait, P., Green, D., and Alexander, L. 2014. Effect of ambient temperature on Australian Northern Territory public hospital admissions for cardiovascular disease among Indigenous and non-Indigenous populations. *International Journal of Environmental Research in Public Health* 11(2): 1942–59.

Chapter 12

Shifting Tides: Climate Change, Migration, and Agency in Tuvalu

Heather Lazrus

> Our sovereignty will not be threatened. . . . Rules would apply to this area and it would remain Tuvalu. Of course, it would be mostly sea, but it would be Tuvalu land area. So we would have our claim maintained on this spot in the Pacific Ocean. —Prime Minister Sopo'aga (Field 2004)

In this chapter an anthropological perspective on migration in response to climate change situates population mobility in cultural and historical contexts, disentangles local views on migration from externally imposed assumptions and discourses, and examines community-based needs and priorities in responses to climate change. In Tuvalu, as in other places facing extreme climate change impacts and creeping environmental problems such as sea-level rise, migration and relocation are often presented as possible responses and adaptations for the affected populations (Lazrus 2012). However, these actions are not considered favorable outcomes in Tuvalu, where climate-driven migration is resisted at the government level and among citizens. The resistance is not a denial of climate change or its effects but rather a historically consistent response to externally imposed ideas about Tuvalu's vulnerability as a small island nation.

Tuvalu and Global Climate Change

The nine islands of Tuvalu, a mix of atolls edged by small *motu* (islets that rise above sea level) and table reef islands, arc across the Pacific Ocean between 5°–10° south and 176°–179° east. The islands are spread across distances of 125 to 150 km, and the combined land area is just 26 square km. The highest point above sea level on the low-lying islands is less than 5 m. From north to south, the islands of Tuvalu are Nanumea, Niutao, Nanumaga, Nui, Vaitupu, Nukufetau, Funafuti, Nukulaelae, and Niulakita, with the last historically uninhabited. The name *Tuvalu* reflects the eight traditionally inhabited islands and means "eight standing

---

Anthropology and Climate Change: From Actions to Transformations (2nd ed.) by Susan A. Crate and Mark Nuttall (Eds.)

together"; *tu* means "to stand," and *valu* signifies "eight." The population of approximately 11,000 now inhabits all nine islands. Each island and community is ecologically and culturally unique, and now, too, each faces unique challenges from the effects of global climate change.

The islands of Tuvalu are geologically very young and dynamic, subject to subsidence, reef growth, and eroding and accreting forces of the sea and weather (Baines, Beveridge, and Maragos 1974). In addition to complex geomorphology, limited technical research contributes to scientific debate about how the islands will be affected by sea-level rise (Hunter 2004; Webb and Kench 2010). Nonetheless, there is broad agreement on the general effects of climate change that include sea-level rise, sea-surface and subsurface temperature increases, ocean acidification and coral bleaching, coastal erosion, increased intensity but decreased frequency of rainfall, and increased frequency of other extreme weather events including drought (Nurse et al. 2014). Already, an upward trend in sea-level rise is discernable at the sea-level gauge on Funafuti, the island that is home to the national capital, of 2 ± 1 mm per year over the period 1950 to 2001. Continued and increasing rates of sea-level rise will lead to increased coastal erosion, flooding, and compromised freshwater resources (Church, White, and Hunter 2006).

The islands of Tuvalu were settled by Polynesian seafarers from elsewhere in the Western Pacific between 800 and 2,000 years ago, when sea levels fell during a small ice age and exposed the atolls (Munro 1982; Chambers and Chambers 2001; Nunn 2007). Land resources are limited owing to sparse alkaline coral sand soils covered only by a thin layer of accumulated vegetable deposit. Freshwater that "floats" in places on the heavier saltwater permeating the porous coral soil, known as the Gyhben-Herzberg lens, is also unevenly distributed and limited (McLean et al. 1986). Narratives, stories, and legends tell of how islanders endure the challenges of daily atoll life and survive droughts and storms. Such stories provide analogies for explaining and coping with contemporary disruptive geological and environmental processes (Cronin and Cashman 2007), although climate change also brings unprecedented challenges and rates of change.

In the 19th and 20th centuries, European interest in the islands went through periods of growth and decline. Whaling and copra production generated short-lived interests, Protestant missionaries gained footholds on each island, and American forces established defenses on Funafuti, Nukufetau, and Nanumea at the frontline of WWII (McQuarrie 1994). In 1889, 40 years after the beginning of sustained European contact, Tuvalu, then named the Ellice Islands, came under British rule. Independence came almost a century later in 1978. Today the national government is a parliamentary democracy with a prime minister and representatives from each of the originally inhabited eight islands, with Niulakita considered a single political entity with Nukulaelae. Tuvalu's economy is now dominated by migration, remittances, aid, and bureaucracy (known as a MIRAB economy) and has proven to be surprisingly robust, contrary to expectations (Knapman, Ponton, and Hunt 2002). The growing importance of a cash-based economy drives heavy urbanization to the capital on Funafuti, where the international airport, wharf, government offices, and a growing retail sector offer economic opportunities. Funafuti's burgeoning economy and increasing population density differentiate the cadence of life in the capital from the outer islands.

Interviews I conducted in 2006 with Tuvaluans who have spent their lives observing and interacting with the environment focused on observations of changes in the sky, sea, and land; on local responses to extreme weather events such as storms, droughts, and floods; and on risk perceptions and expectations of the future (Lazrus 2015).

Migration as a Response to Climate Change?

Since the early 2000s Tuvaluans have been labeled by popular media, non-governmental organizations, and politicians as the world's first "climate refugees" (Farbotko and Lazrus 2012). While the increasing effects of climate change loom on the horizon and cast shadows of uncertainty over the future of the country and its citizens, climate change has not motivated a national conversation about migration in Tuvalu, either within the government or among citizens. Instead, Tuvaluans resist the normative assumption that rising sea levels equate them to being landless refugees (Farbotko and Lazrus 2012; Stratford, Farbotko, and Lazrus 2013; Farbotko, Stratford and Lazrus 2015). Disaster research, and in particular a growing body of theoretical work about the links between climate change and population mobility, is helpful to understand why people in Tuvalu resist the idea of an inevitable future as climate refugees. However, despite an increasing amount of research on the topic, climate-driven migration remains underexplored empirically (Black, Kniveton, and Schmidt-Verkerk 2011; Obokata, Veronis, and McLeman 2014).

Migration is both an adaptive response to environmental change and a driver that has lead humans to populate all areas of the globe. In our globalized world the push and pull factors that motivate population mobility have become increasingly multifaceted, and it is often difficult to disentangle social, political, and economic motivations from environmental ones (Dun and Gemenne 2008; Hartmann 2010). Indeed, a population's vulnerability (often a push factor in migration decisions) to environmental catastrophe is highly complex and socially constructed. Oliver-Smith describes vulnerability as "the conceptual nexus that links the relationship people have with their environment to social forces and institutions and the cultural values that sustain or contest them" (2004:10). Outsider and global discourses about migration in the face of climate change often obfuscate the roots of vulnerability—specifically, the ways that resources, wealth, and security are unequally distributed to systematically reduce people's adaptive options (Oliver-Smith 2013). Hence, when migration is presented as an inevitable outcome of climate change—especially as an emergency response to the effects of climate change but also as a planned adaptation strategy—the root vulnerabilities are not addressed.

In Tuvalu, permanent, temporary, and circular migration are already all key aspects of island life, happening independently of the changing climate. Tuvaluans relocate between islands and internationally to Fiji, New Zealand, Australia, and elsewhere for economic, education, and health care opportunities. Remittances sent home from Tuvaluans living and working overseas pay for children's education, parent's hospital bills, and family's church contributions, among other costs. However, Tuvaluans deeply resist migration in response to climate change, as demonstrated by a lack of migration planning at the national level (Farbotko, Stratford, and Lazrus 2015) and by these representative quotes from Tuvaluans:

> I just don't want to believe that I'll have to leave the country one day if Tuvalu was to sink. . . . I would not wanna hat to leave my home island. I love Tuvalu . . . and I . . . want to be here all my life. (Lina Timala in Horner and le Gallic 2004)

> Can we really allow ourselves to leave our islands? Can we really find another Tuvalu elsewhere? Anyone can uplift us and put us elsewhere, do we really want that to happen? (Community leader speaking during a World Wildlife Funded-lead *Climate Change Impact, Adaptation, and Engaging in International Policy: Awareness Raising Workshop*, February 14, 2006; author's field notes)

> If the worst comes to the worst, I think I would rather just stay here and die. I don't want to leave my country. (Nongovernmental organization officer quoted in Farbotko, Stratford, and Lazrus 2015)

The Importance of Home

Tuvalu's response to climate change is unfolding at home and abroad as national leaders participate in the 21st session of the United Nations Framework Convention on Climate Change (UNFCCC) Conference of Parties (COP) held in Paris in 2015. Tuvaluan Prime Minister Enele Sopoaga emphasized that "Our survival as a nation depends on the decisions we take at this Conference" (Sopoaga 2015). An analogy from Tuvalu's short national history helps explain this abiding connection between home, nation, and cultural integrity even for a population whose past and present are characterized by mobility. Before Tuvalu became an independent nation in 1978 the islands were known as the Ellice Islands and, together with the Gilbert Islands (now Kiribati), was part of the Gilbert and Ellice Island Colony (GEIC). Under colonialism, unease between the Polynesian islanders of the Ellice Islands and their northern Micronesian neighbors in the Gilbert Islands heated relations between the two groups that simmered throughout the 1960s and boiled over in the 1970s, resulting in Tuvaluans' secession from the Colony.

During World War II, school systems were destroyed in the Gilbert Islands and were slow to recover. During the same period Tuvaluans received good education, which placed them in high demand in colonial administrative positions. The widening opportunity gap between the groups planted seeds of dissent. For instance, although Tuvaluans were a small minority they held disproportionately higher numbers of upper-level jobs in the administration. Between 1960 and 1970, Tuvaluans, who made up just a small proportion of the GEIC population, held well over a quarter of the positions in civil service (Macdonald 1975).

A separate political consciousness emerged among Tuvaluans in Tarawa, the post-WWII administrative center of the GEIC. Many feared that they would become a disenfranchised minority in a new county formed from the GEIC. Recognizing that political independence from Britain was approaching, Tuvaluans took a stand of separation before political independence occurred (Macdonald 1975).

In 1974 the United Nations oversaw a referendum that achieved very high participation (88% of estimate eligible voters) and demonstrated almost unanimously (92% of votes) the Tuvaluans' desire for political separation from the Gilbert Islands (Macdonald 1975). A vote in favor of separation was deemed to be implicit consent to the Terms of Separation imposed by, in the words of one of Tuvalu's officials during an interview in 2006, "the scheming bureaucrats in the Foreign and Commonwealth Office in London of Her majesty's Government." The Terms offered the newly formed nation a bare minimum: a second-hand passenger and cargo vessel that was already nearing the end of its lifespan. The sad inadequacy of the Terms indicates that Britain was trying to dissuade Tuvaluans from voting for separation. That Britain did not anticipate the vote for separation reflects not only a severe lack of consultation and understanding before the referendum but also British fear of suffering a "loss of face" and perhaps also apprehension about antagonizing Gilbertese in the late stages of the colony (Macdonald 1975).

The strength of Tuvaluans' desire for a separate country, demonstrated in the nearly unanimous vote and peaceful secession, even at the risk of becoming a new country with only a second-hand transport vessel, reflects a certain set of priorities. International observers accused Tuvaluans of being overly ambitious and taking on an impossible role: once independence was granted, how could such a small, distant, resource-deprived country hope to succeed on the world stage? In contrast, Tuvaluans had conspicuously prioritized their cultural identity and political aspirations over material security and other measurements of legitimacy imposed by Eurocentric expectations. Significantly, Tuvalu's decision for political separation from the Gilbert Islands occurred when proportionally more Tuvaluans lived away from home than ever before or since that time, working in the phosphate mines and colonial administration located in Kiribati.

Following separation, Tuvalu became an independent country on October 1, 1978. At this juncture, nationalism became more than the expression of cultural coherence; it became legalistically inscribed in place. As Gupta describes it: "Nationalism as a distinctively modern cultural form attempts to create a new kind of spatial and mythopoetic metanarrative, one that simultaneously homogenizes the varying narratives of community while, paradoxically, accentuating their difference" (1997: 191). Tuvalu's "mythopoetic metanarrative" is one of perseverance in the face of opposition. As a nation, the individual island communities and cultures have come together across the vast watery distances that separate them physically. This same stand for national identity is again being solidified in opposition to the forces of climate change. The gravitational pull around which national identity is formed is less about the geographic configuration of the islands than it is about shared culture and a home encapsulated in the form of nation-state.

Conclusion

In Tuvalu, migration is seen by most as an option of last resort (McNamara and Gibson 2009) both because of what relocation would mean for loss of identity, culture, and well-being (Mortreux and Barnett 2009), and because of the lack of agency such an outcome represents (Farbotko and Lazrus 2012). Analogies to Tuvaluans' past ways of expressing their desire for a separate home inform understandings of why Tuvaluans are resisting climate-induced migration.

Discourses of climate migration and "climate refugees" fail to accurately represent the perceptions of people affected and the range of adaptation options afforded to them (Orlove 2009). While it may become prudent for Tuvaluans to consider migration among their adaptation alternatives, that realization must evolve in ways that promote Tuvaluans' agency as citizens and as full participants in their culture. Adaptation options that do not fully engage with islander perspectives and priorities risk "adapting" to the very systemic vulnerabilities that render Tuvaluans' futures particularly precarious. Tuvaluans' story illustrates how one of the major challenges of climate change includes imagining new metanarratives of nation, ones in which cultural integrity can be maintained in ways that may not be so tied to "this spot in the Pacific Ocean" (Field 2014).

References

Baines, G. B. K., Beveridge, P. J., and Maragos, J. E. 1974. Storms and island building at Funafuti Atoll, Ellice Islands. *Proceedings of the Second International Coral Reef Symposium* 2: 485–96. Brisbane.

Black, R., Kniveton, D., and Schmidt-Verkerk, K. 2011. Migration and climate change: Towards an integrated assessment of sensitivity. *Environment and Planning-Part A* 43(2): 431–50.

Chambers, K., and Chambers, A. 2001. *Unity of Heart: Culture and Change in a Polynesian Atoll Society*. Prospect Heights: Waveland Press.
Church, J. A., White, N. J., and Hunter, J. R. 2006. Sea-level rise at tropical Pacific and Indian Ocean islands. *Global and Planetary Change* 53: 155–68.
Cronin, S., and Cashman, K. 2007. Volcanic oral traditions in hazard assessment and mitigation, in J. Grattan and R. Torrence (Eds.), *Living under the Shadow: The Archaeological, Cultural and Environmental Impact of Volcanic Eruptions*. Tucson: University of Arizona Press.
Dun, O., and Gemenne, F. 2008. Defining environmental migration. *Forced Migration Review* 31: 10–11.
Farbotko C., and Lazrus, H. 2012. The first climate refugees? Contesting global narratives of climate change in Tuvalu. *Global Environmental Change* 22: 382–90.
Farbotko, C., Stratford, E., and Lazrus, H. 2015. Climate migrants and new identities? The geopolitics of embracing or rejecting mobility. *Social and Cultural Geography*, http://www.tandfonline.com/doi/abs/10.1080/14649365.2015.1089589?journalCode=rscg20.
Field, M. J. 2004. Tuvalu prime minister blames global warming as Funafuti sinks. *Agence France Presse*, February, http://www.tuvaluislands.com/news/archived/2004/2004-02-21b.htm.
Gupta, A. 1997. Song of the nonaligned world: Transnational identities and the reinscription of space, in A. Gupta and J. Fergusson (Eds.), *Culture, Power, Place: Explorations in Critical Anthropology*. Durham, NC: Duke University Press.
Hartmann, B. 2010. Rethinking climate refugees and climate conflict: Rhetoric, reality and the politics of policy discourse. *Journal of International Development* 22: 233–46.
Horner, C., and Le Gallic, G. 2004. *The Disappearing of Tuvalu: Trouble in Paradise*. European Television Center.
Hunter, J. R. 2004. Comment on "Tuvalu not experiencing increased sea level rise." *Energy and Environment* 15(5): 927–32.
Knapman, B., Ponton, M., and Hunt, C. 2002. *Tuvalu 2002 Economic and Public Sector Review*. Manila: Asian Development Bank.
Lazrus, H. 2012. Sea change: Island communities and climate change. *Annual Review of Anthropology* 41: 285–301.
———. 2015. Risk perception and climate adaptation in Tuvalu: A combined cultural theory and traditional knowledge approach. *Human Organization*. In press.
Macdonald, B. 1975. The separation of the Gilbert and Ellice Islands. *Journal of Pacific History* 10: 84–88.
Mclean, R. F., Holthus, P. H., Hosking, P. L., Woodruffe, C. D., and Hawke, D. V. 1986. *Tuvalu Land and Resources Survey: Nanumea*. Auckland: Department of Geography, University of Auckland.
McNamara, K. E., Gibson, C. 2009. "We do not want to leave our land": Pacific ambassadors at the United Nations resist the category of "climate refugees." *Geoforum* 40(3): 475–83.
McQuarrie, P. 1994. *Strategic Atolls: Tuvalu and the Second World War*. Christchurch: University of Canterbury McMillian Brown Center for Pacific Studies, and Suva: University of the South Pacific Institute of Pacific Studies.
Mortreux, C., and Barnett, J. 2009. Climate change, migration, and adaptation in Funafuti, Tuvalu. *Global Environmental Change* 19: 105–12.
Munro, D. 1982. *The Lagoon Islands: A History of Tuvalu 1820–1908*. Ph.D. dissertation, Sydney: Macquarie University.
Nunn, P. 2007. Holocene sea-level change and human response in Pacific Islands: Transactions of the Royal Society of Edinburgh. *Earth and Environmental Sciences* 98: 117–25.

Nurse, L. A., McLean, R. F., Agard, J., Briguglio, L. P., Duvat-Magnan, V., Pelesikoti, N., Tompkins, E., and Webb, A. 2014. Small Islands, in V. R. Barros, C. B. Field, D. J. Dokken, M. D. Mastrandrea, K. J. Mach, T. E. Bilir, M. Chatterjee, K. L. Ebi, Y. O. Estrada, R. C. Genova, B. Girma, E. S. Kissel, A. N. Levy, S. MacCracken, P. R. Mastrandrea, and L. L. White (Eds.). *Climate Change 2014: Impacts, Adaptation, and Vulnerability. Part B: Regional Aspects*. Contribution of Working Group II to the Fifth Assessment Report of the Intergovernmental Panel on Climate Change. Cambridge: Cambridge University Press.

Obokata, R., Veronis, L., and McLeman, R. 2014. Empirical research on international environmental migration: A systematic review, in P. Palutikof, J. van der Linden, and C. E. Hanson (Eds.), *Population and Environment*. Cambridge: Cambridge University Press.

Oliver-Smith, A. 2004. Theorizing vulnerability in a globalized world, in G. Bankoff, G. Frerks, and D. Hilhorst (Eds.), *Mapping Vulnerability: Disasters, Development, and People*. London: Earthscan.

———. 2013. Disaster risk reduction and climate change adaptation: The view from anthropology. *Human Organization* 72(4): 275–82.

Orlove, B. 2009. The past, the present, and some possible futures of adaptation, in W. Neil Adger, I. Lorenzoni, and K. O'Brien (Eds.), *Adapting to Climate Change: Thresholds, Values, Governance*. Cambridge: Cambridge University Press, pp. 131–63.

Sopoaga, E. 2015. Keynote statement delivered by the Prime Minister of Tuvalu, the Honourable Enele S. Sopoaga at the Leaders Events for Heads of State and Government at the Opening of the COP21, November 30, 2015. http://unfccc.int/files/meetings/paris_nov_2015/application/pdf/cop21cmp11_leaders_event_tuvalu.pdf, accessed December 2, 2015.

Stratford, E., Farbotko, C., and Lazrus, H. 2013. Tuvalu, sovereignty, and climate change: Considering fenua, the archipelago, and emigration. *Island Studies Journal* 81: 67–83.

Webb, A. P., and Kench, P. S. 2010. Vulnerability of atoll islands to sea level rise: Multi-decadal analysis of island change in the central Pacific. *Global and Planetary Change* 72: 234–46.

Chapter 13

The Politics of Rain: Tanzanian Farmers' Discourse on Climate and Political Disorder

Michael Sheridan

Some Tanzanian farmers say that "the rain is different now" and that postcolonial leaders and development agency experts cannot "bring rain" the way that colonial chiefs once did.[1] Government officers, expatriate administrators, and Pare farmers agree that average annual rainfall has declined dramatically in recent decades—despite rainfall records that show increasing variability rather than desiccation. The impression that rainfall has declined is a consequence of a particular cultural interpretation of ecology. This cultural model is both a hard-won indigenous knowledge system and a set of moral evaluations of appropriate relationships among people (Shaffer and Naiene 2011). Given this linkage between politics and rainfall, the local narrative of declining rainfall over the 20th century is both a cultural metaphor for changing terms of resource entitlement and the ambiguities of power, morality, and social relations and a description of a geophysical process. Understanding this process requires a closer look at the historical course of social change, the cultural roots of environmental narratives, and the political relationships between powerful institutions (such as governments and development agencies) with rural populations on the periphery of the global economic system. This chapter draws on research into the intersections of culture and power to examine the history of contestation over rainmaking, sacred sites, and climate models in North Pare. After describing these shifting ideologies of power, legitimacy, and value, I conclude that anthropology offers climate change researchers a nuanced vision of "power" and "politics" that can account for how culture affects responses to climate change.

Rain is a political process across much of sub-Saharan Africa. In October 2000 the Spanish government repatriated the remains of a Tswana man who had been on display in a Barcelona museum for over 170 years (Gewald 2001). When the rains failed later that year, many people in Botswana blamed the drought on their government's use of these bones to prevent the rain from falling—and therefore maintain poverty and the dependency of the poor on the ruling party. In July 2002 police in Niger arrested suspected prostitutes in order to bring rain (BBC 2002)—and in 2004 many Tanzanian farmers told me how America's foreign policy in Iraq had angered God and thereby caused that year's erratic rains in East Africa. On the date of a recent American election (November 4, 2008), Barack Obama's patrilateral kin reportedly began celebrating his victory before any election returns were announced, because it had started to rain (NPR 2008). These political meanings of rain have deep roots in African cosmological and political systems. One of the cornerstones of Fortes and Evans-Pritchard's landmark volume on *African Political Systems* was the insight that the authority of leaders in colonial Africa rested, in part, on the performance of rituals to bring rain and ensure the fertility of both land and people (Fortes and Evans-Pritchard 1940: 21). The colonial period did not erase and replace indigenous metaphors and metonyms linking rain and power, however. Instead, bureaucratic notions of accountability and institutional order now coexist and hybridize with cultural models of rain, morality, and metaphysics. This chapter traces the historical trajectory of a particular set of rain discourses in Tanzania and offers some suggestions as to why the politics of rain have continuing salience for the ways that rural Tanzanians evaluate climate change.

The literature on the significance of rain as a metaphor for political power in African societies is too vast (and diverse) to summarize here, but several broad themes stand out (Sanders 2003; Peterson and Broad 2009). First, there is no neat correlation between areas of low (or high) rainfall and the presence of rain-related beliefs, discourses, and social institutions. Rainmakers are as important (albeit in different ways) in the dry Kalahari as they are in the rainforests of the Congo (Schapera 1971; Packard 1981). Second, the politics of rain are deeply interwoven with the politics of kinship, class, ethnicity, and gender (Krige and Krige 1943; Sanders 2014). In Zimbabwe, for example, chiefs become *mhondoro* ("lion" spirits) and "owners of rain" after death, so that current political strategies are also ways to claim this longer-term social status (Ranger 1999). Third, throughout such discourses about rain, the notion that political order leads to ecological order in the form of reliable rain—and that conflict brings drought—functions as the rhythm of political improvisation. These dynamics are, in much of Africa, explicitly gendered, so that both discussions and ritual practices about rain rely on gender categories—for example, the contrast between gentle, fertility-inducing "female" rains and the violence of "masculine" thunderstorms and drought. Rain discourses therefore do not simply reflect social categories and structural conditions but constitute an idiom for conceptualizing, discussing, and sometimes acting on processes of social transformation (for example, Jedrej 1992; Landau 1993; Fontein 2006). In the case of rainmaking in North Pare, Tanzania, although ritual practice has declined in recent years, memories of these older cosmological and political engagements continue to shape farmers' interpretations of social and ecological change at both local and global spatial scales.

Because the effects of climate change on the continent are becoming apparent with daunting implications, the social meanings of rain are more than a cultural curiosity and a recurring theme in African studies (Abdrobo et al. 2014). The task of incorporating sociocultural perceptions of climate into mitigation and adaptation efforts is just beginning. Ethnography is particularly well suited to investigate the ways that environmental change relates to social dynamics, institutional arrangements, and ideological systems (Orlove et al. 2009). Despite the strengths of this qualitative approach, much of the climate change literature still focuses on the quantitative documentation of biophysical change in the atmosphere and biosphere. The emerging anthropology of climate change, however, is injecting social dynamics and cultural complexities into the debate, and the issue of what climate change means cross-culturally is at the cutting edge of this inquiry. As Mike Hulme puts it: "the idea of climate exists as much in the human mind and in the matrices of cultural practices as it exists as an independent and objective physical category" (2009: 28). One implication of this attention to the ideological aspects of climate change is that cultural meanings will affect the course of adaptation.

Rain and Order

North Pare is a mountain block that lies on the Tanzanian-Kenyan border due south of Mt. Kilimanjaro. Like the other highlands in the Eastern Arc Mountains, the 2,000-meter-high peaks rake moisture from the prevailing winds during the area's two rainy seasons, keeping the southeastern slopes of North Pare green and wet. The highest peaks are covered in high-altitude closed-canopy rainforest, while the highland valleys are intensively cultivated with eucalyptus, coffee, bananas, beans, maize, and sweet potatoes. The North Pare ecosystem supports over 400 people/km^2 in some places, which is one of the highest population densities in East Africa. This has meant increasing pressure on the area's resource base, which remains severe, although farmers have taken several measures since the end of colonial rule in 1961, including shifting from a long-fallow system to seasonal fallowing or continuous cropping, incorporating agroforestry into their farming system and overall diversifying the rural economy. In the 1990s the local government worked with several European development agencies to construct community-based natural resource management institutions and to encourage soil and water conservation. The people of North Pare were ambivalent in their responses to these programs, because many of the development agencies' policies and practices violated local understandings of technical efficiency, social equity, and moral order (Sheridan 2002, 2004a).

Throughout my 1997–1998 fieldwork in North Pare, I found many people eager to talk about rain. There had been a severe drought throughout East Africa in early 1997, and the unusually heavy monsoon that arrived in October that year proved to be nearly as damaging as the drought. The sense of crisis reached a crescendo in January and February 1998, when heavy rains persisted over the course of two months that are usually dry. Throughout northern Tanzania, floods washed out roads, ruined bridges, and damaged crops. As my friends learned about the El Niño phenomenon from news reports, they wanted to discuss the drought and strange rains with me. They were not, however, interested in the dynamics of the Pacific Ocean; rather, they focused on the reasons why such odd weather was affecting them. I heard fatalistic musings that the rain was "the will of God" and fervent declarations that the Pare ancestors had sent the rain to punish the living for failing to sacrifice in the area's sacred groves (Sheridan 2009). The common theme running through this discourse was that environmental conditions were fundamentally social and metaphysical matters with the continuing cycle of drought-flood-drought as vivid proof of a deepening crisis. The Pare people had once enjoyed a social, political, and ecological order that peaked in the 1950s, when colonial rule implicitly supported the chiefs' rainmaking functions, but was now lost. The lack of an orderly rainfall regime in the late 1990s was, in the categories used by many Pare residents, a consequence of the collapse of "traditional" institutions (*jadi*) and the ambiguities of "modern" ones (*kisasa*). This narrative is typical of the cultural models I heard in Pare at that time:

> Environmental control was good in the past because of tradition. Now all that is scorned as absurdity, so we get the current environmental problems in our rivers, water sources, and forests. We need to revive tradition so that we don't get the current cycle of drought and flood. Muslims and Christians pray to God for rain, but that troubles God so much that He sends too much to trouble the people in return. It's better to use the ancestors in the sacred forests as intermediaries for rain, so when it comes, it comes carefully.

Rainfall conditions are, for the elderly woman who offered me this ecological analysis, indexed to orderly social, ritual, and moral relationships. Social disorder is most apparent in its ecological effects.

When I returned to North Pare in 2004, I expected more of the local model of blame involving lapsed sacrifices and failed rains. As before, I heard about the ecological functions of tradition. But I also heard new global associations. For example, I learned that in the spring and summer of 2004, George W. Bush was personally responsible for the hot and dry weather. Explaining why his banana plants were starting to show signs of water stress, one elderly farmer said that humans cannot know the reasons for drought for certain but that it was "probably because you Americans had brought war to the world. Drought is God's whip for beating humanity and saying, 'don't do that anymore!'" Another lamented that "we can't depend on the rains this year . . . and it's all because of that Bush and his war. We don't know why God is bringing *us* these problems for *his* mistakes." The African studies literature is replete with meteorological hypotheses linking drought to a loss of tradition and moral order (for example, Eguavoen 2013), but this sort of geopolitical analysis in culture-specific moral and climatological terms appears to be a more recent phenomenon. Today the ongoing conflict in the Middle East clearly looms large in recent East African discourse about rain and morality. In northern Kenya, for example, a village chief told an OXFAM researcher that

> the weather is changing. We used to get heavy rains when the winds came from the west and then came back 2–3 days later with rain. Now the wind comes from the east so it brings little or no rain . . . these are dry winds. I don't know what is causing this. Maybe all this fighting in Iraq and Iran, all this bombing and pollution, bombing the oil fields and all those fumes going into the air . . . and we're not very far from that. (MaGrath and Simms 2006: 6)

The message is clear: local ecological conditions relate to political conditions, and the scale of blame and causation has expanded from the local to the global.

Anthropologists have long explored the cultural logic of the rituals that Africans have used in their attempts to control uncertainty, but we have been less attentive to the logic and meanings of the uncertainties themselves (Whyte 1997: 21). My informants in North Pare did not enact a public ritual to bring reliable rain, nor did they enthusiastically embrace scientific discourse about meteorology and global climatic change. Instead they debated likely causes and appropriate solutions that contrasted their ideas and practices that they considered indigenous and authentic aspects of their cultural repertoire from those of the "outside." Although ecological, political, and ideological processes have rarely, if ever, been both controlled and orderly in North Pare, the simple notion that they existed, *that they had been*, in the late 19th and early 20th centuries continues to shape Pare responses to environmental and social change.

Farmers in North Pare insist that the area's rainfall pattern was predictable until the late 1960s. The rain would begin to fall on certain dates (there are two rainy seasons, in late September and in late February, but different elders gave different dates), and farmers could organize their planting accordingly, confident that there would be appropriate levels of soil moisture at critical phases of plant growth. Now, farmers say, rainfall is so random that they are never sure if their crops will reach maturity. In a 1998 focus group discussion, the farmers' consensus was that "the rain is different now because it rains fewer days, but when it does, it comes in greater volumes. The new rain is more destructive, and now we farm by luck, and we can't count on the rain." These elders agreed that the climate is generally warmer now and that traditional ethnometeorological indicators for gauging the onset and length of the rains (such as the amount of snow on

Mt. Kilimanjaro and the flowering of certain trees) were no longer valid. In fact, their sense that natural processes became disorderly in the 1960s tallies with meteorological observations of Africa's long-term drying trend since 1895, as well as the fluctuations of the 1970s and 1980s (Nicholson 2001). Their assertion that the weather has changed is supported by rainfall data from three rainfall reporting stations in North Pare schools. The annual number of rain days has decreased since the 1950s, and year-to-year variation has declined.[2] Total annual rainfall and the number of rain days for the North Pare highlands do show slight downward trends on average, but these shifts are well within the standard deviation of the data sets. That is, North Pare is indeed drier than it had been in the 1950s, but this "signal" is well hidden within the "noise" of a highly variable seasonal rainfall pattern. If there was no precipitous decline in rainfall, then why do the residents of North Pare envision this so dramatically and as such an important feature of their environmental history? The reasons involve social institutions and cultural changes in the North Pare landscape in addition to biophysical ecological changes such as the declining water retention capacity of the North Pare aquifer (Heckmann et al. 2014; see also Sullivan 2002).

Politics in 19th-century North Pare revolved around rainmaking. The ruling Wasuya and Wasangi clans extracted tribute from subordinate clans in return for bringing rain to Pare and purging the area of disease and pests (Kimambo 1969). Rainmaking rituals took place in sacred groves, which were the nexus of the symbolic oppositions that characterized cultural categories in precolonial North Pare. Contrasts such as hot/cold, wet/dry, male/female, fertile/infertile, and open/closed all featured prominently in domains ranging from initiation ceremonies to irrigation technology. Most important, however, periodic sacrifices to the skulls in these sacred forests "cooled" the "hot" anger and hunger of the ancestors enough for them to bring the peaceful "coolness" of rain. To paraphrase Lévi-Strauss, sacred groves and rain were "good to think" but not in terms of functional integration and social harmony like the rain shrines of southern Africa (Colson 2014). Rain and trees were also good for thinking about conflict and change.

In the early 19th century, the Wasangi clan claimed autonomy from the Wasuya chiefs to the north and brought mercenaries from the Wambaga clan of South Pare to settle the border area between the two chiefdoms. The leader of these new settlers was famous for his rainmaking skills, and the Wambaga were able to collect tribute from farmers for these services, thus giving the Wambaga a crucial foundation for political legitimacy despite their status as recent immigrants. By the late 19th century rainmaking, tribute collection, and guns had made the Wambaga well positioned to challenge their position as clients of the Wasangi (Semvua N.d.; Mbwana N.d.). Rising tensions led to a civil war (ca. 1875–1891) now known in North Pare as Kibonda and remembered for being a period of intense drought and famine. The conflict revolved around Wambaga private rainmaking and Wasangi collective rituals. The rival clans ousted each other's leaders several times before German military intervention in 1891–1894 quelled the dispute. Despite the cultural consensus about the meaning of rain in North Pare, the lesson of this social history is that cultural models of climate are not necessarily functional and adaptive. In the case of North Pare, there were two competing models in the same cultural context, and it was precisely these ideological differences that shaped a bitter dispute that continues to affect politics in North Pare to this day. Political ecology, rather than functionalism, is a more useful approach, because it focuses on both stability and change in both historical and political processes and does not assume that culture somehow "knows" how to adapt.

Rainmaking was also important in 20th-century politics. Shocked at the brutality of the North Pare initiation ceremonies, the British administration banned the Wasangi and Wasuya ritual in 1923 and installed the Wambaga clan leader as chief of Usangi, the southern half of North Pare, in 1925. The Wasangi leaders spent decades asserting that they were the only legitimate political authorities in Usangi because of their control of rainmaking and initiation in sacred groves. They fought this protracted legal battle with British colonial administrators and ultimately through a fruitless series of petitions to the United Nations (TNA 1733/28; UN Trusteeship Council 1955: 7). With their legal options exhausted, the Wasangi turned colonial conservation rhetoric to their advantage in a way that reveals their ongoing preoccupation with the politics of rain. In 1956 the Wasangi convened a summit of Pare chiefs to survey a sacred grove that had been logged by German missionaries and replanted with eucalyptus by the Wambaga chief's bodyguard. The group sent a report to the local government stressing that the forest's conservation value would "bring rain clouds and good weather" (TNA 517/A2/2/88). This was an effort to legitimize Wasangi claims to that forest and by extension, political supremacy. Overall, it was the British colonial policy of indirect rule that supported the continuity of politicized rain discourse in Tanzania (Feierman 1990; Sanders 2014). The British ruled through the chiefs in order to bring progress, but the chiefs' authority rested on far older symbolic foundations. The politics of rain in Pare were maintained by the very administrative system that saw its mission as the creation of rational bureaucratic institutions.

Rain politics continued into the postcolonial era, but ritual-political practice has increasingly given way to frustrated discourse. Climatological models of power were still in effect when the Wasangi regained the chiefdom in 1961. Their Wambaga rivals responded by withholding rainmaking services—and the rain refused to fall. On March 20, 1961, a group of women wrote to the District Commissioner charging four men with causing famine by preventing rainfall, demanding a formal investigation and threatening to "do a Pare dance" that would curse the Wambaga. The somewhat bemused officer responded that only God brings rain, but that if they wanted to perform a ceremony, they should get the appropriate permit from their chief (TNA 517/A2/2/111; TNA 517/A2/2/112). More recently, the GTZ-sponsored TFAP-North Pare Project[3] learned about the politics of rain while collecting environmental information from the local elite. A secondary school teacher told the agency that "people have noticed changes in the rain season [and] change in the amount of rain and its distribution . . . the government should be blamed on this problem" (TFAP-North Pare 1993). In my interviews with North Pare farmers in 1998 and 2004, they repeatedly emphasized that "traditional" environmental management had been orderly and functional but that more "modern" institutions were ambiguous, contradictory, or corrupt (Sheridan 2004b).

In North Pare, the politics of rain form a script that everyone knows but few people perform. The leaders who once made rain (whether they did so through individual supplication or collective sacrifice) no longer govern. The administrative structure is firmly secular and often staffed by men from other parts of Tanzania who do not share Pare cultural models. The key rainmakers of the 1950s and 1960s have either died or moved away. What remains are occasional individual sacrifices and a widespread nostalgia for a lost ecological and social order but no collective ritual practice. This deeply layered frustration is sort of an anti-institution, like the gap of a lost tooth that the tongue cannot help but poke constantly. It is within this gap between ongoing discourse and failed practice that North Pare farmers evaluate climate change and geopolitics as problems of power and meaning.

The Reign of Power

Power is a central theoretical concern in contemporary anthropology, but we are only beginning to develop the vocabulary needed to investigate its usefulness as an analytical concept. Foucault's concept of "biopower" and his insistence that "power/knowledge" are mutually constitutive (yet always in tension), for example, are well suited for the analysis of how rationalizing Western institutions shape behavior and consciousness. The limits and potentials for translating these Foucauldian notions of power into forms that can examine, for example, the dynamics of a society organized largely by kinship and gender remains a key task for anthropology. Yet, as Marshall Sahlins cautions, "power is the intellectual black hole into which all kinds of cultural contents get sucked" (2002: 20). If power explains everything, then it explains nothing. In his final book, Eric Wolf responded to Foucault's challenge by building an alternative typology of power (Foucault 1997; Wolf 1999). Like Foucault, he begins with the assumption that power is inherent in all social relationships and social action and so should never be treated as an essentialized, abstracted, superorganic "thing." If power and the fingerprints that it leaves on people (which we may gloss as "culture") are relational and historically contingent, then power is better conceptualized as a process.

Yet, building on his landmark volume *Europe and the People without History* (1982), Wolf insists that we view the relational power process as socially organized according to the major division of labor (both material and symbolic) in society. Thus Wolf constructed his ethnography of power with case studies of societies based on kinship, tribute, and capitalism, with particular attention to the sort of power that draws on cosmological principles to structure and naturalize the differentiation, mobilization, and deployment of social labor. He calls this "structural power" and illustrates the concept with case studies of three societies—the Kwakiutl of the Pacific Northwest in the 19th century, the pre-Columbian Aztecs of Mesoamerica, and the German Nazi party of the early 20th century—which drew on preexisting elements of "tradition" to construct new ideologies and institutions during periods of rapid social change and political conflict. Elites use ideas of cosmological order and social propriety to relegitimize the threatened social order as a timeless tradition. Yet Wolf's analysis is not an instrumental view of hegemonic elite control of ideology. Those who benefit from the discourses and practices of potlatch, human sacrifice, and genocide both "represent that order . . . [and] enact it in their lives" (Wolf 1999: 224). Thus, as Durkheim taught us long ago, collective representations are coercive, because they categorize and limit individual agency into predetermined channels, using relatively fixed cultural scripts (1954: 29).

Structural power is a useful lens for analyzing rainmaking discourses and institutions in Africa, because the concept helps us to avoid reducing the politics of rain to "tradition." By making rain in sacred groves and extracting farmers' tribute in the 19th century, North Pare elites deployed structural power during a period of rapid social change. As the coastal economics of guns, slaves, and ivory reshaped the East African interior over the 19th century, the structural power of rainmaking became a means of asserting control. Because British colonial rule invested authority in chiefs whose legitimacy rested on local climatological concepts of structural power, colonial rule actually maintained the salience of politicized rainmaking discourse throughout a period of rapid social change. To use a theatrical metaphor, rainmaking remained an "onstage" discourse and a familiar script in North Pare from the 19th well into the 20th century, while the "offstage" practices of making a living and reproducing society underwent drastic agrarian change (from relatively autonomous farming communities to coffee-producing peasants living in an isolated labor reserve). In recent years, however, rainmaking has moved "offstage" as well, while the new institutions that dominate the "onstage" Tanzanian social landscape—authoritarian, intrusive, and often ineffective governments, free markets with arcane and expensive terms of access, and an anemic civil society—are often perceived as absurd, dangerous, and disorderly. In counterpoint to its decline in practice, rainmaking has continued as an ideology in North Pare. What becomes of structural power when its institutions crumble and its cosmological principles become illegitimate? It becomes a kind of deconstructed power that imbues a society with nostalgia for its orderly past, when its values, social organization, and (in this case) its ecology were in perfect harmony. The fact that rainmaking in both the 19th and 20th century North Pare was deeply political and conflict-ridden does not detract from its significance as a metaphor for order.

Colonization, independence, Tanzanian socialism, and, most recently, administrative control by European development agencies have been quite disorderly, and often ambiguous, experiences for the people of North Pare. Talking about rain is a way to express frustration about the contradictions that now pervade matters of kinship, religion, politics, and economics. Is it better to give money to relatives or invest in your business? Is paying respect to the ancestors against Islam or Christianity? Why did the United States, a country that seeks to spread freedom and democracy, practice neocolonial occupation and systemic torture in Iraq? The legitimacy and efficacy of new and old social institutions are questioned in North Pare, so the orderly structural power of what they call "traditional" (*jadi*) looks very appealing compared to the perverse, unpredictable, and contradictory powers of "modernity" (*kisasa*).

Conclusion

Anthropological political ecology is a useful framework for analyzing how politics and power mediate the intersection of human societies and biophysical phenomena. All too often, however, the politics in political ecology appears as a one-dimensional conflict about access to and control over resources (Greenberg and Park 1994; Paulson and Gezon 2005). By following Wolf's expansion of power into discrete forms, political ecology offers more empathetic interpretations and more insightful explanations. Tanzanian rain discourse is the surface manifestation of debates over the efficiency, legitimacy, and morality of social arrangements. In Africa, meteorology is often a political and moral matter, so even weather reports entail structural power and cultural models (Roncoli and Ingram 2002). Climate change is, without a doubt, going to entail disorderly ecologies, societies, and moralities. In much of Africa, these are likely to be unified in both discourse and practice because of the influence of the structural power of rainmaking. Social values, symbolic systems, and cultural models will determine how many Africans evaluate and respond to the risks of climate change and the adaptation choices it necessitates (Adger et al. 2013; Bassett and Fogelman 2013; Eguavoen et al. 2013). Thus, climate change researchers would do well to contextualize rainfall data with social histories of the politics of rain in those parts of Africa where rain is fundamentally political and where nostalgia for order shapes perceptions of disorder and change.

After the failure of the 2009 climate conference in Copenhagen to achieve global reform, climate scientists and policymakers became skeptical about "solving" climate change as they prepared for the COP21 conference in Paris in December 2015. The emerging consensus is that climate change is a "wicked problem" that is complex enough that it cannot be solved with monolithic technological or institutional interventions (Hulme 2014). Instead it requires "clumsy solutions" with diverse, complex, and overlapping efforts that cope with the challenges of our new planet rather than "solving" them (Hulme 2009: 334). The structural power of rain in African societies may indeed make for clumsy climate change policy on the continent, but ignoring the sociocultural meanings of climate change will only contribute to the "wickedness" of disorderly African climates.

Notes

1. Acknowledgments: This is a revised version of an article originally published in the *Journal of Eastern African Studies* in 2012. This research was supported by the Wenner-Gren Foundation and the U.S. Fulbright Program.
2. See Sheridan 2012 for the graphs and charts of rainfall data and perceptions of climate change in North Pare.
3. Respectively, these acronyms stand for Deutsche Gesellschaft für Technische Zusammenarbeit and Tanzania Forestry Action Plan.

References

Abdrabo, M., Essel, A., Lennard, C., Padgham, J., and Urquhart, P. 2014. Africa, in *Climate Change 2014: Impacts, Adaptation, and Vulnerability,* http://ipcc-wg2.gov/AR5/images/uploads/WGIIAR5-Chap22_FGDall.pdf.

Adger, W. N., Barnett, J., Brown, K., Marshall, N., and O'Brien, K. 2013. Cultural dimensions of climate change impacts and adaptation. *Nature Climate Change* 3: 112–17.

Bassett, T., and Fogelman, C. 2013. Déjà vu or something new? The adaptation concept in the climate change literature. *Geoforum* 48: 42–53.

BBC News. 2002. Niger sinners blamed for drought. July 22, http://news.bbc.co.uk/2/hi/africa/2144053.stm.

Colson, E. 2014 [1957]. Rain-shrines of the Plateau Tonga of Northern Rhodesia, in M. R. Dove (Ed.), *The Anthropology of Climate Change*. Chichester: John Wiley and Sons.

Durkheim, E. 1954. *The Elementary Forms of Religious Life*. Glencoe, IL: Free Press.

Eguavoen, I. 2013. Climate change and trajectories of blame in northern Ghana. *Anthropological Notebooks* 19(1): 5–24.

Eguavoen, I., Schulz, K., de Wit, S., Weisser, F., and Muller-Mahn, D. 2013. *Political Dimensions of Climate Change Adaption: Conceptual Reflections and African Examples*. Bonn: ZEF Working Paper Series, no. 120, www.zef.de/fileadmin/webfiles/downloads/zef_wp/wp120.pdf.

Feierman, S. 1990. *Peasant Intellectuals: Anthropology and History in Tanzania*. Madison: University of Wisconsin Press.

Fontein, J. 2006. Languages of land, water and "tradition" around Lake Mutirikwi in southern Zimbabwe. *Journal of Modern African Studies* 44(2): 23–249.

Fortes, M., and Evans-Pritchard, E. E. (Eds.). 1940. *African Political Systems*. Oxford: Oxford University Press.

Foucault, M. 1997. The subject and power, in J. Faubian (Ed.), *Power*. New York: New Press.

Gewald, J.-B. 2001. El Negro, El Niño, witchcraft, and the absence of rain in Botswana. *African Affairs* 100: 555–80.
Greenberg, J., and Park, T. 1994. Political ecology. *Journal of Political Ecology* 1: 1–12.
Heckmann, M., Muiruri, V., Boom, A., and Marchant, R. 2014. Human-environment interactions in an agricultural landscape: A 1400-yr sediment and pollen record from North Pare, NE Tanzania. *Paleogeography, Paleoclimatology, Paleoecology* 406: 49–61.
Hulme, M. 2009. *Why We Disagree about Climate Change*. Cambridge: Cambridge University Press.
———. 2014. *Can Science Fix Climate Change? A Case against Climate Engineering*. Cambridge: Polity Press.
Jedrej, M. C. 1992. Rain makers, women, and sovereignty in the Sahel and East Africa, in L. Fradenburg (Ed.), *Women and Sovereignty*, Edinburgh: Edinburgh University Press.
Kimambo, I. 1969. *A political history of the Pare of Tanzania, c. 1500-1900*. Nairobi: East African Publishing House.
Krige, E., and Krige, J. D. 1943. *The Realm of a Rain-Queen*. London: Oxford University Press.
Landau, P. 1993. When rain falls: Rainmaking and community in a Tswana village, c. 1870 to recent times. *International Journal of African Historical Studies* 26(1): 1–30.
Magrath, J., and Simms, A. 2006. *Africa—Up in Smoke 2*, OXFAM Report, http://policy-practice.oxfam.org.uk/publications/africa-up-in-smoke-2-an-update-report-on-africa-and-global-warming-from-the-wor-121077.
Mbwana, S. N.d. Personal diary. Notebook written intermittently from 1924–1965 in possession of Godwin Ndama, Kiriche, Usangi, Tanzania. Copy in possession of the author.
National Public Radio (NPR). 2008. *In Kenya, Obama Wins Sparks Celebration*, Nov. 5, www.npr.org/templates/story/story.php?storyId=96670528.
Nicholson, S. 2001. Climate and environmental change in Africa during the last two centuries. *Climate Research* 17: 123–44.
Orlove, B., Roncoli, C., Kabugo, M., and Majugu, A. 2009. Indigenous climate knowledge in southern Uganda: The multiple components of a dynamic regional system. *Climate Change* 100(2): 243–65.
Packard, R. 1981. *Chiefship and Cosmology: An Historical Study of Political Competition*. Bloomington: Indiana University Press.
Paulson, S., and Gezon, L. (Eds.). 2005. *Political Ecology across Spaces, Scales, and Social Groups*. New Brunswick, NJ: Rutgers University Press.
Peterson, N., and Broad, K. 2009. Climate and weather discourse in anthropology: From determinism to uncertain futures, in S. Crate and M. Nuttall (Eds.), *Anthropology and Climate Change: From Encounters to Actions*. Walnut Creek, CA: Left Coast Press, Inc.: 70–86.
Ranger, T. O. 1999. *Voices from the Rocks: Nature, Culture, and History in the Matopos Hills of Zimbabwe*. Oxford: James Currey.
Roncoli, C., and Ingram, K. 2002. Reading the rains: Local knowledge and rainfall forecasting in Burkina Faso. *Society and Natural Resources* 15: 409–27.
Sahlins, M. 2002. *Waiting for Foucault, Still*. Chicago: Prickly Paradigm Press.
Sanders, T. 2003. (En)gendering the weather: Rainmaking and reproduction in Tanzania, in S. Strauss and B. Orlove (Eds.), *Weather, Climate, Culture*. Oxford: Berg.
———. 2014 [2008]. The making and unmaking of rains and reigns, in M. R. Dove (Ed.), *The Anthropology of Climate Change*. Chichester: John Wiley and Sons.
Schapera, I. 1971. *Rainmaking Rites of Tswana Tribes*. Leiden: Afrika-Studiecentrum.
Semvua, A. F. N.d. *The Wasangi*. Manuscript written circa 1995. Usangi, Tanzania. Copy in possession of the author.

Shaffer, L. Naiene, J., and Naiene, L. 2011. Why analyze mental models of local climate change? A case from southern Mozambique. *Weather, Climate and Society* 3(4): 223–37.

Sheridan, M. 2002. "An irrigation intake is like a uterus": Culture and agriculture in precolonial North Pare, Tanzania. *American Anthropologist* 104(1): 79–92.

———. 2004a. Development dilemmas and administrative ambiguities: Terracing and land use planning committees in North Pare, Tanzania. *Policy Matters* 13: 186–97.

———. 2004b. The environmental consequences of independence and socialism in North Pare, Tanzania, 1961–1988. *Journal of African History* 45(1): 81–102.

———. 2009. The environmental and social history of African sacred groves: A Tanzanian case study. *African Studies Review* 52(1): 73–98.

———. 2012. Global warming and global war: Tanzanian farmers' discourse on climate and political disorder. *Journal of Eastern African Studies* 6(2): 230–45.

Sullivan, S. 2002. "How can the rain fall in this chaos?" Myth and metaphor in representations of the north-west Namibian landscape, in D. LeBeau and R. Gordon (Eds.), *Challenges for Anthropology in the African Renaissance*. Windhoek: University of Namibia Press.

Tanzania Forestry Action Plan-North Pare (TFAP). 1993. *Opinion Survey on Natural Resources and Land Use*. Archives of TFAP-North Pare, Mwanga. Copy in possession of the author.

Tanzania National Archives (TNA)1733/28. 1923. *Usambara District Annual Report 1923*, Annexure C.

TNA 517/A2/2/88. 1956. *Repoti ya Msitu Mbale Usangi 13 August 1956*, September 18.

TNA 517/A2/2/111. 1961. *Bibi Sefiel Karingo to DC Same*, March 20.

TNA 517/A2/2/112. 1961. *DC Same to Bibi Sefiel Karingo*, April 4.

United Nations Trusteeship Council. 1955. *Official Records of the 15th Session of the Trusteeship Council 1955, Supplement 1, Resolutions January 25–March 28, 1955*. New York: United Nations.

Whyte, S. R. 1997. *Questioning Misfortune: The Pragmatics of Uncertainty in Eastern Uganda*. Cambridge: Cambridge University Press.

Wolf, E. 1982. *Europe and the People without History*. Berkeley and Los Angeles: University of California Press.

———. 1999. *Envisioning Power: Ideologies of Dominance and Crisis*. Berkeley and Los Angeles: University of California Press.

Chapter 14

Cornish Weather and the Phenomenology of Light: On Anthropology and "Seeing"

Tori L. Jennings

Phenomenology as a research method involves the reflective questioning of experiences "as we live through them" (van Manen 2014: 27). For the French philosopher Maurice Merleau-Ponty, phenomenology is furthermore a "style of thinking" about the world in terms of lived meaning (1992: viii). Since 2000 Tim Ingold has written extensively from a phenomenological perspective on the subject of light, weather, and visual perception. Drawing on elements of Merleau-Ponty's phenomenology and James Gibson's ecological approach to perception, Ingold upends many of the implicit assumptions about knowledge and environment that remain rooted in a Western philosophical tradition. Central to Ingold's analysis is the notion that weather is a medium of perception, and perception is tantamount to experience (2000b, 2007). Thus we do not observe weather in a positivist sense but instead inhabit weather "as the experience of light itself" (Ingold 2005: 97). Despite the methodological relevance of light, anthropologists interested in weather, climate, and culture have paid remarkably little attention to this work. Why this might be the case is not altogether apparent since, as Bille and Sørensen assert in their survey of anthropology and luminosity, "light affects everything we experience, in obvious or subtle ways" (2007: 265).

The eclectic title of this chapter references a largely uncharted area of study within anthropology. However, in the spheres of philosophy, art, and history, Nolen Gertz (2010), among others, has probed if it is possible to have something like a phenomenology of light. To that end, Gertz explores the "peculiar preoccupation with light" shared by philosophers Husserl, Heidegger, and Levinas (2010: 41). Gertz further attempts to expound the relationship between light, sight, and meaning while simultaneously asking, what *is* light?—subjects Ingold explores in a series of essays on perception and environment (2000a). More recently, Samuel Galson historicizes phenomenology by using the "supposedly unique quality of light in Greece" to examine phenomenological claims themselves (2013: 249). Galson transcends landscape as a way of seeing to envisage instead ways of seeing vision itself. Given that phenomenology intends to overturn Cartesian nature-culture dualisms by investigating insight gained from everyday lived experience, and anthropologists are well acquainted with knowledge *as* lived experience (for example, Nadasdy 2003), it seems that anthropology can offer its own insights into the possibility of a phenomenology of light. But why should such an enquiry matter beyond the realm of philosophy or art history? What anthropological insights might be gained from a phenomenology of light?

To address these questions, I consider how light informs our way of thinking about weather and even climate science itself. This chapter attempts to foreground an elusive yet fundamentally important set of ideas that place *light* at the center of an ontological discussion about human engagement in a "weather-world" (Ingold 2005). I begin by taking a critical look at so-called Cornish light and its historical relation to the art and tourism industry of Cornwall, England. I then analyze how light brings lived experience to reflective awareness and conclude with ethnographic examples from my fieldwork to illustrate this point. This study builds on my dissertation field investigation conducted in Cornwall from 2003 to 2005 and on return trips there annually.

Cornish light

The aesthetics of light has long been the purview of artists, architects, and historians. The "painter's way" of seeing things, in the words of Merleau-Ponty (1964), provides useful insight into a phenomenological orientation. In short, phenomenology is a method of qualitative enquiry that makes accessible a "prereflective" experience of the world that exists before becoming an object of thought—that is, before we theorize about it through language and semiotic means (Merleau-Ponty 1992; Duranti 2010; Desjarlais and Throop 2011; van Manen 2014). This enigma, often described as being-in-the-world, is sometimes brought into visible existence by way of the "painter's gaze," which is itself a phenomenological modification since the artist shows how meaning reveals itself through light or color (Merleau-Ponty 1992). In less prosaic terms, prereflectivity as conceived by Merleau-Ponty is "that world which precedes knowledge, of which knowledge always *speaks*, and in relation to which every scientific schematization is an abstract and derivative sign-language, as is geography in relation to the countryside in which we have learnt beforehand what a forest, a prairie or a river is" (1992: ix). Paradoxically, phenomenology (rather than science) concretely explores the "raw reality" of a nonconstructed world becoming constructed (van Manen 2014). A principal aim of phenomenology, and justification for making it a topic of this chapter, is its ability to reveal unexamined assumptions and abstractions embedded in scientific practice. Specifically, the subjective "feel" and phenomenon of light that so fascinate painters have a bearing on scientific approaches to weather and climate that researchers often take for granted but cannot account for. An obvious place therefore to begin a discussion about the aesthetics and experience of light in Cornwall and elsewhere is among the artists and theorists who have contemplated light over the centuries.

The rural coastal region of Cornwall in southwest England was first noted as an artistic location during the Romantic revival from 1830–1870. During the interwar years of the 20th century, Cornwall was the most popular location outside London for panoramic illustrations, and, according to Patrick Laviolette (2011), nowhere else in Britain have artists paid more attention to light than in Cornwall. The majority of these illustrations incorporate the atmospheric qualities of natural light to generate romantic and sublime landscapes along with scenes of Cornish working life made famous by the Newlyn School painters (c. 1880–c. 1940). Artists and writers alike have commented on the "peculiar clarity of light" associated with Cornwall, and many take advantage of this quality to promote their work (Laviolette 2011: 36; 1999: 113). Even the media has capitalized on Cornish light. In 2008, for instance, BBC Two invited neuroscientist Beau Lotto of University College London to carry out sensor tests to determine "what makes light so special in Cornwall" for the television episode, *St. Ives: The Quality of the Cornish Light*. In August 2013 BBC Radio Four *Open Book* presenter, Mariella Frostrup, interviewed leading contemporary novelist Patrick Gale, who claimed that the Cornish landscape is peculiar because of "the light." Using Cornwall as a setting for many of his novels, Gale explained: "It's a bit like East Anglia, it has enormous horizons . . . even in bad weather there is drama." While Cornwall and East Anglia probably have little in common beyond horizons, the analogy itself illustrates the author's attempt to explain Cornish light via the "drama" of weather.

From the 19th century to the present, Cornish light and weather have played both an essential and perplexing role in the region's tourism industry. Historians observe that mass tourism in Cornwall began with the arrival of the railway in 1859 (Payton 2004). It was the Great Western Railway's (GWR) Cornish Riviera campaign that transformed Cornwall into a commercial asset by convincing the public that the Cornish peninsula was Britain's own Riviera, taking inspiration from Romantic ideas about the health-giving properties of sunshine, refreshing air, and sea-bathing that date back to ancient Greece (Kevan 1993). Despite the realities of Cornwall's frequent rain, severe gales, and sometimes deadly storms all too familiar to local sailors, farmers, and fishermen, the myth of the Cornish Riviera proved a robust marketing scheme. The GWR developed aesthetically pleasing advertising techniques including guide books and posters to promote a Mediterranean version of Cornwall, where "the sun always shines but never proves harmful, where it is always warm but never enervating, where we may bathe in the winter and take active exercise in the summer" (Mais 1934: 1). The GWR transformed historic fishing villages such as Newquay into holiday resorts and produced an image of Cornwall that lingers to this day and still dominates the region's economy. While light has been implicated in the art, literature, and economics of the Cornish Riviera, the phenomenon of light is largely absent from scholarly discussions on tourism in Cornwall.

The Meaning of Light

Although assumptions about light play figuratively into Cornwall as a peculiar and special place, Ingold suggests that similar associations about light and place are made elsewhere—for example, the northeast of Scotland, where light has long been an obsession of Nordic landscape painters (pers. comm. 2012). This observation invites the question: is Cornish light actually unique? Farther afield, Galson (2013) explores the luminous quality of light in Greece expanding yet again the list of places where painters and architects find light incomparable. However, Galson observes that "Greek light" as a supposed "natural" feature of the Athenian landscape "emerges at a particular historical moment and under particular cultural forces" (2013: 250). Tracing landscape phenomenology to its eighteenth century origins, Galson argues that Greek light begins with the German polymath Johann Wolfgang von Goethe (1749–1832), whose critique of Enlightenment empiricism and the shortcomings of Newton's theory of light opened up a phenomenological way of seeing. So-called Greek light, explains Galson, is the result of a phenomenological engagement with light in the nature-culture hybrid of landscape. Even though Greek light is taken for granted as a physical feature of Greece, the historical "invention of Greek light" reflects the process whereby landscape painters and architects transposed Goethe's phenomenological study of light and color onto an Athenian landscape (2013: 256). Galson reveals how the early Romantics created a new concept of light to negotiate concerns over Newtonian science and by extension provide insight into ways that Cornish light might be examined and questioned.

Thus to understand Cornish light, or know if such a thing exists, one must consider a more complicated and ambiguous account of what we mean by *light*. Although light as metaphor is used widely in art and literature, theorizing the meaning of light or even "seeing" has been only vaguely investigated in social studies (Galson 2013). Instead, scholars in material culture studies, history, and science and technology studies explore light and luminosity in relation to health, religion, agency, and social life in general. Frequently light is taken as given, and what follows are questions such as those presented by Bille and Sørenson: "How do people use light, and what does light *do*?" (2007: 280; see also White 2012; Winther 2013). Elsewhere, Simon Carter (2007) explores our ambivalent attitudes to sunlight and the way it mediates health, pleasure, the body, race, and class from ancient to modern times. While these are indeed fascinating studies, they often reveal unexamined assumptions about light as a material object. In the field of architecture, for example, light is regarded as a "building material" (Bille and Sørensen 2007: 270). This impulse to objectify light presupposes a physics of light that requires some explanation.

In general, light is conceptualized in two opposing ways resulting in an enduring philosophical paradox articulated in René Descartes' *Dioptrics*. On the one hand, the light of physics (*lumen*) consists of radiant energy that produces physical impulses in the retinal nerves and parts of the brain responsible for sight. Light in this sense is a physical phenomenon that can be measured and quantified and exists outside the body. On the other hand, the light of illumination and awareness of things (*lux)* is constructed within the mind. Light in this second sense is subjective, interior, and a matter of mental speculation (Ingold 2000b). Given these distinct views, where then does vision occur? Does light, asks Ingold, "shine in the world or in the mind?" (2000b: 256). Do we actually see light or merely the surfaces of objects that light illuminates? Merleau-Ponty writes that we do not so much see light as live *in* it, we are "immersed in it," he declares (1964: 178). He then makes a distinction between seeing and vision. Whereas seeing presumes a differentiation between a perceiver and objects, things and phenomena, the process of vision is attached to bodily movement in a world of constant flux—vision embodies our very involvement in an intersubjective world. Vision, says Merleau-Ponty, has the "fundamental power of showing forth more than itself" (1964: 178).

The subject-object, mind-body binary of light that so puzzled Descartes goes to the very heart of the nature-culture relation we at once critique and struggle against in anthropology. However, as scholars have noted before, this dualist epistemology is itself a by-product of our own Western categories of thought, and despite efforts to the contrary we remain tethered to its formulation. The significance of a phenomenology of light is one of ontology, since it concerns where "life" is situated. Ingold is emphatic on this point:

> Questions of the meaning of light must be wrongly posed if they force us to choose between regarding light as *either* physical *or* mental. . . . Thus the phenomenon known as "light" is neither on the outside nor on the inside, neither objective nor subjective, neither physical nor mental. It is rather immanent in the life and consciousness of the perceiver as it unfolds within the field of relations established by way of his or her presence within a certain environment. It is, in other words, a phenomenon of *experience*. (Ingold 2005: 99)

Acknowledging light as a two-way process of engagement between individuals and their environment has significant implications for weather, climate, and culture studies. First, reiterating Ingold's assertion that weather enters our awareness as the multisensory experience of light, scientific methods that transform weather to data (temperature, precipitation, climate, and so on) distill out lived experiences while simultaneously obscuring the cultural features responsible for this distillation.[1] Largely absent from the climate and culture literature are ethnographic analyses that examine what it means to be *in* light. Second, a phenomenology of light provides a resourceful and dynamic framework for addressing practical concerns in research—namely, how do researchers "see" and why. Third, light and therefore weather transcend the nature-culture binary that presents itself as a recurring problem in human-environmental research. Ideologies of knowledge that contrast "folk" or "indigenous" environmental knowledge with "scientific" knowledge are but one example (Scott 1996; Nadasdy 2003). To illustrate, I use two examples from my research that show how light brings lived experience to reflective awareness.

"You Know It When You See It, but You Can't Explain It"

Since 2003 I have carried out fieldwork with farmers, granite quarrymen, fishermen, and descendants of these same traditional tradespeople while living in Blisland, a 12th-century parish located on the edge of Bodmin Moor in north Cornwall. My research involves Cornish people who once made their living directly from the land and sea, and my interests include understanding how these individuals experience and interpret the natural world. Marjorie Webber, now among the parish's aging inhabitants, is one whose practical knowledge of subsistence farming, animal husbandry, and the moorland environment I rely on. Several times I have noted Mrs. Webber's reference to a "brassy sky" in the context of weather turning for the worse. I never understood the meaning of a brassy sky until when, on a mild spring afternoon, I took the 2-mile walk along the narrow hedge-lined lane from the village of Blisland to Bodmin Moor, something I had done routinely for almost two years. As I crested the hill that day, I took in the expansive view and found myself caught up in a remarkable stillness. Cottony white clouds hung motionless in a cobalt sky. The late afternoon light transformed the colors of the landscape into their richest hue of red, green, yellow, and blue. I then perceived, in the stillness of the moment, that everything seemed to shine, as if gold dusted the landscape and faintly hovered in the air. The spectacle made me almost giddy. As my thoughts returned to me, I ascribed the vividness of the encounter to an especially good mood and continued my ramble across the moor. Returning to the village the next morning, I mentioned my impression of the moor to Mrs. Webber who without hesitation exclaimed approvingly: "That's right, it was a brassy sky. Everything shines like!" Only then did the association of lustrous polished brass make any sense at all. However, I did not "see" a brassy sky as much as felt immersed in a particular kind of glimmering light.

In 1939 the significance of ambient light in forming perceptions about landscape, climate, and weather was observed by meteorologist, L. C. W. Bonacina, who argued that scenic or "pictorial impressions" of weather events have scientific value for their "question-raising power" (1939: 485).

Writing for the journal of the Royal Meteorological Society, Bonacina associates landscape meteorology with "sense-impressions" that we store in memory and also photograph, paint, and describe (1939: 485). He contends that the "artistic side of meteorology comprises those scenic influences of sky, atmosphere, weather and climate which form part of our natural human environment" as given in everyday experience. Meteorologists do not pay attention to these "pictorial impressions," he acknowledged, because "they have conceived meteorology to be a *science*—only concerned with scientific methods of investigation" (1939: 485). Bonacina proceeds to describe the "quality of light" that distinguishes the seasons, and, while the author takes a material approach to landscape in general, he anticipates a cultural understanding of weather and climate. Applied meteorologist, John Thornes (2003), aptly refers to this approach as "cultural climatology." Bonacina's attempt to merge the subjective and objective by way of light and color seems obvious. At the same time he is aware that perceptual experience reified in landscape art presents a challenge to the atmospheric sciences. Bonacina's insights reinvest meteorology with the artist's conscious and subconscious awareness of subtle environmental detail and should be taken seriously (Thornes 1999). For instance, Hans Neuberger's (1970) intriguing study of climate in art from 1400 to 1967 determined that meteorological features in art do in fact correspond with quantitative climate differences in various regions over time. My goal differs in that I aim to understand how the experience of light affects the way people do things.

The second example from my research is suggestive on this last point. Fox hunting with hounds, despite its controversies, is extremely popular in Cornwall and is an activity that crosscuts all socioeconomic groups. The hunt master of the North Cornwall Hunt once explained to me how he uses the hounds to understand, in his words, "nature." For example, he systematically discussed how weather affects both the hound's noses and the condition of scent itself, which together determine the success or failure of the hunt. In effect, the hunt master reads the hounds to understand details about the physical environment and anticipate the behavior of his quarry—animals that in turn have their own awareness of how to manipulate scent to confuse the hounds and avoid attack. As I sat listening to the hunt master in his kitchen, he looked out the window and paused for a moment. "Sometimes there is a blue haze," he said. "I don't know what causes it, but when you see it there's hardly any scent. . . . If you see the hounds rolling, it means the blue haze is about." What to "see" means in this instance is somewhat obscure, because the blue haze was apprehended indirectly through the hounds. I suggest that the "blue haze" is part of the hunt master's watchful engagement with the natural world and a linguistic expression he used to convey an experiential account of light and color that goes beyond words. "You know it when you see it," he thought out loud, "but you can't explain it." Weather it would seem—is *our being in it*—and light is the "ground of being" from which the meaning of weather surfaces (Ingold 2000b: 265).

Conclusion

Light is what renders to weather its felt meaning, and people act on meaning in culturally specific ways. The quality of light that formed the "brassy sky" for Mrs. Webber or the "blue haze" for the hunt master is not a mere isolated example of metaphorical use of light or commonplace folk prognostications. Instead, these linguistic expressions reflect different kinds of perception. Ingold has criticized anthropology in the way it follows a Western ontological tradition of separating vision and light (2005). How we regard vision, argues Ingold, is the outcome of a specific historical trajectory, one that prejudices us against a variety of perceptual experiences that exist in the world. I follow Ingold in advancing a phenomenological engagement with light as a means of expanding our awareness and sensitivity to the broader context in which people experience their world. Since experience itself is the bearer of meaning, we have much to gain by adopting this painterly view of things. Above all, a significant challenge confronting climate researchers is making meaningful connections between abstract data and the weather-world that people inhabit. In demonstrating the social and environmental effects of climate change, the science-driven solution to public indifference and political wrangling has often been more data. The focus on light returns us to meaningful accounts of weather and the practice of living.

Acknowledgments

This research was supported by a grant from the University of Wisconsin-Stevens Point and a generous fellowship from the International Dissertation Field Research Program of the Social Science Research Council (SSRC), with funds provided by the Andrew W. Mellon Foundation. I would like to thank Susan Crate for her continued interest in my research, and Paul Nadasdy, Samuel Galson, Karin Fry, Christian Diehm, Gwyn Howells, and Rachel Hunt for their helpful comments on early versions of this chapter.

Note

1. Much discourse has taken place regarding the distinction between climate and weather. Constructionist critiques regard climate as an abstraction, an accumulation of recorded numerical data. Despite its many quantitative features, weather, by contrast, is sensed and experienced.

References

Bille, M., and Sørensen, T. F. 2007. An anthropology of luminosity: The agency of light. *Journal of Material Culture* 12: 263–84.

Bonacina, L. C. W. 1939. Landscape meteorology and its reflection in art and literature. *Quarterly Journal of the Royal Meteorological Society* 65(282): 485–97.

Carter, S. 2007. *Rise and Shine: Sunlight, Technology, and Health*. Oxford: Berg.

Desjarlais, R., and Throop, J. C. 2011. Phenomenological approaches in anthropology. *Annual Review of Anthropology* 40: 87–102.

Duranti, A. 2010. Husserl, intersubjectivity, and anthropology. *Anthropological Theory* 10(1): 1–20.
Galson, S. 2013. “The Singular Light”: Phenomenology in its landscape. *Cultural History* 2(2): 247–61.
Gertz, N. 2010. On the possibility of a phenomenology of light. *PhaenEx* 1: 41–58.
Ingold, T. 2000a. *The Perception of the Environment: Essay on Livelihood, Dwelling, and Skill*. London: Routledge.
———. 2000b. Stop, Look, and Listen! Vision, Hearing, and Human Movement, in *The Perception of the Environment: Essay in Livelihood, Dwelling and Skill*. London: Routledge, pp. 243–87.
———. 2005. The Eye of the storm: Visual perception and the weather. *Visual Studies* 20(2): 97–104.
———. 2007. Earth, Sky, Wind, and Weather. *Journal of the Royal Anthropological Institute* N.S.: S19–38.
Kevan, S. M. 1993. Quests for cures: A history of tourism for climate and health. *International Journal of Biometeorology* 37: 133–24.
Laviolette, P. 1999. An Iconography of landscape images in Cornish art and prose, in P. Payton (Ed.),*Cornish Studies: Seven*. Exeter: University of Exeter Press, pp. 107–29.
———. 2011. *The Landscaping of Metaphor and Cultural Identity: Topographies of a Cornish Pastiche*. Frankfurt am Main: Peter Lang.
Mais, S. P. B. 1934. *The Cornish Riviera* (3rd ed.). London: Great Western Railway Company.
Merleau-Ponty, M. 1964. Eye and mind, in M. Merleau-Ponty and J. Edie (Eds.), *The Primacy of Perception: And Other Essays on Phenomenological Psychology, the Philosophy of Art, History, and Politics*. Evanston, IL: Northwestern University Press, pp. 159–90.
———. 1992. *Phenomonology of Perception*. London: Routledge.
Nadasdy, P. 2003. *Hunters and Bureaucrats: Power, Knowledge, and Aboriginal-State Relations in the Southwest Yukon*. Vancouver: UBC Press.
Neuberger, H. 1970. Climate in art. *Weather* 25(2): 46–56.
Payton, P. 2004. *Cornwall: A History*. Fowey: Cornwall Editions Ltd.
Scott, C. 1996. Science for the West, myth for the rest? The case of James Bay Cree knowledge construction, in L. Nader (Ed.), *Naked Science: Anthroplogical Inquiry into Boundaries, Power, and Knowledge*. New York: Routledge, pp. 69–86.
Thornes, John E. 1999. *John Constable's Skies*. Birmingham: University of Birmingham Press.
Thornes, John E., and McGregor, G. R. 2003. Cultural climatology, in S. Trudgill and A. Roy (Eds.), *Contemporary Meanings in Physical Geography*. London: Arnold, pp. 173–97.
van Manen, M. 2014. *Phenomenology of Practice: Meaning-Giving Methods in Phenomenological Research and Writing*. Walnut Creek, CA: Left Coast Press, Inc.
White, L. 2012. The impact of historic lighting, in M. Forsyth and L. White (Eds.), *Interior Finishing & Fittings for Historical Building Conservation*. Oxford: Wiley-Blackwell, pp. 143–62.
Winther, T. 2013. Space, time, and sociomaterial relationships: Moral aspects of the arrival of electricity in rural Zanzibar, in S. Strauss, S. Rupp, and T. Loue (Ed.), *Cultures of Energy: Power, Practices, Technologies*. Walnut Creek, CA: Left Coast Press, Inc., pp. 164–76.

Chapter 15

Making Sense of Climate Change: Global Impacts, Local Responses, and Anthropogenic Dilemmas in the Peruvian Andes

Karsten Paerregaard

This chapter explores how Andean people make sense of climate change via the process of climate ethnography. However, rather than applying a multisited approach and investigating climate change as simultaneous global and local phenomena, one of several possible approaches described by Crate (2011), my aim is to make a strategically situated ethnography. Such an ethnography can be thought of as a foreshortened multisited project that attempts to understand something broadly about the world system and current globalization processes in ethnographic terms by understanding them in context of the local and its local subjects (Marcus 1998: 95). More specifically, a strategically situated ethnography identifies places that are of pertinent relevance to the chosen topic of research and that allows the researcher to draw on and use local insights to shed light on issues of global importance, such as climate change. To these ends, an Andean climate ethnography must document how the people of the Andes experience, interpret, and respond to such environmental change as melting glaciers, unusual temperature fluctuations, irregular precipitation, and growing water scarcity.

The Peruvian Andes are a powerful context for such a strategically situated climate ethnography. Peru is counted as one of the world's countries most vulnerable to climate change, with the effects causing water conflicts in both urban and rural environments (Carey 2010; Paerregaard 2013a). Peru contains 70% of the world's tropical glaciers, providing a majority of the water used for irrigation and consumption in the country's rural and urban areas (Vuille et al. 2008). It has been calculated that within the coming 15 years all the glaciers below 5,500 m above sea level (m.a.s.l.) are bound to disappear, a worrying scenario that represents a threat in a country where 90% of the population lives in dry, semidry or subhumid areas (Oré et al. 2009: 56–57). However, the mountains and glaciers are

---

not only the sources of water for the physical life in the Andes; they are also critical elements in the cosmology and cultural memory of Andean people (Rhoades et al. 2008; Bolin 2009; Paerregaard 2013b, 2014). Since Inca times mountain deities have been the objects of adoration. To this day in many parts of the Andes people make annual offerings to the nearby mountains that supply them with fresh water (Besom 2013). Imagined as gifts that humans offer to the supernatural powers who control their lives, such offerings constitute a critical means to regulate the relation between society and nature, between humans and their environment (Gose 1994; Paerregaard 2014). But what happens when the nature that Andean people have known for centuries changes and begins to behave in unrecognizable and alarming ways? More specifically, what happens when the glaciers and the ice caps that cover the mountains of the Andes start to melt noticeably and rapidly and the mountain deities no longer respond to the offerings humans make to ask them for water? In other words, how do Andean people account for and respond to current global climate change and its effects on their surroundings and daily lives? My extensive fieldwork shows that, although people are increasingly concerned about the impact of climate change and have adopted its global vocabulary, they do not blame humans in other parts of the world for causing their environmental problems. Although this fact may make the people with whom I have worked look like the companions of conventional climate skeptics, this chapter argues that their skepticism is of a very different kind.

Water Management in Tapay

I illustrate this point by reviewing ethnographic data based on my long-term fieldwork in Tapay, an Andean community located in the Colca Valley in Peru's southern highlands (Figure 15.1). I made my first study of Tapay in 1986, and I have returned recurrently to conduct somewhat shorter periods of fieldwork (Paerregaard 1997). My 1986 study of Tapay consisted of a detailed account of the community's social organization, agricultural specialization, and irrigation system. Geographically, Tapay is a vertical community. Dispersed among 247 households, the villagers, who in 2011 numbered 671 (INEI 2007), live on different altitudes ranging from 2,250 m on the banks of the Colca River to 5,400 m at the top of Mount Seprigina (Figure 15.2), home to the water source that supplies the community for irrigation and human consumption. Within this ecological diversity and these vertical extremes, the villagers grow fruit, plant crops, and raise animals such as cattle, sheep, alpacas, and llamas, which provide them a variety of products that they consume domestically and also barter or sell for other products at markets in the neighboring villages of Cabanaconde and Chivay and in more remote places, such as Yauri in the province of Cusco.

Mount Seprigina and a few other nearby mountains are critical to the villagers' economy and their future survival. Thawing snow from these

Figure 15.1 Village of Tapay

Figure 15.2 Mount Seprigina of Tapay

mountains assures a constant flow of water that feeds a number of rivers and a multitude of springs in the community and that allows the villagers to operate various independent irrigation systems. Thus a total of 53 water sources of which 2 are springs and 32 are off-takes in the forms of 6 rivers and streams supply the community with water. Moreover, the villagers have constructed a total of 21 reservoirs in Tapay that allow the water users to capture the melt water overnight. For centuries, the villagers have maintained and managed the canals and the reservoirs without external support or interference just as they have a long tradition of forming their own irrigation and drinking-water committees and selecting their own authorities to allocate water, settle water disputes, and collect water tariffs (Paerregaard 1997). Two *regidores* ("water allocators") are elected in each irrigation cluster during annual cleaning and water ritual celebrations. The *hatun regidor* ("principal water allocator") is in charge of water allocation from July until January, and the *huch'uy regidor* ("supplementary water allocator") holds office during the rainy season from January to April.

The *hatun regidor* is a mandatory task at which each water user takes a turn once in his lifetime. Appointed by the local water users of the irrigation cluster at an annual assembly, this person has historically been an elderly male villager with the required experience. Today, however, it is not uncommon for the *hatun regidor* to delegate the role either to a relative, who could be female, or to another villager for pay. In addition to allocating irrigation water, the responsibility of *hatun regidor* includes several ritual duties. At the start of his or her term, the *hatun regidor* is expected to organize an offering to Mount Seprigina and at the end of the term (Figures 15.3 and Figure 15.4) is responsible for arranging the *yarqa aspiy* (canal cleaning), an event that is both a mandatory workday for cleaning the canals and reservoir and a ritual to welcome the next *hatun regidor* (Figures 15.5 and 15.6). Another ritual duty is to pay tribute to the mountain deities believed to control the water flow in Tapay (Paerregaard 1994a). In the eyes of the villagers, maintaining good relations with these powers and, in particular, the deity of Mount Seprigina by making offerings is crucial to ensure the water supply and the future life of Tapay. It is also a common belief that the harvest of fruits and crops and the well-being of the animals hinge on these ceremonies intended to appease the spiritual powers and request water.

Climate Change in Tapay

Even though local lifeways and activities in Tapay have changed little since I first arrived in 1986, the outside world is increasingly present. For example, the effects of the national economy are evident. In the past decade Peru's national economy has tripled, not only generating a boom in the country's cities but also precipitating a trickle-down effect in marginal places such as Tapay. Although the long-promised road to the community

Figure 15.3 The *hatun regidor* and his assistants conducting the annual offering ceremony at Mount Seprigina in Tapay

is still under construction, the villagers now have running water, electricity, and telephones. Moreover, Tapay receives substantial support from both the central and regional governments. In the last few years government inputs have worked to improve agricultural production and to build a new high school and a health center in the community.

The effects of global change are present. The snow that falls and lies on Seprigina and other mountains lasts only for short periods of time and, as I and local people observe, produces less melt water than previously. Few Tapeños have detailed knowledge or understanding of what causes these striking environmental changes. However, during 2011 interviews with villagers, many expressed concern that now it is only during the rainy season that Mount Seprigina is ice or snow covered, when before ice and snow lasted most of the dry season.[1] A majority of interviewees also confirmed that the weather had changed since I did my first fieldwork in 1986, pointing to temperature fluctuations and irregular precipitation as the most notorious signs of that change. The vast majority of those I interviewed also claimed that water today is scarcer than before, suggesting that there is a broad consensus among Tapeños that climate change is seriously affecting their lives.

These changes in climate, weather, and water access have not come alone but in the context of other changes from the outside world. For

Figure 15.4 Women serving homemade *chichi* ("corn beer") at the annual offering to Segrigina in Tapay

example, new opportunities have emerged providing the villagers with alternative livelihoods and introducing them to different ways of seeing the world. Since Tapay became a backpacker haven a decade ago, many villagers have established hostels and restaurants for the dozens of tourists who visit the community on a daily basis and provide them with welcome extra income. In addition to bolstering household incomes, these visitors also bring stories of modernity and globalization. In the process Tapeños have been introduced to Western ideas and have appropriated worldly terms and concepts, including global warming and climate change.

Even though I had been back to Tapay several times between 1986 and 2010 I was struck when I returned in 2011 by the villagers' use of a new terminology to observe and interpret environmental change and by their talk of the climate as a phenomenon posing serious threats to their lives. It was even more surprising that the villagers indicated a variety of causes for environmental change, and, while several of them adeptly used such terms as "global warming" and "contamination" to account for these, they disagreed when I suggested the change perhaps could be seen as a problem caused by industrialized countries. Some of the interviewed argued that rising temperatures are due to the introduction of modern lifestyles in Tapay and Peru's contaminating mining industry. Others suggested that climate change is a cyclical phenomenon that people in the Andes have

Figure 15.5 Ritual specialist preparing the offering to an irrigation canal in Tapay

Figure 15.6 Men taking a pause during the mandatory work day to clean the irrigation canals in Tapay

known for centuries that is related to natural disasters such as earthquakes, hunger, and plagues. One man said: "In Europe you're worried about climate change, but we've known it for a long time. Today water is scarce, but things will change again." Another villager pointed out to me that climate change is a phenomenon that occurs locally, not globally, and that global warming is caused by the encounter between warm water coming up from the earth and the cold water falling as rain from the sky.

Although Tapay inhabitants were reluctant to recognize environmental change as a global and irreversible phenomenon, many expressed doubts about the mountain deities they and their predecessors had believed in. During my field research in the late 1980s and early 1990s, a small group of *hermanos* ("brothers") who had converted from Catholicism to Protestantism campaigned against not only the Catholic Church but also the villagers' Andean customs and ceremonial practices such as the offerings to the mountain deities. In particular, the *hermanos* repeatedly tried to obstruct the villagers' ritualization of the annual cleaning of the water reservoirs and canals in Tapay. Even so, the vast majority of the villagers continued to participate in these events, and some *hermanos* actually gave up their struggle and reconverted to Catholicism (Paerregaard 1994b). In 2011 I noticed a change in not only the way the *hermanos* view the community's ritual practices but also how the rest of the villagers perceive the cosmology. A growing number of villagers now had their doubts about the effectiveness of the offerings and the many rituals associated with irrigation in Tapay, and although they continued to comply with them many said they merely do it because it is a custom. These doubts were particularly evident among the *hermanos,* who now were even more radical in their rebellion against Catholic and Andean rituals compared to what I observed in 1986. Indeed, on a previous visit I paid to Tapay in 2010 the *hatun regidor*, who was *hermano*, refused to make the traditional offering to Mount Seprigina on November 1. Some villagers claimed that the *regidor*'s neglect could bring misfortune on him and his family, but others found the issue of little importance. Curiously, in 2011 several villagers reported to me that more rain had fallen that year than in the past 15 years. Nonetheless, when I pointed this out to the *regidor* who was responsible for the ritual the following year, he told me that he intended to carry out the ritual as planned, despite the seemingly unpredictable climate and the mountain deities' dubious response. "Not because I believe it'll make any difference but because it's a custom," he said. Some villagers approved of his decision to continue the offerings, but as one pointed out to me: "We do it not because there'll be more water but to elude the anger of the mountains."

Although the 15 villagers I interviewed in 2011 interpreted the cause of climate change in various ways, they agreed it is to be found locally rather than globally and that they therefore must assume the responsibility for its consequences. Moreover, the interviews showed that although the villagers concurred that the offerings no longer have the effect they used to have, few dared to neglect them altogether. While raising doubts whether the rituals any longer serve as a useful means to establish a relationship of exchange with nonhuman beings, the majority of villagers interviewed still believed that these beings maintain the power to punish them for their misdeeds. According to several of these villagers, misdeeds include the extent to which they have acquired a modern lifestyle and consumer habits that contribute to the increased pollution of the local environment. Yet, despite how the majority interviewed recognizes their impact on the environment in this way, they believe it is the mountain deities who control the water flow and the climate. While five of the interviewed claimed it was the *hermanos* and their stubborn resistance to conduct the offering ceremonies that has resulted in the anger of the nonhuman beings and their holding back the water, three suggested that climate change is one of many ongoing changes of nature that create water shortages. In the eyes of these villagers, even though rising temperatures are inevitable, they are perceived as only temporary, and, like other natural disasters including earthquakes, they will end and be followed by better times. Rather than accepting the Western notion of climate change as a global irreversible phenomenon caused by people elsewhere, then, the villagers I interviewed and spoke with attributed the environmental changes Tapay is experiencing and undergoing to their own acts and the will of the mountain deities.

In the context of global change, other kinds of change are on inhabitants' minds. Outside interests are planning to mine gold in Tapay, which has added fuel to the uncertainty many villagers feel about the future water supply in the community. In 2011 several villagers reported to me that engineers from the Peruvian mining company had visited Tapay on several occasions offering them information about how the exploitation of gold will affect the community. Few believed the engineers when they explained that their mining activities would not affect the villagers' access to fresh and clean water. The same year the company made a study of the mine's environmental impact on Tapay and concluded that effects on local water would be insignificant. A group of villagers then visited the compound where the company was carrying out preliminary explorations to obtain more knowledge about its activities and to engage in negotiations with its leaders. While the visit affirmed the distrust many of the elder villagers felt for the company, it served to sway several younger villagers in favor of the mine, since it could provide them with well-paid jobs in a situation where water scarcity increasingly threatens traditional livelihoods such as agricultural production. Others expressed worries that the mine's need for water would worsen the community's environmental degradation, pointing to the fact that some of Tapay's springs had recently dried up. They suggested that the company should finance the construction of a new channel to direct water from a neighboring river to Tapay in return for the community's approval of the mine. Yet others found it futile to engage in negotiations with the company or to protest against it concluding that there is nothing Tapay can do to prevent it from operating in the community.

Conclusion

Even though the villagers of Tapay are experiencing and beginning to suffer the consequences of global warming, they are reluctant to endorse the discourse on climate change and the idea that it is caused by the developed world. While recognizing that industrialization and modern consumption lead to rising temperatures, many attribute this to pollution caused by people locally, regionally, and, to some extent, nationally rather than globally. Some even claim that climate change is a cyclical phenomenon related to other natural disasters such as earthquakes. The villagers thus interpret climate change very differently, some attributing it to nature itself and others to human agency. Still, most agree that climate change is produced in their own surroundings and not somewhere else in the world. The majority also perceives the climate as an essential element of nature and the nonhuman forces they believe regulate humans' relation to the environment. From this perspective the claim brought forward in the developed world that it is human activity that causes climate change is met with skepticism given that humans are viewed as one among many living agents in the world. Although changes in the environment indirectly can be related to human behavior—for example, when humans neglect to make offerings to the nonhuman beings—ultimately, it is these forces and not humans who control nature and therefore cause environmental change.

Anthropologists can make an important contribution to climate change research by exploring the human perspective on environmental change and thus correcting the reductionism of the climate change models natural scientists create. As I have demonstrated, such studies not only help with understanding the many ways people experience, perceive, and respond to climate change, but they also bring to the fore immanent tensions in the discourse on global warming and the concept of the Anthropocene. Inherent in this discourse and the idea of anthropogenic climate change is the idea that human activity is the main cause of global warming and that humans all over the world must act as a global community to save the planet from their own misdeeds. This view is gaining ground in the developed world, but in marginal places such as the Peruvian Andes people are critical of the notion of a global anthropogenic world. While many are aware of and deeply concerned about the impact of climate change in their communities they see it as something they can do little to counteract and as part of a broader and longer process of change that implies new problems and opportunities.

Note

1. In 2011 I also conducted a survey on climate perceptions with one-fourth of the households in Tapay's two most populated hamlets on such issues as land holding, livelihoods, irrigation, and community participation. I also conducted informal interviews on climate change with members of 15 households.

References

Besom, T. 2013. *Inka Human Sacrifice and Mountain Worship: Strategies for Empire Unification*. Albuquerque: University of New Mexico Press.

Bolin, I. 2009. The glaciers of the Andes are melting: Indigenous and anthropological knowledge merge in restoring water resources, in S. Crate and M. Nuttall (Eds.), *Anthropology and Climate Change: From Encounters to Actions*. Walnut Creek, CA: Left Coast Press, Inc., pp. 228–39.

Carey, M. 2010. *In the Shadow of Melting Glaciers: Climate Change and Andean Society*. Oxford: Oxford University Press.

Crate, S. 2011. Climate and culture: Anthropology in the era of contemporary climate change. *Annual Review of Anthropology* 40: 175–94.

Gose, P. 1994. *Deathly Waters and Hungry Mountains: Agrarian Ritual and Class Formation in an Andean town*. Toronto: University of Toronto Press.

INEI. 2007. *Censos nacionales 2007 XI de población y VI de vivienda*. Lima: Instituto nacional de estadística y informática, http://censos.inei.gob.pe/cpv2007/tabulados/#)

Marcus, G. 1998. *Ethnography through Thick & Thin*. Princeton, NJ: Princeton University Press.

Oré, M. T., del Castillo, L., Van Orsel, S. and Vos, J. 2009. *El Agua, ante nuevos desafíos: Actores e iniciativas en Ecuador, Perú y Bolivia, Agua y Sociedad*. Lima: Instituto de Estudios Peruanos.

Paerregaard, K. 1994a. Why fight over water? Power, conflicts and irrigation in an Andean village, in M. William and D. Guillet (Eds.), *Irrigation at High Altitudes: The Social Organization of Water Control in the Andes*, Washington, D.C.: Society for Latin American Anthropology and the American, Anthropological Association, pp. 189–202.

———. 1994b. Conversion, migration, and social identity: The spread of Protestantism in the Peruvian Andes. *Ethnos* 59(3-4): 168–86.

———. 1997. *Linking Separate Worlds: Urban Migrants and Rural Lives in Peru*. Oxford: Berg.

———. 2013a. Governing water in the Andean community of Cabanaconde: From resistance to opposition and to cooperation (and back again?)" *Mountain Research and Development* 33(3): 207–14.

———. 2013b. Bare rocks and fallen angels: Environmental change, climate perceptions, and ritual practice in the Peruvian Andes. *Religions* 4(2): 290–305.

———. 2014. Broken cosmologies: Climate, water, and state in the Peruvian Andes, in K. Hastrup (Ed.), *Anthropology and Nature*. London: Routledge, pp. 196–210.

Paerregaard, K., Stensrud, A., and Andersen, A. O. In press. Water citizenship: Negotiating water rights and contesting water culture in the Peruvian Andes. *Latin American Research Review* (2016).

Rhoades, R. Zapata, X., Rios, Z., and Ochoa, J. A. 2008. Mama Cotacachi: History, local perceptions, and social impacts of climate change and glacier retreat in the Ecuadorian Andes, in B. Orlove, E. Wiegandt, and B. Luckman (Eds.), *Darkening Peaks: Glacier Retreat, Science, and Society*. Berkeley and Los Angeles: University of California Press, pp. 216–25.

Vuille, M., Francou, B., Wagnon, P., Juen, I., Kaser, G., Mark, B. G., and Bradley, R. S. 2008. Climate change and tropical Andean glaciers: Past, present and future. *Earth-Science Reviews* 89: 79–96.

Chapter 16

Climate Change Beyond the "Environmental": The Marshallese Case

Peter Rudiak-Gould[1]

That climate change is unambiguously an "environmental issue" is often taken for granted, sometimes even within the field of anthropology. This assumption must be questioned, because the notion of an "environmental issue" is irretrievably caught up in Western notions of nature and carries with it all the liabilities and limitations of that category. In this chapter I illustrate these liabilities through my fieldwork on climate change perceptions and responses in the Marshall Islands. I do so by applying the environmental frame to this topic to show how it undervalues the human costs of the threat and ignores Marshallese views of climate change as a cultural crisis. I argue that framing climate change as "environmental" predisposes anthropologists to adopt either an ecological-anthropology paradigm, which, in the Marshallese case, overlooks the influence of science communication on public understandings of climate change, or a political-ecology paradigm, which misrepresents Marshallese narratives of climate change causes and responsibility. The framing of climate change as "environmental" is not wrong, but it is incomplete, and anthropologists of climate change must remain mindful of the blind spots that it generates.

Climate Change and the Environmental Frame

"Climate change is merely the latest northern environmental issue experienced by Inuit," writes environmental studies scholar Timothy Leduc (2011: 110). "Global warming is the most serious environmental problem of our time and a major issue of environmental justice," writes sociologist Kari Marie Norgaard (2006: 347). "The anthropogenic contribution to the greenhouse effect is an environmental problem for which we still cannot envisage any credible solution," write a group of science educators (Schreiner, Henriksen, and Kirkeby 2005: 36). None of these commentators could be accused of taking a narrow or uncritical view of climate change. Leduc rejects dominant scientific understandings of climate change and articulates an indigenous counter narrative. Norgaard challenges self-serving national discourses of carbon innocence. Schreiner and colleagues advocate climate change pedagogy as a form of grassroots empowerment rather than simply another branch of science education. But there is one assumption that these critical scholars leave largely unchallenged: the assumption that climate change is an "environmental issue."

That assumption needs to be put under the microscope. The category "environmental issue" is hardly timeless or universal, and its definition is a slippery one. It is far from obvious what constitutes the common denominator of (to pick four randomly chosen entries from Wikipedia's "list of environmental issues" [Wikipedia contributors N.d.c]) whaling, landslide, Agent Orange, and overpopulation in companion animals. Wikipedia defines "environmental issues" as "harmful aspects of human activity on the biophysical environment" (Wikipedia contributors N.d.b), which in turn is defined as "the biotic and abiotic surrounding of an organism or population" (Wikipedia contributors N.d.a). But these definitions are evasive. They do not specify what constitutes a "harmful" impact (was the eradication of smallpox an environmental harm?), where the line between "environmental" harms and "non-environmental" harms should be drawn (is vandalism an environmental issue because it damages people's built "environment"?), and why these issues should be singled out in the first place.

More helpful than a definition of *environmental issue* is a historical account of how the term has been invented and deployed by various actors. Sociologist Bronislaw Szerszynski's work provides exactly this. As he writes, issues such as deforestation, pesticide use, nuclear energy, and so forth "were not self-evidently 'environmental.' It was the task of the [environmentalist] movement to . . . develop an emerging discourse of the 'environment' to which a developing portfolio of issues, shaped by political opportunity and public resonance, could be linked" (Szerszynski 1993: 42). The rise of this environmental frame in the 1960s and 1970s was the result of a New Left that inherited the older left-wing view that "society—the system —was sick and needed a radical reorientation if it was to survive" (ibid.) and then "broaden[ed] and deepen[ed] this framework by inserting into it the representations of nature as threatened, and nature as a moral source . . . representations of nature [that] were firmly embedded in industrialised societies" (ibid.). In a less sympathetic account of the genesis of the environmental frame, anthropologist Mary Douglas and political scientist Aaron Wildavsky argue that the 1960s counterculture invented an "environmental" crisis because its preference for small-scale, egalitarian, acephalous social organization made it difficult to "hold [its] members together without coercion or overt leadership" (Douglas and Wildavsky 1982: 11). To prevent schismogenesis and defection, the counterculture organizations needed to "invok[e] danger. Either the backlash of God or the backlash of nature is an effective instrument for justifying membership . . . [and] in a secular civilization nature plays the role of grand arbiter of human designs more plausibly than God" (Douglas and Wildavsky 1982: 127). The "environmental issue" was born.

In other words, an "environmental issue" means a dysfunctional relationship between humans and "nature." The anthropological critique now becomes inevitable: if the environmental frame is irretrievably caught up in a Western nature-culture dichotomy, then using it to understand climate change results in the usual modernist sins. It divorces humans from the world in which they live (Ingold 2008 [1993]) and thus widens the very schism it aimed to bridge (Latour 2004). It adds fuel to spurious debates about "balancing" ecology with economy, as though they were separate entities. It imposes a Western ontology onto other ways of life (Smith 2007: 200–01).[2] It reduces climate change to one dimension, discouraging the kind of cross-sectoral policy-making that is necessary for successful local adaptation (Baker et al. 2012). And last, it fails as a tool of public communication: various studies show that framing climate change mitigation measures as "environmental" actions is less effective than framing them in terms of national security or economic thriftiness (Zang 2009; Lockwood 2011; Gromet et al. 2013).

All of this may seem obvious to anthropologists, trained as we are to cast a skeptical eye on dominant categories and paradigms. But such is the clout of the environmental frame in today's world that, even in anthropology, we find it hard to escape it. Those of us engaged in issues of climate, resource extraction, ethnobotanical knowledge, and so forth often call ourselves "environmental anthropologists" rather than just anthropologists. We use such terms as *eco-colonialism*, *environmental justice*, and *political ecology* rather than simply *colonialism*, *justice*, and *politics*. We join the Anthropology and Environment section of the American Anthropological Association, which gives out awards for specifically "environmental" studies in anthropology, invites sessions on "environmental" issues, and so forth. When I meet anthropological colleagues at conferences and tell them that I study climate change, they often suggest other "environmental" panels that I should attend. When I started a teaching position in the Department of Anthropology at McGill University, it was noted that I conduct fieldwork on the human dimensions of climate change, so I was asked to teach a class on Environment and Culture. And so forth.

The environmental frame dies hard. But my aim in this chapter is not to deliver the death blow but rather to show the limitations of the environmental frame for climate change anthropology and to illustrate them using my research in the Marshall Islands.

The Environmental Frame in the Marshall Islands

The Republic of the Marshall Islands is a sovereign nation in eastern Micronesia composed of 29 coral atolls and 5 single coral islands, totalling over 1,000 individual islets (Figure 16.1). With an average elevation of 2 m above mean sea level and a maximum elevation of 10 m, this country's

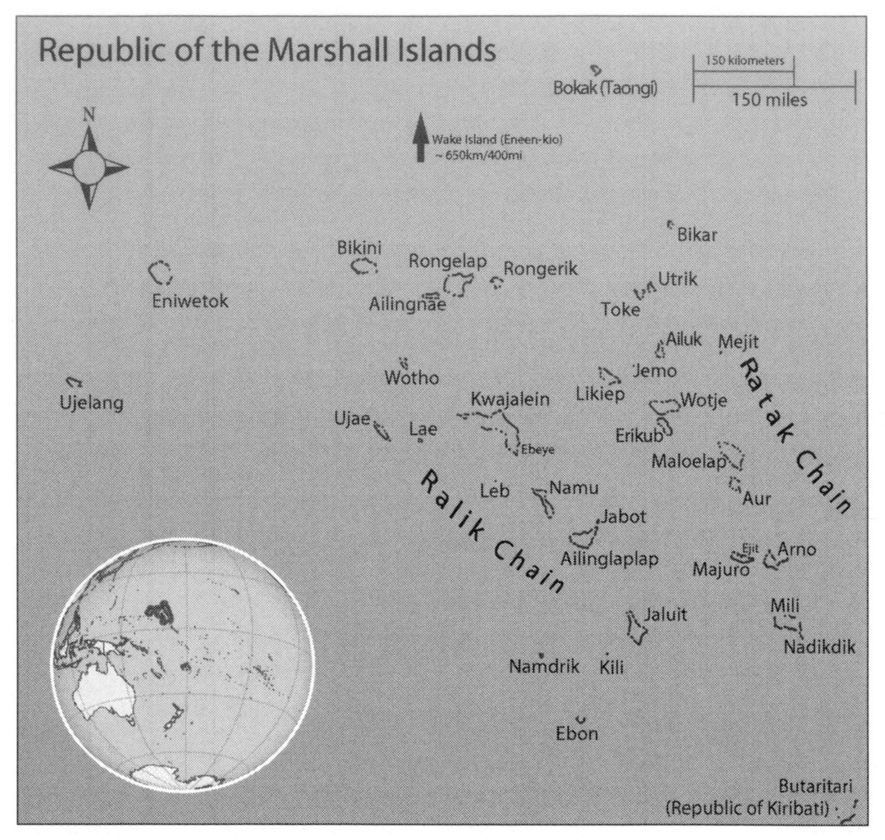

Figure 16.1 Map of the Republic of the Marshall Islands

most obvious climate-related danger is sea-level rise. Universalizing statements about the country's vulnerability—"All the country's islands . . . lie only a few metres above sea level and could be engulfed by rising oceans" (The Economist 2013)—ignore the intimately local nature of climate change vulnerability—for example, the fact that the heavily developed shorelines of the urban centers are far more exposed to sea-level rise than outer island communities (Spennemann 1996) and that certain somewhat higher outer islands actually have little to fear from 0.5 m sea-level rise scenarios (Kench et al. 2011). Nonetheless, the danger that sea-level rise, intensifying storms, coral bleaching, and changing precipitation patterns will make large swatches of the country uninhabitable in the medium term cannot be denied (Barnett and Adger 2003; Nunn 2013).

Already we see a liability of the "environmental" frame for understanding climate change in the Marshall Islands: its potential to cheapen the problem. In the Marshall Islands, climate change is not so much an environmental threat as it is an existential one. If Americans became convinced that their entire country was facing a condition of uninhabitability by the end of the century, would President Obama address this issue in the same breath as oil drilling in the Arctic National Wildlife Refuge? Would the editors of *The New York Times* place articles on this issue in the Science-Environment section? Would Fox News pundits argue that the United States must prioritize its economy over its inhabitability? Would Greenpeace, in order to reach the public, emblematize charismatic megafauna (other than humans!) that face extinction when the United States disappears? It is hard to imagine that they would. In other words, the issue would probably not be framed as "environmental"; and anyone framing it as such would appear to be belittling the issue.

Appropriately, Marshall Islanders do not usually speak about the existential threat of climate change within an environmental frame. It is spoken of as a crisis of tradition not a crisis of nature; the cause is understood to be the local adoption of foreign, untraditional, polluting artifacts such as air-conditioning units and gas-powered motorboats, and the consequence is understood to be the increased difficulty (and eventual impossibility) of sustaining traditional livelihoods. Marshall Islanders do share with Western environmentalists the idea of a pristine past spoiled by human greed and short-sightedness—but for them that "past" is not a natural past but a *cultural* past, the idyllic precontact era of untainted Marshallese tradition (*m̧antin m̧ajeļ*) (for an extended discussion see Rudiak-Gould 2013). Indeed the environmental frame barely exists in the Marshallese language, with no word that can be translated as "nature," and the closest equivalent to the English "environment" is *peļaak*, which simply means surroundings. Marshallese climate educators translate English "climate" as Marshallese *mejatoto*, a word that refers to prevailing social *and* meteorological conditions. A local NGO, the Marshall Islands Conservation Society (MICS), brands itself as environmental: in its Mission Statement it emphasizes "raising environmental awareness and building capacity for conservation, sustainable use of resources and protection of biodiversity in the Marshall Islands" (Rare Planet website N.d.), which is quintessential environmental language. However, this framing is primarily a way of marketing itself to outside donors who understand the language of "biodiversity" and "conservation" better than the language of *kōjparok m̧anit* ("maintaining tradition") and *lale doon* ("taking care of one another"). In its local activities, partnering with outer island communities to safeguard livelihoods against cultural and climatic change, the NGO frames its work as protecting culture rather than protecting nature.

Other liabilities of the environmental frame become apparent when one considers the informational outlets that shape Marshallese understandings of climate change. A 1996 Marshallese government document summarized the threat of climate change to the country climate change as follows: "this threat of a potential sea-level rise remains only speculative and far in the future. In any case, there is nothing that the atolls can do about it" (RMI Government 1996: 19–20). Seventeen years later, attitudes had clearly changed. The country's president, Christopher Loeak, wrote an open letter to U.S. Secretary of State John Kerry, declaring,

> [M]y country . . . is in the eye of a gathering climate storm. In May, I was forced to declare a state of disaster as our northern islands wilted under the effects of a climate-driven drought. A mere six weeks later, a king tide and rising oceans topped the sea walls in our capital, Majuro, flooding the airport runway and many neighborhoods, including my own. Climate change has arrived in the Marshall Islands. . . . We are fighting a war for the survival of my country. (Loeak 2013)

Among Marshallese citizens the change in attitude was nearly as dramatic. In 2003–2004, during my first field trip to the Marshall Islands, the idea of burgeoning climatic devastation was almost never discussed, and no local NGO was tackling the issue in any serious way. By 2009 a survey I conducted suggested that climate change issues had become the number-five concern among Majuro residents (Rudiak-Gould 2013a: 90), and two local NGOs (MICS and Women United Together Marshall Islands) had enthusiastically adopted climate change as one of their major raisons d'être.

What brought about this sea change in concern? The obvious answer—and the answer suggested by the environmental frame—is that concern increased because local environmental impacts intensified. Indeed, during the time when climate change concern was burgeoning in the Marshall Islands, the country was hit by a number of climate-related extreme events, including the cocktail of disasters described in President Loeak's letter. But this is not the whole story. Experiencing a climate-related disaster strengthens concern about climate change only if people perceive the disaster to be unusual, unprecedented, and indicative of a trend. Marshall Islanders, living in an archipelago that has been prone to climatic disturbance for centuries, do not always perceive climate-related disasters with such anxiety (Rudiak-Gould 2013b). People have to *learn* to perceive floods, droughts, and so on as part of a wider and longer trend with a particular name: "climate change." Only then does concern about climate change coalesce. My fieldwork suggests that the process of bundling events into a single conceptual entity ("climate change") was not as much due to a changing "natural environment" as it was to the result of a changing institutional arrangement, in which a number of NGOs, media outlets, educational institutions, and governmental bodies intensively marketed scientific narratives of "climate change" to the Marshallese public starting in 2009 and continuing to the present day. Quantitative evidence strongly indicates that Marshall Islanders' reports of local sea-level rise, changing seasonality, rising temperature, and decreasing rainfall are a result of their *reception* of scientific narratives of climate change, not just their observation of local change (Rudiak-Gould 2013c).

Framing climate change as environmental does not necessarily exclude these considerations of science communication, but neither does it highlight them. As I have argued elsewhere, anthropologists have been more comfortable conducting studies of the local observation of climate impacts than the local *reception* of climate science (Rudiak-Gould 2011). I argue that anthropologists' and other researchers' persistence in employing the environmental frame, which directs us to think of climate change as a thing happening in the "natural environment" that people observe and respond to, is one of the main culprits.

To put this differently, when we approach climate change anthropology as a kind of "environmental anthropology," we default to one of environmental anthropology's dominant theoretical paradigms. One of these go-to frameworks is ecological anthropology—associated with Julian Steward's investigations of environmental influences on cultural traits, Roy Rappaport's studies of humans as part of ecosystems, and Harold Conklin's documentation of non-Western systems of environmental knowledge. Applying such an ecological-anthropology approach to climate change in the Marshall Islands would mean focusing on topics such as pre- and post-contact Marshallese adaptation to the climatic conditions of coral atolls; Marshallese meteorological terminology and folk theories; Marshallese observation of local effects of global climate change; Marshallese livelihood strategies in response to these perturbations; and so on. These are certainly valid topics. But they fail to consider an affected community's reception of scientific narratives of "climate change," "greenhouse gases," "sea-level rise," and so forth, which I have found crucial to understanding Marshallese attitudes toward climate change.

The other dominant paradigm in contemporary environmental anthropology is political ecology. This is a Foucauldian or poststructuralist approach to "environmental" issues that is primarily concerned with the operation of power and discourse. It tracks the historical invention of such categories as "nature" and "the environment," the dissemination of hegemonic environmental discourses by powerful elites (for example, REDD programs that greenwash land grabs), the unequally distributed environmental hazards and benefits that these discourses justify and obscure, and the emergence of emancipatory environmental counter-narratives among marginalized groups. Taking a political-ecology approach to climate change in the Marshall Islands would mean studying the colonial origins of the country's vulnerability to climate change (Spennemann 1996); the problematic eco-colonialist framing of low-lying island countries as "canaries in the coal mine" (Farbotko 2010); the environmentally unjust mismatch between the country's tiny crime (about three millionths of the world's total greenhouse gas emissions) and huge punishment (possible nationwide uninhabitability) (Barker 2008); and so forth.

But Marshall Islanders themselves rarely talk about climate change in these terms.[3] Rather than pointing fingers at the United States or other egregious carbon offenders, they tend to place blame for climate change on all human beings, with a special focus on Marshallese people themselves. The prevailing response is therefore local mitigation rather than foreign protest. The political-ecology perspective might overlook these tendencies. Worse, it could delegitimize them—since Marshallese narratives of climate-change blame are not counter-narratives in any obvious way, the political-ecology paradigm is forced to understand them instead as false consciousness, denial, ignorance, or insincere performance. As I have argued in detail elsewhere, these sorts of explanations fail to shed much light on the causes or consequences of Marshallese climate-change blame narratives, which are more brightly illuminated by the literature on risk (Rudiak-Gould 2014). Political ecology's insights are many, but they are not all-encompassing.

Conclusion

I am not calling for the abolishment of the environmental frame. It does, at times, provide a serviceable platform for climate-change action (Bodansky 1993: 463–64; Attari et al. 2009). It helps us to connect our work to that of other scholars, to foster dialogue across disciplines and to communicate our ideas to a wider world using a familiar language. These are all laudable goals—and most of us use the frame with some awareness of its limits and pitfalls. Humans, in any case, *need* frames. "An individual cannot look in all directions at once," write Mary Douglas and Aaron Wildavsky, and as a result "social life demands organization of bias" (1982: 9). The environmental frame is one such organization of bias, and it is no more "wrong" than a camera lens, or a culture, can be "wrong."

It *is*, however, partial. When we put some elements in the center (cultural understandings of "nature"; firsthand observation of climatic change; adaptation to impacts) we relegate others to the peripheries (cultural understandings of *culture*; second-hand reception of climate science; mitigation of carbon emissions). The depth of focus is reduced, and some objects appear sharply only because others have been blurred. Climate change *is* an environmental issue, but as anthropologists we must realize, and help others to realize, that it is so many other things besides.

Notes

1. I wish to acknowledge the generosity and hospitality of the many Marshall Islanders who shared their thoughts and assisted the research process in innumerable ways. An affiliation with the Department of Anthropology at the University of Toronto provided an academic home during the revision of this article. Earlier phases of research and fieldwork were supported by Dr. Alun Hughes; the Andrew W. Mellon Foundation; Jesus College, All Souls College, St. Hugh's College, Oxford; and the Institute of Social and Cultural Anthropology, Oxford.

2. These and other liabilities were explored in a panel entitled "Green power? On the Uses, Abuses, and Limits of an Environmental Frame," organized by Emily McKee and Karen Rignall at the American Anthropological Association's 2013 annual meeting.
3. The only significant exception to this is the statements made by Marshallese government officials when airing the country's climatic plight abroad.

References

Attari, S. Z., Schoen, M., Davidson, C. I., DeKay, M. L., de Bruin, W. B., Dawes, R., and Small, M. J. 2009. Preferences for change: Do individuals prefer voluntary actions, soft regulations, or hard regulations to decrease fossil fuel consumption? *Ecological Economics* 68(6): 1701–10.

Baker, I., Peterson, A., Brown, G., and McAlpine, C. 2012. Local government response to the impacts of climate change: An evaluation of local climate adaptation plans. *Landscape and Urban Planning* 107(2): 127–36.

Barker, H. M. 2008. The inequities of climate change and the small island experience. *Counterpunch*, November 4.

Barnett, J., and Adger, W. N. 2003. Climate dangers and atoll countries. *Climatic Change* 61: 321–37.

Bodansky, D. 1993. The United Nations Framework Convention on Climate Change: A commentary. *Yale Journal of International Law* 18(2): 451–558.

Douglas, M., and Wildavsky, A. B. 1982. *Risk and Culture: An Essay on the Selection of Technical and Environmental Dangers*. Berkeley and Los Angeles: University of California Press.

Farbotko, C. 2010. Wishful sinking: Disappearing islands, climate refugees and cosmopolitan experimentation. *Asia Pacific Viewpoint* 51(1): 47–60.

Gromet, D. M., Kunreuther, H., and Larrick, R. P. 2013. Political ideology affects energy-efficiency attitudes and choices. *Proceedings of the National Academy of Sciences* 110(23): 9314–19.

Ingold, T. 2008 [1993]. Globes and spheres: The topology of environmentalism, in M. R. Dove and C. Carpenter (Eds.), *Environmental Anthropology: A Historical Reader*. Malden, MA: Blackwell Publishing, pp. 462–69.

Kench, P. S., Owen, S., Resture, A., Ford, M. R., Trevor, D., Fowler, S., Langrine, J., Lometo, A., Alefaio, S., Kitala, T., Latasi, P., Penivao, F., Tanielu, S., and Pese, T. 2011. *Improving Understanding of Local-Scale Vulnerability in Atoll Island Countries: Developing Capacity to Improve In-Country Approaches and Research*. Final Report to the Asia-Pacific Network for Global Change Research.

Latour, B. 2004. *Politics of Nature: How to Bring the Sciences into Democracy*, C. Porter, Trans. Cambridge, MA: Harvard University Press.

Leduc, T. B. 2011. *Climate, Culture, Change: Inuit and Western Dialogues with a Warming North*. Ottawa: University of Ottawa Press.

Lockwood, M. 2011. Does the framing of climate policies make a difference to public support? Evidence from UK marginal constituencies. *Climate Policy* 11(4): 1097–12.

Loeak, C. J. 2013. Dear Secretary Kerry, join the Pacific fight against climate change. *Huffington Post*, July 24, www.huffingtonpost.com/christopher-jorebon-loeak/dear-secretary-kerry-join_b_3642230.html?utm_hp_ref=tw.

Norgaard, K. M. 2006. "We don't really want to know": Environmental justice and socially organized denial of global warming in Norway. *Organization and Environment* 19(3): 347–70.

Nunn, P. D. 2013. The end of the Pacific? Effects of sea level rise on Pacific island livelihoods. *Singapore Journal of Tropical Geography* 34(2): 143–71.

Rare Planet website. N.d. Marshall Islands Conservation Society, www.rareplanet.org/en/organization/marshall-islands-conservation-society (accessed July 16, 2014).

RMI Government. 1996. *A Situation Analysis of Children and Women in the Marshall Islands 1996*. The Government of the Marshall Islands with the assistance of UNICEF.

Rudiak-Gould, P. 2011. Climate change and anthropology: The importance of reception studies. *Anthropology Today* 27(2): 9–12.

———. 2013a. *Climate Change and Tradition in a Small Island State: The Rising Tide*. New York: Routledge.

———. 2013b. Memories and expectations of environmental disaster: Some lessons from the Marshall Islands, in M. Davies and F. Nkirote (Eds.), *Humans and the Environment: New Archaeological Perspectives for the 21st Century*. Oxford: Oxford University Press. 231–44.

———. 2013c. The influence of science communication on indigenous climate change perception: Theoretical and practical implications. *Human Ecology* 42(1): 75–86.

———. 2014. Climate change and accusation: Global warming and local blame in a small island state. *Current Anthropology* 55(4): 365–86.

Schreiner, C., Henriksen, E. K., and Kirkeby Hansen, P. J. 2005. Climate education: Empowering today's youth to meet tomorrow's challenges. *Studies in Science Education* 41(1): 3–49.

Smith, H. A. 2007. Disrupting the global discourse of climate change: The case of indigenous voices, in M. E. Pettenger (Ed.), *The Social Construction of Climate Change: Power, Knowledge, Norms, Discourses*. Aldershot: Ashgate Publishing Limited, pp. 197–215.

Spennemann, D. H. R. 1996. Non-traditional settlement patterns and typhoon hazard on contemporary Majuro Atoll, Republic of the Marshall Islands. *Environmental Management* 20(3): 337–48.

Szerszynski, B. 1993. *Uncommon Ground: Moral Discourse, Foundationalism, and the Environmental Movement*. Unpublished doctoral thesis, Lancaster University.

The Economist. 2013. Sea change. www.economist.com/news/asia/21584396-gathering-pacific-leaders-worries-about-climate-change-sea-change, August 31.

Wikipedia contributors. N.d.a. List of environmental issues. *Wikipedia, The Free Encyclopedia*, http://en.wikipedia.org/wiki/List_of_environmental_issues (accessed July 14, 2014).

———. N.d.b. Environmental issue. *Wikipedia, The Free Encyclopedia*, http://en.wikipedia.org/wiki/Environmental_issue (accessed July 14, 2014).

———. N.d.c. Environment (biophysical). *Wikipedia, The Free Encyclopedia*, http://en.wikipedia.org/wiki/Biophysical_environment (accessed July 14, 2014).

Zang, D. 2009. From environment to energy: China's reconceptualization of climate change. *Wisconsin International Law Journal* 27(3): 543–74.

Chapter 17

"This Is Not Science Fiction": Amazonian Narratives of Climate Change

David Rojas[1]

At a 2010 conference on climate change I attended in Brasília a senior Brazilian scientist stated that the combined effects of deforestation and global warming could disrupt rain regimes and transform parts of the Amazon rainforest into a savanna-like ecosystem. He added that, as a result of these macro-ecological changes, agricultural and ranching operations in the basin would suffer, and Amazonian populations would find themselves living in environments radically different from those to which they were accustomed. Following his presentation, I asked one of the conference organizers (an environmentalist who has worked in Amazonia for decades) how it was for him to work on the premise that unprecedented socioecological futures are coming. He responded by explaining that the basin had a long history of rapid and profound socioenvironmental changes. Some populations living in Amazonia's agricultural regions, he added, had already lived through rapid ecological and economic shifts that were analogous to transformations associated with climate change; for them such futures would not be unprecedented. "This is not science fiction," he claimed.

By "science fiction" the Amazonian environmentalist meant stories about people facing situations unlike any they have experienced before (consider *Star Trek*'s motto: "to boldly go where no man has gone before"). This understanding of the genre echoes the suggestion by Slusser and Rabkin that conventional science fiction narratives depict futures that, although envisioned as radically different from the present, are mastered through enhanced technologies (Slusser and Rabkin 1987; see also Stover 1973). From this perspective, a science fiction narrative of climate change would offer reassurance that technological investments can re-create previous socioenvironmental orders even after the planet is transformed by human actions. In contrast, the environmentalist invited me to consider views on climate change that, in two respects, are far less reassuring. First, profound socioenvironmental disruptions are not completely new to Amazonian populations who have contributed to transforming large swaths of forests into pastures and farmlands. Second, the futures that some Amazonian populations describe do not include the re-creation of vanished socioenvironmental orders by technology. The environmentalists at the conference argued, to the contrary, that Amazonian populations anticipated that farming and ranching technologies were likely to continue to destabilize socioecological dynamics in the foreseeable future.

This advice helped me to examine narratives of climate change I heard from people in Amazonia. Take the story of Luíz, a rancher I met in the southern Amazonia town of Butantá.[2] Like other landholders in the region, Luíz talked about climate change dynamics as part of the history of development efforts to which he contributed and that drove previous socioecological shifts. Luíz explained that decades ago he was part of a group of land speculators who built several new cities in Amazonia, the names of which translate to English as "New World," "New Planet," "New Satellite," and "New Earthly Body." These settlements, he argued, added to the Brazilian military government's (1964–1985) project to "occupy" the basin and halt the perceived United States plans to control the region. More recently, Luíz and other local politicians strived to transform these cities into technology-intensive agroindustrial landscapes housing export-oriented enterprises capable of competing with Northern agribusiness. Based on his experience in past world-making efforts, Luíz discussed rain-regime disruptions as a challenge not unlike other problems he had already faced in Amazonia's socioenvironmental history. Although he recognized the disruptive potential of climate change dynamics, he suggested *engaging with* bewildering environmental futures rather than trying to avoid them, reflecting lessons he had learned when facing extremely uncertain situations in the past.

The conference organizer who preceded my encounter with Luíz was right: Luíz's narrative of climate change (and the narratives of others in similar situations as Luíz) is not *conventional* science fiction. Rather, he and others like him offer histories wherein human interventions have created a succession of unfamiliar worlds that have already tested and familiarized ranchers with potentially inhospitable situations. What Luíz's story challenges us to ponder is the extent to which he and other ranchers may be open to climate change based on their historical *familiarity with the unfamiliar*.

In this chapter I examine Amazonian accounts that reflect familiarity with the strange worlds of climate change using science fiction as an analytic. Even if, as my interlocutor at the conference argued, climate change is not like *most* science fiction, anthropologists have shown that the tenor of climate change dynamics is analogous to that of a science fiction subgenre that focuses on the "extreme" (Valentine, Olson, and Battaglia 2012), with the word *extreme* pertaining to worlds and situations that are radically different from the conditions in which humans have learned how to live

(Helmreich 2012; Masco 2010, 2012; Oreskes and Conway, 2014). As philosophers of science have also pointed out, this particular science fiction subgenre (also called *extro-science fiction*) challenges readers to consider that humans may find themselves navigating worlds unbounded by invariable laws (Meillassoux 2010, 2015; Morton 2013). This is to say, extreme science fiction brings to the foreground the possibility that environmental patterns may suddenly shift in ways that create profoundly uncertain situations (Harman 2012; Meillassoux 2015). Framing climate change narratives of Amazonian populations in terms of extreme science fiction allows me to understand the actions of ranchers who, foregoing normal and familiar worlds, embrace extreme environments as an opportunity and prepare themselves to live on inhospitable landscapes.

I first examine how ranchers like Luíz discuss the emerging worlds created by climate change as representing an "extreme" situation wherein mysterious and sudden changes of behavior in organic and inorganic entities have undermined familiar traits of the worlds that had been built by ranching operations. In the second section I show how these unfamiliar worlds were not new to people such as Luíz, who succeeded economically and politically in the similarly "extreme" land-speculation environments of Amazonia. I conclude by arguing that such familiarity with the unfamiliar in ranching areas in Amazonia makes it easier to understand how Luíz and others like him respond to climate change by investing in "terra-forming projects"—agricultural operations that remake parts of the world into inhospitable environments.

Changing Climate, Sudden Death

I first visited Butantá in 2010, a year in which the Amazon basin suffered through one of the most severe droughts in its history and the region revealed itself as an "extreme" setting. As in other parts of Southern Amazonia, in Butantá the dry season extends from July through September—after which rains are an almost daily occurrence. While the 2009–2010 rainy season was particularly strong, during the dry months of 2010 rains were unusually scarce, and significant precipitation arrived more than two months later than usual, in November. However, more than just the timing of the rains was wrong; their quality had also changed. Some people in Butantá claimed that the rains of recent years had fallen in particularly short and strong downpours—unlike the long and mellow showers characteristic of the region when the city was founded a few decades back. These 2010 "rains out of control," or *chuva descontrolada*, as some locals called them, came after the 2005 drought that scientists described as a once-in-a-century drought. Furthermore, the 2005 drought had occurred just a few years after a particularly harsh dry period in 1997, the severity of which led some scientists to argue that such an event would reoccur only after some decades. Environmental scientists who specialize in Amazonia argue that the increasingly strange Amazonian weather is likely associated with climate change and may be taken as an omen of extreme futures that are devoid of key macroecological traits that make our worlds familiar (Lewis et al. 2011; Davidson et al. 2012; Brando et al. 2014).

For ranchers like Luíz, the droughts of 1997, 2005, and 2010 most significantly revealed disconcerting aspects of the worlds he and his associates had built while struggling to make Butantá's ranching operations into global enterprises. The scarcity of rains in 2010, for example, seemed to have contributed to an extreme agricultural condition known as the "sudden death of the pastures" (*a morte súbita das pastgens*). People talked about the sudden death when parts of their pastures rapidly dried out "as if you had applied herbicide," as ranchers often described it. At the affected sites, seemingly healthy pastures died in circular patches that in some cases expanded to engulf most of the property. The sudden death affected one of the main technologies that had enabled Butantá's ranching economy to thrive: a grass variety known as *Brachiaria brizantha*. Since its introduction from Africa in the 1960s *brizantha* had been the subject of much agronomic research, becoming one of the best-performing grasses in Southern Amazonia's acidic soils and long dry seasons. As *brizantha* pastures collapsed with the expansion of the sudden death, ranchers were forced to destroy their grass and plant another variety (*mombaça*), which demanded a costly and lengthy operation (if they replanted *brizantha* the sudden death would reappear). The grass was not, however, the only problem. Some ranchers pointed out that the sudden death was particularly aggressive toward *brizantha* that was growing on degraded soils, while others remarked that it appeared where cattle had compacted the soil and therefore undermined the grass's root systems. Many pointed out that the unusually strong droughts made the pastures even more vulnerable to the sudden death—which expanded rapidly across tens of millions of hectares seeded with *brizantha*. At any rate, it was clear that, mysterious as it was, the sudden death thrived in a human-shaped ecology composed of mechanized soils, engineered pastures, and carefully bred cattle varieties.

Although the ranchers understood the sudden-death phenomenon as a result of the region's change in ranching ecology, they had no comprehensive theory or knowledge of the phenomenon. For example, ranchers could not say whether the sudden death was caused by a fungus, an insect, soil conditions, or a combination of all these factors. I interpreted their descriptions of the phenomenon to echo extreme science fiction stories wherein a sudden turn of events places characters into worlds that are radically different from a familiar law-bounded cosmos. Like the science fiction worlds studied by philosophers Graham Harman and Quentin Meillasoux (Harman 2012; Meillasoux 2015), Butantá's ranching environments could not be explained by human understanding and therefore remained withdrawn, exterior to human capacities for explanation and control.

Ranchers in Butantá clarified their role as being more than breeding cows, describing their efforts as enterprises that brought entire worlds into existence. And, like the heroes (and antiheroes) of extreme science fiction narratives, they persevered in their terra-forming activities unmoved by signs that announced the self-destructiveness of their actions. In what follows, I explain how the unfamiliar worlds created by phenomena such as out-of-control rains and the sudden death were not unprecedented for people like Luíz and thereby how Butantá elites are willing to face climate change dynamics owing to their previous political and economic success in navigating the inhospitable worlds of land speculation in Amazonia.

Building New Worlds

As mentioned, in the decades before the drought Luíz had built several cities in Southern Amazonia, and these experiences had prepared him to explore, build, and rebuild unfamiliar and inhospitable worlds. Born outside the basin, Luíz moved to the region in his mid-20s to try his luck as a pilot flying provisions and machinery into illegal gold mines (*garimpos*) and taking Amazonian gold to sell in southern Brazil. Working as an intermediary between small mining centers and large cities, Luíz amassed considerable wealth, and in a few years his reputation had reached the ears of land speculators who were building Butantá—and who promised him considerable economic rewards in exchange for his help.

Luíz explained to me that Butantá was built by what is known in Amazonia as a "private colonization company." These were enterprises that were granted legal ownership over large swaths of land by state institutions. In turn, recipients were expected to create new agricultural regions in which rural commodities were produced and sent to international markets or Brazil's cities. The company that built Butantá was awarded millions of hectares, in which it built roads, laid out the basic urban infrastructure of a new city, and attracted immigrants into the region to whom it sold lands. The final goal of this project, Luíz told me, was *ocupar para não entregar*: to occupy Amazonia in order to avoid giving it away to Northern powers (in particular to the United States, which was perceived to be trying to turn Amazonia into a gigantic natural park or agricultural area of its own). Effectively operating as an arm of state institutions, Butantá's elites (as in several other private colonization projects at the time) often used force to implement their projects (Schmink and Wood 1984). Luíz was called to Butantá precisely to further the elites' capacity to exercise such power in response to profound crises that were ailing the young city.

Butantá's first agricultural operations failed after a few years owing to a mixture of economic, political, and ecological problems that are too complex to explain here in detail (for an analysis of a broadly analogous case, see Bunker 1985). Suffice it so say that, against the plans of its founders, Butantá's economy came to rely on illegal gold mining, an activity that was not under the control of the local elites. Instead, these prosperous mining operations were carried out by a decentralized and highly mobile population, mostly young and male, who followed rumors of promising gold pits to prospect. Those who owned the land in which gold was found could receive a cut of the profits for allowing the exploitation of the pit. However, those who refused faced expulsion from their lands and sometimes death. Thanks to their numbers and fearless tactics, the miners (*garimpeiros*) trumped the elite's plans to contain them, and, as Butantá's population doubled, the city became a place in which large fortunes were quickly made through ruthless means and were spent even more quickly in a booming illegal economy. Poor neighborhoods, with their dire sanitary conditions, had an explosion in malaria cases, while alcohol and mining rivalries fueled fights that translated into very high homicide rates. Watching their city slip from their control, Butantá's elites hired Luíz and his partners to build new cities nearby in order to relocate the *garimpeiros* and thereby lower their influence on Butantá.

Luíz and his partners literally had to step out into unknown worlds to build these new cities. The first, a city fittingly named New World, was built through a series of steps. First Luíz hired a gold prospector to locate an area with promising gold pits to base a new mining settlement. The prospector covered hundreds of miles by foot and boat, venturing deep into unknown forests for weeks at a time. Once he located the desired spot, he returned to Butantá and led an expedition of about two dozen men and one woman (in charge of cooking) to the site, entailing a grueling effort to manually clear the forest and build a camp and an airstrip. Once this work was completed, Luíz landed planes loaded with machines and materials to build permanent structures and mining operations. As the gold started flowing more people arrived bringing more machines, clearing more forests, and building houses, shops, taverns, and brothels around the airstrip.

However, Luíz and his partners did not settle down in New World's increasingly urban environment to enjoy the small fortune they made for themselves. Instead they invested their new wealth in building a series of cities including New Planet, New Satellite, and New Earthly Body. Their celestial names accurately conveyed the unfamiliar worlds they brought into existence. Soon Luíz and his partners were controlling flows of provisions, alcohol, medicines, gold, and sex workers into a vast region in Amazonia. As the head of a sizable fleet of planes, Luíz reigned over, by his count, more than 4,000 people who were distributed across a constellation of illegal mining centers that were rapidly becoming villages and cities. Luíz's dominant position in this process of rapid socioecological change was based to a great extent on his capacity to navigate situations lacking a sense of lawfulness.

During my fieldwork I worked with peasants who had been *garimpeiros* on Luíz's lands. They described the cities he built as extremely unruly and uncertain—places in which wealth and power relations could change overnight. For example, fortunes changed hands when someone had a stroke of luck finding a particularly productive area. Force also led to quick fortunes for those who were capable of taking control of a rival's site. The capacity to effect large changes through extralegal means often led to violence against weak individuals who became prey to particularly violent entrepreneurs. However, extremely powerful men met their deaths at the hands of collectives who rose against perceived excesses of power. Luíz and his partners were often arbiters in violent conflicts delivering old-testament justice that matched the ways of the more violent players in the *garimpo*.

Although Luíz's success was built partly on his strong reputation, even more important was his capacity to shift between diverging world-making endeavors. Many of those who took part in land speculation enterprises from positions similar to Luíz's (including his closest partner) were killed by rivals, while others found themselves entangled with law enforcement agencies or saw their fortunes disappear in Amazonia's famous cyclical economic crises (Cleary 1993; Campbell 2015). Luíz, however, was the type of land speculator who avoided being linked to only one economic sector or single kind of political power. While he was transforming forests and riverbeds into cities and mines he was at the same time buying lands in which he cleared forests to introduce pastures. When the gold market collapsed he shifted the core of his operations to ranching and reinvented himself as a "rural producer," successfully deploying technologies such as expensive cattle breeds and seeds. He was one of the few large landholders I met who were willing to discuss the links between the sudden death, deforestation, and the global ecological impacts of ranching operations. Furthermore, he was vocally supportive of some minor environmental regulations that were designed to limit deforestation in Amazonia. He thought of environmental policies as a medicine. "No one likes to take his medicine," he told me, "but it is something you have to do." He explained that he was willing to place his ranching site under a federal monitoring system that would use satellite imagery to make sure he did not cut forests on his property.

In a sense, then, Luíz's story can be examined as a piece of extreme science fiction that deals with how persons and groups built and inhabited extreme environments (Valentine, Olson, and Battaglia 2012: 1014–15)—zones in which forests fell rapidly, wealth was hastily accumulated and expended, and human lives were often lost to aggressive economic competition. Luíz was one of those who succeeded by placing himself in a position from which he benefited from ever-changing worlds whose capacity to provide a ground for desirable futures was never assured.

Conclusion

I met Luíz at a time when he was cultivating the persona of a wise old man. The relatively upscale atmosphere of the office in which we talked made it difficult for me to see the same man in Luíz that my friends described in fear. Nevertheless, I caught a glimpse of the unforgiving airman when he described why he supported environmental policies. With great enthusiasm he explained that once his operations were certified as environmentally sound he would access new lines of credit, purchase novel technologies, and sell his products in environmentally conscious European markets. Not for a moment had he entertained the possibility of halting the expansion of the ranching economy in order to preserve Butantá's familiar worlds. His dreams were of a new region teeming with confined animal farm operations, mass-scale meatpacking factories, and export-oriented soy plantations replacing extensive ranching. He expected that Amazonia's biophysical traits would continue to change, offering surprise after surprise, imposing hardships on him similar to those he had faced at previous stages in his life. From his perspective, not even the sudden death or the ever-stranger weather were reasons to halt his projects. Quite the opposite: climate change dynamics represented an opportunity to renew his commitment to break away from known worlds and engage in the construction of new ones.

We may shiver when we hear scientists (such as the speaker I mentioned at the beginning of the chapter) warn about how dire Amazonian futures are likely to be in the context of climate change. In our trembling we may be tempted to ask: Why do landholders refuse to support environmental policies capable of halting the very economic activities driving climate change? Why do they ignore evidence showing how climate change futures will negatively affect their livelihoods? A response to these questions, at least in Butantá, may be that not all ranchers ignore the socioenvironmental changes of climate change. If we analyze Amazonian narratives of climate change by underlining the elements of extreme science fiction in them, it becomes possible to acknowledge landholders' rather surprising willingness to march with heads held high into worlds that, in their inhospitality and uncertainty, withdraw from human control. As we closely listen to the stories of rogue entrepreneurs such as Luíz and his collaborators we may start to understand that they fashion an unsettling familiarity with the unfamiliar, a strange capacity to work toward social and ecological worlds that are expected to collapse.

Notes

1. I want to thank Daena Funahashi, Andrew Johnson, and Ryan Adams for their insightful comments on early drafts. Susan Crate and Mark Nuttall offered valuable suggestions that improved the manuscript.
2. I have altered the names of persons and places in order to protect my interlocutor's identities.

References

Brando, P., Balch, J., Nepstad, D., Morton, D., Putz, F., Coe, M., Silvério, D., Macedo, M., Davidson, E., Nóbrega, C., Alencar, A., and Soares-Filho, B. 2014. Abrupt increases in Amazonian tree mortality due to drought–fire interactions. *Proceedings of the National Academy of Sciences* 111(17): 6347–52.

Bunker, S. G. 1985. *Underdeveloping the Amazon: Extraction, Unequal Exchange, and the Failure of the Modern State*. Urbana: University of Illinois Press.

Campbell, J. *Conjuring Property: Speculation and Environmental Futures in the Brazilian Amazon*. Seattle: University of Washington Press.

Cleary, D. 1993. After the frontier: Problems with political economy in the modern Brazilian Amazon. *Journal of Latin American Studies* 25(2): 331–49.

Davidson, E., de Araujo, A. C., Artaxo, P., Balch, J., Brown, Bustamante, M. M., Coe, M., DeFries, R., Keller, M., and Longo, K. 2012. The Amazon Basin in transition. *Nature* 481(7381): 321–28.

Harman, G. 2012. *Weird Realism: Lovecraft and Philosophy*. Zero Books: http://www.zero-books.net/home.html.

Helmreich, S. 2012. Extraterrestrial relativism. *Anthropological Quarterly* 85(4): 1125–39.

Lewis, S., Brando, P. M., Phillips, O. L., Van Der Heijden, G. M., and Nepstad, D. 2011. The 2010 Amazon drought. *Science* 331(6017): 554.

Masco, J. 2010. "Sensitive but unclassified": Secrecy and the counterterrorist State. *Public Culture* 22(3): 433–63.

———. 2012. The End of Ends. *Anthropological Quarterly* 85(4): 1107–24.

Meillassoux, Q. 2010. *After Finitude: An Essay on the Necessity of Contingency*. London: Bloomsbury Publishing.

Morton, T. 2013. *Hyperobjects: Philosophy and Ecology after the End of the World*. Minneapolis: University of Minnesota Press.

———. 2015. *Science Fiction and Extro-Science Fiction*. Minneapolis: Univocal.

Oreskes, N., and Conway, E. M. 2014. *The Collapse of Western Civilization: A View from the Future*. New York: Columbia University Press.

Schmink, M. C., and Wood, C. H. 1984. *Frontier Expansion in Amazonia*. Gainesville: University Press of Florida.

Slusser, G. E., and Rabkin, E. S. 1987. *Aliens: The Anthropology of Science Fiction*. Carbondale, IL: SIU Press.

Stover, L. E. 1973. Anthropology and science fiction. *Current Anthropology* 14(4): 471–74.

Thacker, E. 2011. *In the Dust of This Planet: Horror of Philosophy*. Zero Books: http://www.zero-books.net/home.html.

Trexler, A., and Johns-Putra, A. 2011. Climate change in literature and literary criticism. *Wiley Interdisciplinary Reviews: Climate Change* 2(2): 185–200.

Valentine, D., Olson, V. A., and Battaglia, D. 2012. Extreme: Limits and horizons in the once and future cosmos. *Anthropological Quarterly* 85(4): 1007–26.

Part 3: Refining Anthropological Actions

Chapter 18

Fostering Resilience in a Changing Sea-Ice Context: A Grant Maker's Perspective

Anne Henshaw

In the mainstream media, images of polar bears perched on pans of melting ice continue to be emblematic of climate change and its immediate threats. However, what is increasingly clear is that the image is not particularly effective in motivating nations to mitigate greenhouse gas emissions nor does it adequately reflect the political milieu that defines the Arctic today. As the multiyear sea ice continues to show remarkable decline and the ocean becomes more accessible, communities face increased pressures from industrial-scale development and a regulatory environment not well matched with the pace and scope of the changes taking place. The Arctic has increasingly become a place of competing visions, environmental agendas, geopolitical claims of sovereignty, and multinational corporate frontier operations. So where do the rights of indigenous peoples, particularly Inuit, fit into the complex array of actors and agendas?

Since the first edition of this volume appeared in 2009, the narratives and actions of Inuit operating at the regional and national level have tended to place greater emphasis on staking claim to resource development opportunities as part of their right to self-determination over the threat posed by climate change (see also the chapters by Hastrup, Johnson, and Nuttall in this volume). At the same time, local communities voice concern over the posed risks of such large-scale development projects on the health of the resources they depend on and for their own safety. Within this context, social science has kept pace with these changes by placing increasing attention to political developments and applying social science methods in addressing some of the current challenges both in terms of climate impacts and the social and environmental implications of large-scale resource development. Drawing on my experiences as a program officer for a private foundation, this chapter, a version of which was originally published in the journal *Polar Geography* in 2013, provides an overview of these recent developments, as well as the role philanthropy can play in advancing self-determination and resilience in a changing Arctic context.

Introduction

> The Tuaq [shorefast ice] used to be very thick, and it froze a long distance from shore. Nowadays our ocean doesn't freeze far from shore, and our tuaq and rivers become unsuitable for hunting because they are too thin and dangerous. And last year, we really couldn't go out seal hunting [in Kongiganak and Kwigillingok], because the shorefast ice was too thin. (John Philip, Yupik Elder from Kongignak, 2005, as quoted in Fienup-Riordan and Rearden 2010: 317)

Sea ice is arguably one of the most sensitive indicators of a changing Arctic. It is also a feature of the environment, as Yup'ik elder John Philip describes, that is inextricably linked with the livelihoods and identities of people in coastal communities who rely on landfast sea ice for travel and to harvest ice-associated species including birds, fish, seals, whales, and walrus. As sea-ice conditions become increasingly dynamic and unpredictable, the food security of many Arctic communities is coming into question (White et al. 2007; Duhaime and Bernard 2008; Wesche and Chan 2010). Compounding this threat is the fact some coastal communities, especially in Alaska, are vulnerable to erosion, because the ice no longer protects their shorelines from powerful ocean storms (Jones et al. 2009). Moreover, as the ocean becomes more accessible, communities face increased pressures from industrial-scale development and a regulatory environment not well matched with the pace and scope of the changes taking place (Young 2009).

In many ways, arctic communities face a "double exposure"—whereby they are simultaneously experiencing the effects of climate change and globalization (O'Brien and Leichenko 2000). In the context of the changing sea-ice conditions in the Arctic, this double exposure makes communities vulnerable, because it threatens the ability of people to harvest renewable resources safely, especially ice-associated species, which form the cornerstone of their livelihoods and cultural identities (Ford et al. 2007; Krupnik et al 2010). Exacerbating these effects, most arctic communities represent a mix of market and subsistence-based economies affected by changes in the globalized market place (Wenzel 2000). Many Inuit youth today, educated in school systems largely imported from the south, are not gaining the experience they need to understand the inherent dangers of Arctic travel on the land and the sea. In addition, rapid social change can be related to a whole host of social ills including high suicide rates, domestic abuse, and health problems seen in many northern communities (Einarsson et al. 2009). As Udloriak Hanson, an Inuit leader from Nunavut reminds us: "Inuit in the north have moved from igloos to the internet in one generation" (Hanson 2009: 500).

The extent to which communities in the Arctic can become less vulnerable, more resilient, and better prepared for the social and ecological transformations underway will be seen in the coming decades. Chapin and colleagues have argued that one of the key mechanisms to foster resilience and the sustainable use of resources in times of rapid change is the practice of ecosystem stewardship, the central goal of which is "to sustain the capacity to provide ecosystem services that support human well-being under conditions of uncertainty and change" (2009: 242). This approach is considered a paradigm shift, because it moves away from steady-state resource management and instead recognizes that people are integral components of social-ecological systems. I would further argue that it is an approach closely linked with research in common property, traditional ecological knowledge, environmental ethics, political ecology, environmental history, and ecological economics that have "bridged" natural and social science research in recent decades, as well as linked with emerging discussions in domestic and international law (Berkes 2004: 624; Baker and Mooney 2012). From a human geographic perspective, the effects of diminishing sea ice represent a crisis point for resource-dependent communities on the one hand and an opportunity to innovate in ways that enhance community and ecosystem resilience during a period of profound change on the other.

In this chapter, I explore how approaches to resilience building are being implemented in a philanthropic context. Drawing from my own research experience as an anthropologist who transitioned into more applied work as a program officer with the Oak Foundation, I discuss how philanthropy as a field is well suited to catalyze policy changes that embrace community solutions to the socioecological challenges of a melting environment. Specifically, I argue that making direct investments in Native communities and organizations most affected by the changes taking place represent one important way to implement the linkages between resilience theory and the practice of ecosystem stewardship. While philanthropic investments make up a small percentage of the total funding available to arctic communities, I conclude that they can nonetheless play an important role in fostering resilience by giving communities a chance to take ownership and drive policy processes that will help them to shape their own futures positively.

Philanthropy in Practice

In 2007 I moved from working as an anthropologist in academia, teaching and conducting research in the Canadian Arctic, to serving as a program officer at the Oak Foundation, a private family foundation that commits its resources to address issues of global social and environmental concern, particularly those that have a major impact on the lives of the disadvantaged. The Oak Foundation adheres to six funding principles: to (1) target the root causes of problems; (2) be replicable either within a sector or across geographical locations; (3) include plans for long-term sustainability; (4) secure cofunding; (5) strive to collaborate with like-minded organizations; and (6) value the participation of people and communities. How these funding principles get translated into grant-making practice provides the basis for the methods employed in this chapter. These same principles also lay the groundwork for applying anthropological approaches to fostering self-determination, a core value in anthropology, in real-time contexts today.

Within the field of philanthropy, grant-making practices vary depending on the foundation size, staffing level and professional backgrounds, and the interest of individual donors. In many ways philanthropy runs along a continuum from classical models of charitable giving to one defined by comprehensive theories of change that are strategic and evidence-based (McCully 2008). Within this continuum, grant making can best be understood as a community of practice whereby professionals in the field are involved in collective learning in a shared endeavor (grant making) as a way to improve their performance and to make a difference (Wenger 1998). For example, I engage with a specific subset of social and environmental funders interested in the Arctic to help learn about investment opportunities and challenges, to gain in regional experience, and to engage in collaborative grant making where our interests align. As a community, grant makers also share a common framework for discussing their work in the context of theories of change, capacity building, evaluation, and impact and are dedicated to growing the field by documenting best practices and sharing them with others through philanthropic affinity groups and support organizations, such as Center for Effective Philanthropy, Grant Craft, and Grantmakers for Effective Organizations.

In my role as a program officer, I draw from the collective experience of this community of practice to develop a strategic framework that builds on the Foundation's funding principles to guide investments in Arctic marine conservation around three key areas: to improve ocean governance in ways that value community-based stewardship of ocean resources; to mitigate impacts of large scale industrialization on local communities; and to reduce overfishing. Each year I invite grant applications from nonprofit organizations that align with our strategic goals and address one or more of these three broad areas either on an individual basis or as a cluster of activity, with Alaska being the program's primary geographic focus. Increasingly, our funding strategy has focused on the indigenous nonprofit sector as a way to build in regional capacity and to amplify the voices of indigenous peoples in ocean-related policy decisions that directly affect their communities. Oak evaluates its grant-making portfolios through a variety of means. The passage and implementation of policies represent one key way the foundation tracks the impact of projects focused on protecting marine resources and the communities who depend on them. For example, Oak funds the Alaska Eskimo Whaling Commission's Conflict Avoidance Agreement (CAA) process, an innovative approach to protect Bowhead whale habitat and the subsistence hunt from the effects of offshore drilling. While these represent private agreements between the AEWC and the different companies operating in the Alaskan Arctic, they are increasingly used as a way for National Oceanic and Atmospheric Administration to fulfill its obligations under the Marine Mammal Protection Act (LeFevre 2012). Therefore, for the Oak Foundation, policy becomes an important measure to track progress and gauge the impact of investments.

Changing sea ice can be considered a bellwether for understanding how philanthropic investments might better position communities to cope and innovate through a period of transformative change. Such innovations will no doubt come in a variety of forms, but from a policy perspective, creating a regulatory environment that enables people and marine resources, especially ice-associated species, to respond and adapt effectively to changing circumstances is critically important. By directly supporting community-based organizations, people who live in the Arctic can play an active role in shaping their own futures by representing themselves and their interests in the policy making arena in ways that can help to ensure their needs are heard. This approach to grant making is a direct outgrowth of the paradigm shift that recognizes communities as integral components of social-ecological systems. It is important because it provides a rationale for creating a more diversified and equitable approach to funding projects and organizations that represent the needs and priorities of the communities experiencing change most directly. In the context of Alaska, such approaches to direct grant making include a diverse set of Native nonprofit, Native government entities (tribal or hamlet), and pooled funds to Native-led initiatives that have a genuine incentive to ensure the future well-being of the communities they represent and the ecosystems they depend on for the livelihood and identities.

Policy Dimensions of a Melting Environment

As arctic communities face increasing environmental and social uncertainty, they are also confronting the pressures that a more accessible ocean brings as sea-ice seasons grow shorter and multiyear ice conditions decrease in thickness and extent (NSIDC 2011). With increasing pressure and impacts from oil and gas development, shipping, tourism, and industrial fishing, the arctic seas and sea ice are becoming increasingly crowded with multiple users with competing goals. While multilevel governance and models for co-management currently being developed and tested in the Arctic hold some promise in practice, they are not addressing the current needs of communities who rely on these resources (Siron et al. 2008). In many cases current regulations are hampering community resilience because of a variety of problems including the fact that policies developed in the 1960s and 1970s, especially in the United States, lag behind the current pace and magnitude of change taking place today (Lovecraft and Meek 2011). While much has been written on the challenges facing the regulatory environment in Alaska (Robards and Lovecraft 2010; Meek et al. 2011), I argue there is little socioecological fit between the scales in which sociopolitical and ecological processes function (Young 2002). Figure 18.1 illustrates the patchwork of arbitrary political boundaries that divide the large marine ecosystems of Alaska that occur at a scale inconsistent with community use and the ecosystems themselves. As the figure shows, two different federal agencies, the Department of Commerce (DOC) and the Department of the Interior (DOI), have overlapping jurisdictions within the U.S. Exclusive Economic Zone (EEZ). Each federal agency has its own mandates that

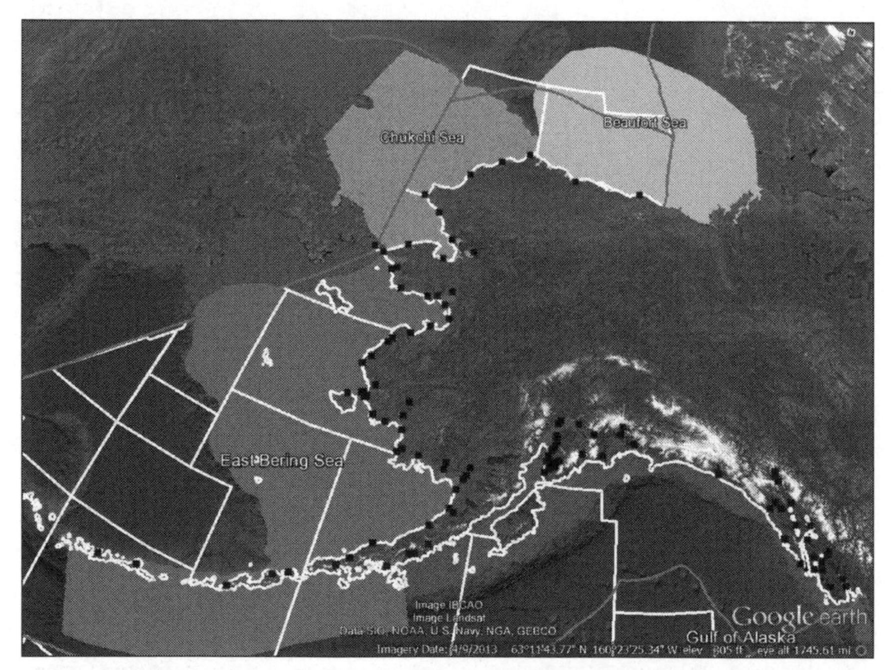

Figure 18.1 Political boundaries and large marine ecosystems of the U.S. Arctic (EEZ = grey line; DOI planning areas are in white)
Source: Google Maps

often work independently of the other: DOI regulates offshore oil and gas development, while DOC governs fisheries through the North Pacific Fisheries Management Council. One of the fundamental problems with this kind of fragmented approach to regulation and management is that it fails to address the cumulative effects of human activity on marine ecosystems both within and across national borders over time (Young et al. 2007).

In addition, arctic communities that rely on marine mammals are subject to the provisions of the 1972 Marine Mammal Protection Act (MMPA) and Section 119, which allow for subsistence take by Alaska Natives. The MMPA takes a single-species approach to co-management and is overseen by two different federal agencies—the Fish and Wildlife Service under DOI and the National Marine Fisheries Service under the DOC—both of which have different organizational cultures that do not consistently value community input (Meek 2009). Taken as a whole, the ability of coastal communities to engage effectively and proactively to steward subsistence resources and related habitats is clearly being compromised by the bureaucratic maze of rules governing renewable and nonrenewable resources and also the arbitrary planning boundaries that do not match up well with the changing ecosystems they regulate.

Therefore, to create a better socioecological fit in the policy arena, one must invest in strengthening the ability of communities to play an active role in shaping and framing the way ocean resources and sea ice are governed. While the new National Ocean Policy approved in 2010 by the Obama Administration is an attempt to address some of the inadequacies of a fragmented approach to ocean governance through initiatives like Coastal and Marine Spatial planning, the fact that the State of Alaska is less than supportive of federal mandates and recently closed its own Coastal Zone Management Program in state waters[1] means it will take some time before meaningful change comes about—time that communities can no longer afford. Concrete strategies that foster greater community resilience are needed now.

Resilience Theory and Philanthropic Approaches to Funding in a Changing Arctic

Resilience is a concept that refers to the capacity of a social-ecological system to absorb a spectrum of shocks or perturbations while sustaining its fundamental function, structure, identity, and feedbacks as a result of reorganization in a new context (Chapin et al. 2009; see Oliver-Smith, this volume, for an extended discussion). As a theory it has gained significant traction in recent years, not only in academic circles (Gunderson et al. 2001; Berkes et al. 2003; Walker et al. 2004; Plummer and Armitage 2007; Folke 2006; Folke et al. 2010) but also in a variety fields, including philanthropy. Philanthropy is a field that includes a broad suite of initiatives by private actors serving the public good that, as Joel Fleishman would argue, "powers innovation and diverse experimentation in the civic sector"

(2007: 59). It is also a field that increasingly encourages transparency, embraces risk, focuses on ideas instead of problems, and is action-oriented (Somerville 2008). In a philanthropic context, fostering resilience is well placed, because it requires all of these attributes.

A good example of a foundation that has adopted resilience as a basis for its grant making programs is the Christensen Fund. The Christensen Fund is designed to promote biocultural diversity, long-term sustainability, and human well-being. As Director Ken Wilson explains:

> The term "resilience" is central at Christensen. We are not preservationists seeking to maintain diversity just how it is, pickled behind real or conceptual fences. Instead we recognize that cultures, species and landscapes are inevitably and necessarily in motion, replete with aspiration, indeed that it is the evolutionary interactions between innovation, selection and changing context that itself creates diversity. . . . Resilience can be defined as a property of complex dynamic systems, namely their capacity to self-organize, learn and adapt and thus absorb shocks and change while maintaining structure and function and avoiding qualitative change. Resilience thinking becomes a way to protect systems from collapse, to help them recover from breakdown, to escape certain undesirable "traps," and to embrace the possibilities of adapting to a new world (e.g., climate regime) by building upon and reworking heritage. (Christensen Fund 2011)

From an arctic sea-ice perspective, resilience theory is a particularly powerful concept for grant makers in light of the profound social, psychological, and ecological changes that define the region today. However, its application requires a culturally appropriate, cross-scale, and diversified approach, particularly as it relates to the hotly contested politics around environmental sustainability and stewardship in Alaska (Eicken and Lovecraft 2011: 694–99).

One way for grant makers to begin fostering resilience is by implementing some of the basic sustainability approaches embodied in the concept of ecosystem stewardship. Specifically, Chapin and colleagues (2009) defined three over-lapping approaches that have direct bearing on how funding could be channeled, including (1) reducing vulnerability to expected changes, (2) fostering resilience to prepare for and shape uncertain change, and (3) transforming to potentially more favorable trajectories (Chapin et al. 2009: Boxes 2–4). These three approaches have been broken down further into specific strategies, some of which the philanthropic communities as well as government agencies such as the National Science Foundation are ideally suited to fund. In terms of grant making, one way to foster the implementation of these ecosystem strategies is by making direct investments to Alaska's Native nonprofits that have a stake in resource management decisions. Figure 18.2 combines the approaches identified by Chapin and colleagues (2009) with specific philanthropic interventions that could, in part, help to meet the current needs of communities in coping with rapid change.

Such investments are important, because they enhance local participation in framing policy and are critical for maintaining access to and stewardship

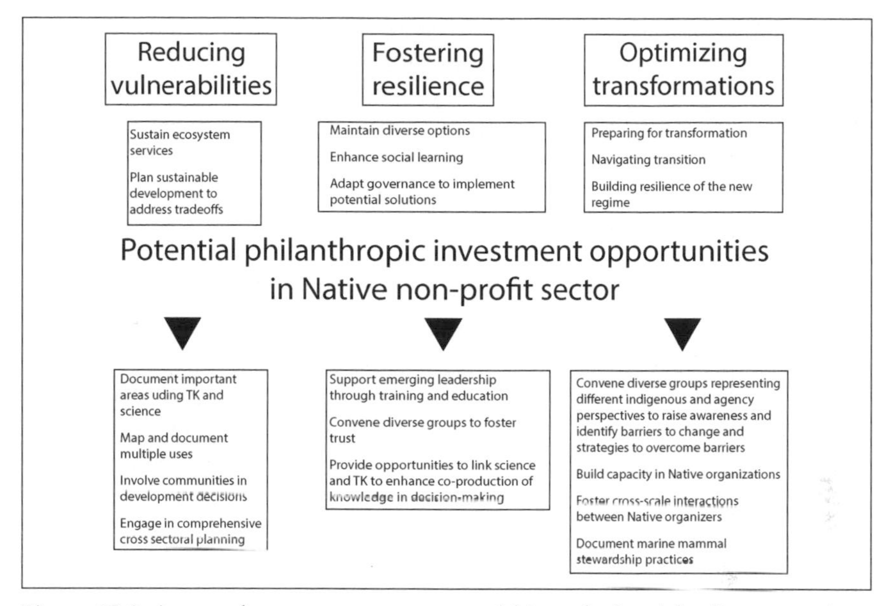

Figure 18.2 Approaches to ecosystem stewardship and related funding strategies (adapted from Chapin et al. 2009)

over marine mammals so critical to Native's livelihoods and identity. Table 18.1 shows a selection of key Alaska Native nonprofit organizations across the state that have a clear stake in natural resource management decisions. The list includes a diverse group of organizations with different origins, missions, constituents, language groups, and priorities that represent the civil society sector tied most closely with Arctic communities and living resources (Huntington and Pungowiyi 2008). One feature they all hold in common is the fact that each provides a channel through which community needs, collectively and on an individual basis, can be voiced in the policy arena. Investing in a diverse set of institutions, such as those that make up Alaska Native organizations, also promotes vertical and horizontal linkages and networks important to address various scale-appropriate solutions to social-ecological resilience building (Ostrom 1990; Ostrom et al. 1999; Young 2002; Ostrom 2008). For example, the regional nonprofits and the communities they represent encompass areas that are consistent with ecosystem scale processes that match the large marine ecosystems of the region (Figure 18.3).

Regional nonprofits also represent a mosaic of indigenous voices that tend to get homogenized by oversimplified definitions of *community* in locally based conservation work (Berkes et al. 2003). In Alaska, for example, views of how best to address the increasing pressure from offshore oil and gas development vary among tribes, boroughs, marine mammal commissions, and Alaska Native corporations, but that diversity gets lost

Table 18.1 Alaska Native organizations that have a stake in natural resource management

Type	*Organizations*
Statewide organizations	Alaska Federation of Natives
	Alaska Inter-Tribal Council
	Rural Alaska Community Action Program
	Alaska Native Science Commission
	Indigenous Peoples Council for Marine Mammals
Regional Nonprofits	Inupiaq Community of Arctic Slope
	Arctic Slope Native Association
	Maniilaq Association, Kawerak, Inc.,
	Association of Village Council Presidents
	Bristol Bay Native Association
	Aleutian Pribilof Islands Association
	Kodiak Area Native Association
	South Central Foundation,
	Copper River Native Association
	Tanana Chiefs Conference
	Chugachmuit, Inc.,
	Central Council Tlingit Haida
Native Tribes	200 Federally Recognized Tribes
Co-Management Organizations	Alaska Eskimo Whaling Commission
	Eskimo Walrus Commission
	Alaska Migratory Bird Co-Management Council
	Alaska Beluga Whale Committee
	Alaska Sea Otter and Steller Sea Lion Commission
	Alaska Native Harbor Seal Commission
	Alaska Nanuuq Commission
	Ice Seal Committee
	Aleut Community of St. Paul Island
	Aleut Marine Mammal Council
	Bristol Bay Marine Mammal Commission
	Sitka Marine Mammal Commission
	Traditional Council of St. George Island
	Central Council Tlingit & Haida Indian Tribes of Alaska
	Cook Inlet Marine Mammal Council
Alaska Permanent Participant Organizations to Arctic Council	Inuit Circumpolar Council
	Aleut International Association
	Arctic Athabaskan Council
	Gwich'in Council International

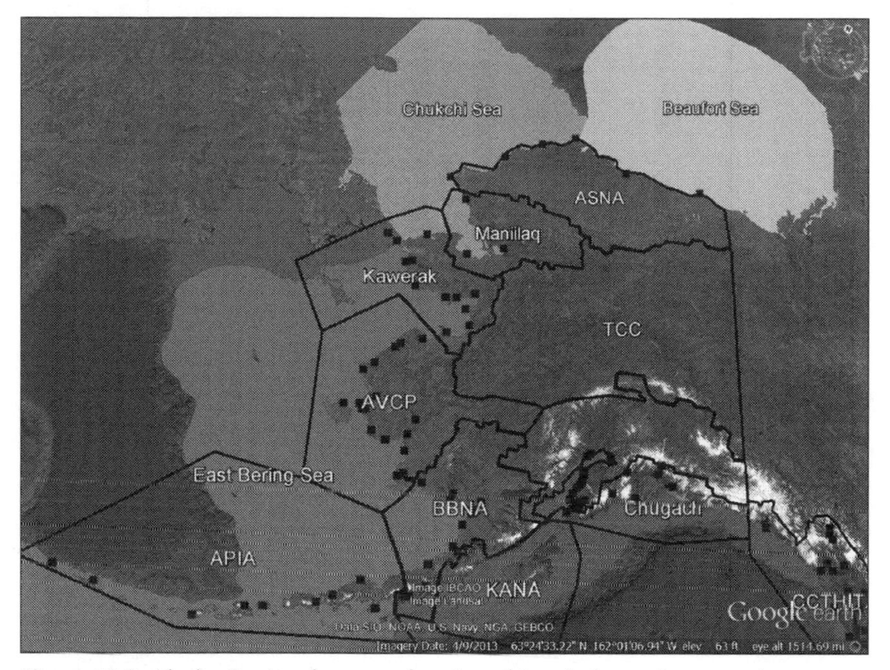

Figure 18.3 Alaska Regional nonprofits plotted in relation to large marine ecosystems of the U.S. Arctic
Source: Google Maps

in the way these views are expressed through the media or in litigation proceedings. Native nonprofit boards represent an important mix of perspectives that include these various entities and are well positioned to build consensus on highly controversial topics, such as offshore oil and gas development. In the end, many Arctic communities seek to strike a balance between economic development on the one hand and the protection of and access to renewable resources critical for their livelihoods on the other. The United States national office of the Inuit Circumpolar Council (ICC) in Anchorage provides a good example of an Alaska Native nonprofit that illustrates many of these features through the makeup of its board, including representatives from tribes, Native corporations, marine mammal commissions, and regional nonprofits. It is also an organization that fosters both vertical and horizontal linkages by bringing Inuit perspectives from different communities together and raising diverse concerns in Alaska and international policy forums such as the Arctic Council.

In the conservation funding community, directly investing in such groups is largely a departure from the norm in philanthropy generally, and in Alaska in particular. Grants to indigenous communities are largely funneled through intermediaries, and in the world of conservation this generally means large national and international environmental groups. Such practices came under criticism by anthropologists who recognized the power disparity between indigenous communities and the large NGOs that dominate the world's conservation agenda (Chapin 2004; Igoe 2005; Dove 2006). A recent report titled *Foundation Funding for Native American Issues and Peoples,* published by the Foundation Center, captures this trend all too clearly (Mukai and Lawrence 2011). The report synthesizes U.S. foundation giving that explicitly targets Native Americans (including Alaska Natives) from 2000–2009 and shows that the share of overall grant dollars benefiting Native Americans represents just 0.5 to 0.3% of total giving ($22.1 billion in 2009). Of that fraction just 7.7% (or $4,418,138) is going toward environmental issues (Mukai and Lawrence 2011: 1–3). These findings support trends from earlier decades reported by Brescia (1990) and Hicks and Jorgensen (2005), who found that less than 1% of foundation giving went to support Native American causes, and a significant portion of those grants went to non-Native controlled organizations. In part, the lack of direct investment is because "donors are failing to recognize existing capabilities and . . . the bureaucracy they impose often limits or even prevents the engagement or even use of that capacity" (Adamson 2011: 22). Adamson (2011: 23) goes on to note that foundation culture has a lot to learn about the values of indigenous peoples, traditional mechanisms for decision making, and engagement in culturally diverse paradigms of giving that are more holistic in nature and often not issue-specific.

Philanthropic funding targeting Alaska largely mirrors what is happening on the national scene, where Alaska Native organizations receive only a fraction of total giving by private foundations. In a survey by Philanthropy Northwest (2010), an affinity group representing social and environmental private foundations from the west coast, giving to Native populations constituted 1.2% (or $16.4 million) of total dollars granted in 2008. This amount represents a significant decline from $22 million reported in 2006. Of the $16.4 million granted in 2008, 44% went to Native populations in Alaska. The same survey reports that between 2006 and 2008 the percentage of dollars granted to Alaska nonprofits by foundations grew from 74% to 81%. However, most of those funds were directed to education and health and human services, not toward environmental stewardship issues also important to communities (Philanthropy Northwest 2010: 10). Because each foundation operates under a different set of mandates and priorities of its principle donors and boards, it is difficult to assess cause-and-effect relationships that can account for these trends. That said, among more progressive conservation-funders in Alaska there is a general reluctance to invest in a Native nonprofit sector they are less familiar with and where the perceived risk to investment is higher. Also, many U.S. foundations are rooted in Western-based ideologies of environmentalism that do not easily accommodate diverse cultural approaches to conservation (Igoe 2005).

Although there remains a long way to go in terms of providing more resources directly to Alaska Native nonprofits, a paradigm shift is beginning to take shape at least within a small subset of funders in Alaska and Canada. In Canada, the Walter and Duncan Gordon Foundation has led the way in working directly with northern Canadian communities through its Arctic program. For more than 25 years the Walter and Duncan Gordon Foundation has invested over 17 million dollars in developing programs with people in the North, with grant making focused on three main areas: emerging leadership, equity in resource development, and indigenous self-determination (Walter and Duncan Gordon Foundation 2011). In the Marine Conservation Initiative at the Gordon and Betty Moore Foundation, multimillion-dollar investments have been made directly to Coastal First Nations in support of their community-led efforts to conduct comprehensive marine spatial planning (GBMF 2012). In Alaska, the Alaska Conservation Foundation recently launched a new Alaska Native Fund (ANF), a granting mechanism designed to

> advance Alaska Native priorities for protecting our land and sustaining our ways of life and to provide an indigenous framework for impacting critical environmental issues while promoting innovative strategies to strengthen the capacity of Alaska Native organizations and communities. The goals of the fund are to: (1) leverage more foundation and donor resources for Alaska Native organizations; (2) support Alaska Native strategies and solutions on environmental issues; and (3) build relationships that will grow and strengthen the conservation movement in Alaska. (Alaska Conservation Foundation 2011)

Most important, the ANF is led by an Alaska Native Steering Committee that defines its program areas, criteria for funding, and decisions on the approval of grants. Similar approaches are being led by a host of funder affinity groups such as the International Funders of Indigenous Peoples (IFIP), Native Americans in Philanthropy (NAP), and the Circle on Philanthropy and Aboriginal Peoples in Canada. Each of these groups recognizes the value of investing directly in the capacity of Native peoples focused on sustaining their communities and stewarding resources that have sustained their health and well-being for generations.

Direct investing into Alaska Native nonprofits has also been on the rise in the Arctic marine conservation program at the Oak Foundation. Figure 18.4 illustrates the dollars invested by the Oak Foundation into the Alaska Native nonprofit sector involved in marine conservation between 2007 and 2011. The data indicate that in 2007, 0% of Oak's Arctic grant-making dollars ($1.5 million) went to Alaska Native nonprofits, whereas in 2011, 47% of the total Arctic marine program budget ($4.1 million) was invested in such groups—and not just to build capacity but to recognize and value the role indigenous people play in stewarding resources. Over the last five years, 16 separate grants were made to a total of 12 different organizations. The Oak Foundation also made an additional $500,000 in grants to Alaska

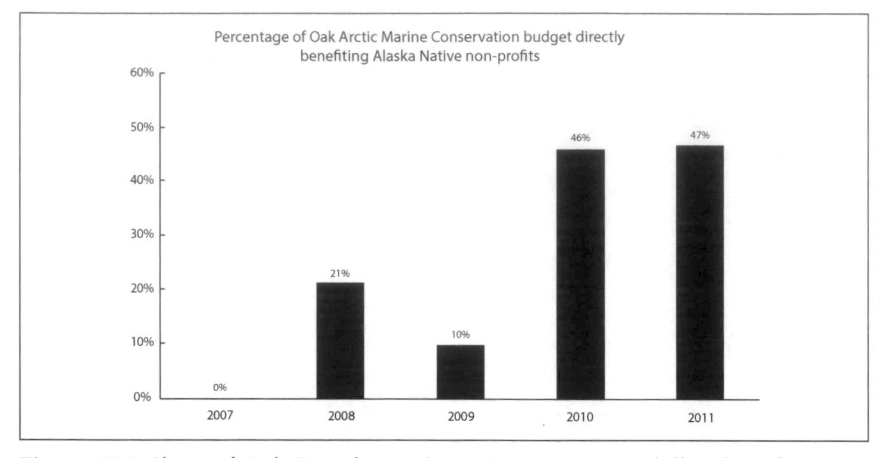

Figure 18.4 Share of Oak Foundation Arctic Program grant dollars benefiting Alaska Native nonprofit organizations (Oak Foundation 2011)

universities in support of higher education of Alaska Native students in the fields of marine science and co-management. More important often than the number of grants or dollars invested are the types of work or support that is given. Grant makers typically differentiate capacity-building grants (for example, grants that support long-term sustainability and organizational effectiveness) from those that are project-specific (for instance, grants that support a specific program or campaign). For some Alaska Native nonprofits involved in environmental stewardship or resource management, capacity varies and funders need to be flexible enough to understand where their resources can have the most influence. In the case of Oak, a mix of capacity building and project-based funding enables Native organizations to map and plan for their future and the opportunity to engage in time-sensitive projects that advance their own program goals. Although Oak continues to work with intermediary organizations that redistribute funds (for example, regranting) to indigenous communities and organizations, the foundation is clearly making a strategic effort to level the playing field so that more funds flow directly to community-led initiatives and institutions.

In developing and implementing this grant-making strategy, I have learned that grant making is more than a financial transaction or furthering a donor-driven agenda; it is about developing relationships with people, being open to Native worldviews about how funds can best serve communities, developing tailor-made administrative and evaluation mechanisms, and sharing lessons learned with colleagues in ways that will grow a community of practice centered on indigenous grant making. Specifically, there is an element of trust that must develop between donor organizations and their grantees whereby their priorities align so that true partnership can develop.

Discussion and Conclusion

This chapter describes how coastal communities in Alaska, affected by the effects of changing sea-ice conditions, are becoming increasingly vulnerable to the compounding effects of globalization and environmental change. It also highlights how an anthropological approach, especially as it relates to advancing self-determination in the Arctic, can be applied outside traditional academic research contexts. With a marginalized voice in natural resource decision making at the state and federal level, communities are faced with uncertainty not just from biophysical and economic changes but also from a regulatory environment that is not designed to value local input or foster adaptive management approaches necessary for accommodating the enormous changes they confront. One way to address these challenges and advance self-determination is by investing resources in the Native nonprofit sector in a way that fosters resilience using an ecosystem stewardship framework. Figure 18.2 illustrates how such a framework translates into grant making—specifically, by investing in the capacity of Native nonprofits to

- map important subsistence and ecological areas;
- engage in comprehensive planning;
- train younger generations as Native scientists, co-managers, lawyers, policy specialists, rights advocates, and nonprofit leaders;
- partner with other organizations and institutions where their priorities align; and
- participate in decision-making processes across different levels of government.

All have the potential to foster greater ecosystem stewardship and ultimately social-ecological resilience.

Although private philanthropic investments represent a very small piece of the overall funding pie in the north (largely dominated by public government agencies), they have the advantage of being nimble and able to experiment with ideas and novel approaches that are sometimes challenging in other public sector contexts. If being proactive and innovative are important to reducing vulnerabilities, fostering resilience, and optimizing transformation for people and the ecosystems they depend on, then philanthropy has the potential to play a catalytic role in helping communities define for themselves how they will fare in a new and transformed Arctic. In this regard, public-private-public partnerships with progressive agencies within the government could leverage larger public investments toward innovative northern solutions designed to help Arctic peoples meet the complex set of challenges facing their communities today. Specifically, funding programs that value traditional and scientific knowledge, build institutional capacity, create partnerships and collaboration, and foster the intergenerational transmission of knowledge would go a long way toward creating the enabling conditions for communities to secure a healthy and sustainable future (Henshaw 2010).

However, it would be naïve to assume that this kind grant making is not without significant challenges. For example, many conservation policy-related grants deal with hotly contested political issues, such as offshore oil and gas development. These are often highly divisive issues within communities, and even families. First, grant making has to be sensitive to the diverse needs and opinions of groups in the region, and this requirement is often a very delicate arena to navigate. Second, as indigenous use of the marine environment, including sea ice, becomes better documented, how the information is used and by whom is something that Alaska communities need to address. NSF human subjects' policy and data management plans provide some guidance but in the area of applied research such questions are all the more pressing. Third, while the Alaska nonprofit sector represents a key piece of the state's economy, it is one that is experiencing a funding crisis owing to an overcrowded nonprofit "ecosystem" and downward trends in revenue streams. To sustain the nongovernmental organizations groups must increasingly cooperate and integrate their programs where possible (Foraker Group 2011).

Ultimately, resilience building is a multidimensional, highly complex endeavor for Alaska Native communities faced with rapid social and environmental change. While sea-ice loss represents just one dimension of these changes, it provides a catalyst for private foundations to (re)think their giving practices in ways that better recognize people of the Arctic as integral parts of ecosystems and the political processes that govern the region. Such giving practices have to do more than ensure that community voices are heard in policy decisions that affect them. The real challenge for arctic philanthropy moving forward is to embrace cultures whose knowledge of, dependence on, and respect for living resources engenders a strong sense of stewardship and a vital voice for conservation. Without Arctic community engagement, marine resources, including ice-associated species, lose this constituency and lose the leadership necessary to sustain these resources into the future. That said, the Native nonprofit sector cannot afford to work in isolation, and grant makers need to recognize the importance of building diverse portfolios that help to foster partnerships and strategic collaborations with other civil society actors. Such partnerships, however, can take place only if private foundations are willing to help build the in-depth organizational capacity for communities and institutions that have the most at stake in terms of securing a sustainable, equitable, and resilient future.

Acknowledgments

This chapter derives in part from an article published in *Polar Geography* 2012, www.tandfonline.com/doi/full/10.1080/1088937X.2012.724460#.VLF98Xva1kk.

I am grateful for the support I received from my colleagues at the Oak Foundation for providing me an opportunity to develop a portfolio of grants that builds on the approach described in this chapter. I would like to acknowledge the helpful feedback I received on the *Polar Geography* manuscript from Leonardo Lacerda, Kathleen Cravero, Amy Lovecraft, Susan Kaplan, Betsy Baker, Meaghan Calcari Cambell, Jenny Grimm, Dave Secord, Chanda Meek, and James Stauch. I also want to thank Susan Crate and Mark Nuttall for their comments and feedback on this chapter and for encouraging me to submit a piece for this volume. I particularly thank the two anonymous reviewers for their constructive comments, which led to substantial improvements of this piece.

Note

1. An extension of Alaska's Coastal Zone Management Program failed to pass in the Alaska Legislature in 2011, leaving no formal mechanism for local input into federal projects, including offshore oil and gas development.

References

Adamson, R. 2011. Learning to see "invisible" capacity. *Alliance* June 16(2), http://www.alliancemagazine.org/, accessed November 8, 2011.

Alaska Conservation Foundation. 2011. Alaska Native Fund, http://alaskaconservation.org/grant-opportunities/alaska-native-fund/, accessed January 7, 2012.

Anderies, J., and Janssen, M. 2011. The fragility of robust social-ecological systems. *Global Environmental Change* 21: 1153–56.

Baker, B., and Mooney, S. 2012. The legal status of Arctic sea ice in the United States and Canada. *Polar Geoegraphy* 36(1-2): 86–104.

Berkes, F. 2004. Rethinking community-based conservation. *Conservation Biology* 18(3): 621–30.

Berkes, F. Colding, J., and Folke, C. (Eds.). 2003. *Navigating Social-Ecological Systems: Building Resilience for Complexity and Change*. Cambridge: Cambridge University Press.

Brescia, W. 1990. *Philanthropy and the American Indian*. Lumberton, NC. Native Americans in Philanthropy.

Chapin, F. S. III, Carpenter, S. R., Kofinas, G. P., Folke, C., Abel, N., Clark, W. C., Olsson, P., Smith, D. M. S., Walker, B., Young, O. R., Berkes, F., Biggs, R., Grove, J. M., Naylor, R. L., Pinkerton, E., Steffen, W., and Swanson, F. J. 2009. Ecosystem stewardship: Sustainability strategies for a rapidly changing planet. *Trends in Ecology and Evolution* 25(4): 241–49.

Chapin, M. 2004. A challenge to conservationists. *World Watch* 17(6): 17–31.

The Christensen Fund. 2011. *Christensen Mission Explication,* http://www.christensenfund.org/wpcontent/uploads/2010/12/Christensen_Mission_Explication.pdf, accessed January 7, 2012.

Circle on Philanthropy and Aboriginal Peoples in Canada (CIRCLE). 2011. *About Us,* http://philanthropyandaboriginalpeoples.ca/about/about-us/, accessed January 12, 2012.

Dove, M. 2006. Indigenous people and environmental politics. *Annual Review of Anthropology* 35: 191–208.

Duhaime, G. and Bernard, N. (Eds.). 2008. Arctic food security. *Occasional Paper 58*. Edmonton: Canadian Circumpolar Institute Press.

Eicken, H., and Lovecraft, A. L. 2011. Planning for northern futures: Lessons from socio-ecological change in the Alaska region, in A. L. Lovecraft and H. Eicken (Eds.), *North by 2020: Perspectives on Alaska's Changing Social-Ecological Systems*. Fairbanks: University of Alaska, pp. 681–700.

Einarsson, N., Larsen, J., Nilsson J., and Young, O. 2009. *Arctic Human Development Report*. Akureyri: Stefannson Arctic Institute.

Fienup-Riordan, A., and Reardan, A. 2010. The ice is always changing: Yup'ik understandings of sea ice past and present, in I. Krupnik, A. Aporta, S. Gearheard, G. J. Laidler, and L. K. Holm, *SIKU: Knowing Our Ice—Documenting Inuit Sea Ice Knowledge and Use*. New York: Springer, pp: 295–320.

Fleishman, J. 2007. *The Foundation: A Great American Secret*. New York: Public Affairs.

Folke, C. 2006. Resilience: The emergence of a perspective for social-ecological systems analysis. *Global Environmental Change* 16(3): 253–67.

Folke, C., Carpenter, S. R., Walker, B., Scheffer, M., Chapin, T., and Rockström, J. 2010. *Resilience Thinking: Integrating Resilience, Adaptability, and Transformability: Ecology and Society* 15(4): 20, http://www.ecologyandsociety.org/vol15/iss4/art20/, accessed November 8, 2011.

Foraker Group. 2011. http://www.forakergroup.org/layouts/forakergroup/ files/documents/resources_research/FOR_ER_11.pdf.

Ford, J., and Smit, B. 2004. A framework for assessing the vulnerability of communities in the Canadian Arctic to risks associated with climate change. *Arctic* 57(4): 389–400.

Ford, J., Pearce, T., Smit, B., Wandel, J., Allurat, M., Shappa, K., Ittusujurat, H., and Qrunnut, K. 2007. Reducing vulnerability to climate change in the Arctic: The case of Nunavut, Canada. *Arctic* 60(2): 150–66.

Gordon and Betty Moore Foundation (GBM). 2011. Grants Awarded, http://www.moore.org/init-grants-awarded.aspx?init=112, accessed January 23, 2012.

Gunderson, L. 2000. Ecological resilience: In theory and application. *Annual Review of Ecology and Systematics* 31: 425–39.

Gunderson, L., Holling, C. S., and Light, S. (Eds.). 2001. *Panarchy: Understanding Transformations in Human and Natural Systems*. Washington D.C.: Island Press.

Hanson, U. 2009. The next generation, in F. Abele, T. Courchene, and S. F. Seidle (Eds.), *Northern Exposure: Peoples, Powers, and Prospects*. Montreal: The Institute of Research and Public Policy, pp. 389–94.

Henshaw, A. 2009. Sea ice: The sociocultural dimensions of a melting environment in the Arctic, in S. A. Crate and M. Nuttall (Eds.), *Anthropology and Climate Change: From Encounters to Actions*. Walnut Creek, CA: Left Coast Press, Inc., pp. 153–65.

———. 2010. Partnerships in policy: What lessons can we learn from IPY SIKU? in I. Krupnik, A. Aporta, S. Gearheard, G. J. Laidler, and L. K. Holm, *SIKU: Knowing Our Ice—Documenting Inuit Sea Ice Knowledge and Use*. New York: Springer, pp. 427–44.

Hicks, S., and Jorgensen, M. 2005. *Large Foundations' Grantmaking to Native America. Kathryn M. Buder Center for American Indian Studies, Washington University, and the Harvard Project on American Indian Economic Development*, http://hpaied.org/images/resources/publibrary/LargeFoundationsGrantmakingtoNativeAmerica.pdf, accessed January 7, 2012.

Huntington, H., and Fox, S. 2005. The changing Arctic: Indigenous perspectives, *Arctic Climate Impact Assessment*. Cambridge: Cambridge University Press, pp. 62–98.

Huntington, H., and Pungowiyi, C. 2008. *Briefing Paper for Oak Foundation on Alaska Native Organizations and Environment Issues*. Unpublished manuscript on file with author.

International Funders for Indigenous Peoples (IFIP). 2011. *Mission, Vision, and Values*, http://www.internationalfunders.org/english/about-ifip/ifip-values/, accessed January 7, 2012.

Igoe, J. 2005. Global indigenism and spaceship earth: Convergence, space, and re-entry friction. *Globalizations* 2(3): 377–90.

Jones, B., Arp, C., Jorgenson, M., Hinkel. K., Schmutz, J., and Flint, P. 2009. Increase in the rate of uniformity of coastline erosion in arctic Alaska. *Geophysical Research Letters 36*. L03503, DOI:10.1029/2008GL036205.

Krupnik, I., Aporta, C., Gearheard, S., Laidler, G. J., and Holm, L. K. (Eds.). 2010. *SIKU: Knowing Our Ice: Documenting Inuit Sea Ice Knowledge and Use*. Berlin: Springer Science & Business Media.

Larsen, J., Schweitzer, P., and Fondahl, G. 2010. *Arctic Social Indicators*. Copenhagen: Nordic Council of Ministers.

LeFevre, J. 2012. *The Alaska Eskimo Whaling Commission's Open Water Season Conflict Avoidance Agreement: A Case Study in the Design of Process and Law Supporting Ecosystem-Based Multi-Use Ocean Management* (manuscript under review). *Environmental Law Reporter*.

Lovecraft, A. L., and Meek, C. 2011. The human dimensions of marine mammal management in a time of rapid change: Comparing policies in Canada, Finland, and the United States. *Marine Policy*: 427–29.

Lovecraft, A. L. and Eicken, H. (Eds.). 2011. *North by 2020: Perspectives on Alaska's Changing Social-Ecological Systems*. Fairbanks: University of Alaska Press.

McCully, G. 2008. *Philanthropy Reconsidered: Private Initiative, Public Good, Quality of Life*. Bloomington, IN: AuthorHouse.

Meek, C. 2009. *Comparing Marine Mammal Co-Management Regimes in Alaska: Three Aspects of Institutional Performance*. Ph.D. dissertation. Fairbanks: University of Alaska, http://www.uaf.edu/rap/students/alumni/.

Meek, C. L., Lovecraft, A. L., Varjopuro, R., Dowsley, M., and Dale, A. T. 2011. Adaptive governance and the human dimension of marine mammal management: Implications for policy in a changing North. *Marine Policy* 35: 466–76.

Mukai, R. L. S. 2011. *Foundation Funding for Native American Issues and Peoples. New York: The Foundation Center*, http://foundationcenter.org/gainknowledge/research/pdf/ff_nativeamerican.pdf

Mukai, R., and Lawrence, S. 2011. *Foundation Funding for Native American Issues and Peoples. The Foundation Center*. New York, http://www.issuelab.org/resource/foundation_funding_for_native_american_issues_and_peoples.

Native Americans in Philanthropy (NAP). 2011. *Vision, Mission, and History*, http://www.nativephilanthropy.org/about, accessed January 7, 2012.

National Snow and Ice Data Center (NSIDC). 2011. http://www.nsidc.org/arcticsea ice-news, accessed January 5, 2011.

Naylor, R. L. 2006. Directional changes in ecological communities and social-ecological systems: A framework for prediction based on Alaskan examples. *The American Naturalist*, 168, Supplement: 36–49.

Oak Foundation. 2011. *Annual Reports 2007–2011*, http://www.oakfnd.org/library, accessed January 7, 2012.

O'Brien, K., and Leichenko, R. 2000. Double exposure: Assessing the impacts of climate change within the context of economic globalization. *Global Environmental Change, Part A: Human and Policy Dimensions* 10(3): 221–32.

Ostrom, E. 1990. *Governing the Commons: The Evolution of Institutions for Collective Action*. Cambridge: Cambridge University Press.

———. 2008. The challenge of common pool resources. *Environment* 50(4): 8–20.

Ostrom, E. Brunger, J., Field., R., Norgaard, R., and Policansky, D. 1999. Revisiting the Commons: Local Lessons, Global Challenges. *Science* 284: 278–82.

Plummer, R., and Armitage, D. 2007. A resilience-based framework for evaluating adaptive co-management: Linking ecology, economics, and society in a complex world. *Ecological Economics* 61: 62–74.

Philanthropy Northwest. 2010. *Trends in Northwest Giving: A Comprehensive Report on Giving in our Region,* http://www.philanthropynw.org/s_pnw/bin.asp?CID=6393&DID=17242&DOC=FILE.PDF, accessed January 7, 2012.

Ribot, J. 2011. Vulnerability before adaptation: Toward transformative climate action. *Global Environmental Change* 21: 1160–62.

Robards, M., and Lovecraft, A. L. 2010. Evaluating comanagement for socio-ecological fit: Indigenous priorities and agency mandates for Pacific walrus. *Policy Studies Journal* 38(2): 257–79.

Siron, R., Sherman, K., Skjoldal, H. R., and Hiltz, E. 2008. Ecosystem-based management in the Arctic Ocean: A multi-level spatial approach. *Arctic* 61, Supplement: 86–102.

Somerville, B. 2008. *Grassroots Philanthropy: Fieldnotes of a Maverick Grantmaker.* Berkeley, CA: Heyday Books.

Walker, B. H., Holling, C. S., Carpenter, S. R., and Kinzig, A. 2004. Resilience, adaptability, and transformability in social-ecological systems. *Ecology and Society* 9(2): 5, http://www.ecologyandsociety.org/vol19/iss2/art5.

Walter and Duncan Gordon Foundation. 2011. Arctic Program, http://gordonfoundation.ca/programs/arctic-program, accessed January 7, 2012.

Wegner, E. 1998. *Communities of Practice.* Cambridge: Cambridge University Press.

Wenzel, G. W. 2000.Sharing, money, and modern Inuit subsistence: Obligation and reciprocity at Clyde River, Nunavut, in G. W. Wenzel, G. Hovelsrud-Broda, and N. Kishigami, *The Social Economy of Sharing: Resource Allocation and Modern Hunter-Gatherers*, Senri Ethnological Studies no. 53, pp. 61–87.

Wesche, S., and Chan, H. M. 2010. Adapting to the impacts of climate change on food security among Inuit in the western Canadian Arctic. *Ecohealth* 7: 361–73.

White, D. Gerlach, S. G., Loring, P., Tidwell, A., and Chambers, M. 2007. Food and water security in a changing Arctic climate. *Environmental Research Letters* 2(4): 1–4.

Young. O. 2002. *The Institutional Dimensions of Environmental Change: Fit, Interplay, and Scale.* Cambridge, MA: MIT Press.

———. 2009. The Arctic in play: Governance in a time of rapid change. *The International Journal of Marine and Coastal Law* 24: 423–42.

Young, O., Osherenko, G., Ekstrom, J., Crowder, L., Ogden, J., Wilson, J., Day, J., Douvere, F., Ehler, C., McLeod, K., Halpern, B., and Peach, R. 2007. Solving the crisis in ocean governance. *Environment* 49(4): 21–32.

Chapter 19

Is a Sustainable Consumer Culture Possible?[1]

Richard Wilk

This chapter takes an approach to consumer culture that is quite different from my chapter in the first edition. The original chapter, "Consuming Ourselves to Death," argued that anthropologists have a wealth of knowledge about consumer cultures that should be applied to the task of making consumption more sustainable. I gave examples of some relevant kinds of anthropological work, particularly among the many anthropologists who now work in business schools and market research. I urged caution in making sweeping moral arguments about good and bad forms of consumption, which have a tendency toward ethnocentrism and "blaming the victim." This chapter takes a different approach, questioning the category of *consumption* itself, suggesting that the term is profoundly cultural and is structured by metaphor. The tools we anthropologists use for cultural analysis are therefore crucially important in the quest for sustainability. As many of the contributors to this new collection have pointed out, cultural constructions of climate change are a necessary part of our toolkit in crafting solutions that will actually work.

Consumer culture is so well established that it has become unremarkable and taken for granted across most of the world. We measure standards of living by measuring how much people consume, and when overall consumption goes down the economy is considered sick or stagnant. When the economy is growing and healthy, consumption rises. As the "economic miracle" spreads to hugely populous countries such as Brazil, China, and India in the next decades, we should expect a rapid increase in the greenhouse gas (GHG) emissions associated with making, moving, and using goods. China is already building coal-fired electrical plants at such a rate that the central government has not been able to keep track of their numbers.

At the time I wrote my chapter for the first edition, few anthropologists had taken part in cross-disciplinary scientific discussions of the causes, or "drivers," of climate change beyond traditional topics such as deforestation and desertification. That chapter was really a wake-up call, urging anthropologists to question what we mean by *standard of living* and pointing out the need for more research on the global middle class. I cautioned against a simple definition of overconsumption and warned about the consequences of blanket messages telling people to consume less to save the planet. Many of the causes of high-level consumption are structural, legal, and social; we cannot solve the climate change problem by convincing billions of people to change their buying habits and live simply, while the comparatively wealthy continue to gobble up far more than their share of energy and resources. If people need to drive cars to get to work because their city has grown through endless sprawl, asking people to carpool, share cars, or ride bicycles is not going to reduce GHG emissions very much. Many studies also show that increases in efficiency just lead to higher levels of consumption, leaving total consumption about the same (Owen 2012). These "rebound effects" remain controversial, but so far very few rich countries have been able to stabilize their GHG emissions while their economy is growing. The growing economic inequality in the United States and other countries just makes the problem worse by adding a sense of injustice to peoples' aspirations. Many studies show how mass media has raised expectations, and they provide high-consumption "reference groups," which spark new needs and aspirations among viewers (Benson and Reilly 2012).

In this chapter I question some of the basic assumptions of consumer culture. This kind of critique can be difficult, because consumption is so deeply embedded in fundamental categories of kinship, gender, individualism, ethnicity, and nationality. As George Orwell said: "To see what is in front of one's nose needs a constant struggle" (2000: 122). Even fundamental concepts of wealth and poverty, health and well-being, are so deeply connected with consumption that in everyday life people assume that the poor are miserable and the rich are happy, despite abundant evidence that these connections are simply not true (Frey and Stutzer 2010).

According to many critics, consumer culture also has a strong and pernicious effect on deeply held political and cultural values, because it substitutes freedom of choice for other important political and intellectual freedoms. Cohen calls the contemporary United States "the consumer republic," where comfort and convenience are the supreme values (Cohen 2003, see also Markus and Schwartz 2010). The use of the term *consumer* instead of *citizen* or *worker* is an obvious example of the salience of this particular set of values. So far anthropologists have not entered this debate, even though economic and ecological anthropology offer important tools for questioning folk-models of consumption, standards of living, and ideologies of freedom. As a discipline we have worked extensively with people and cultures that *do not* exalt consumption over other values. We can easily imagine a Yanomami from the rainforest of Venezuela asking an American how we can possibly tolerate such a profoundly unfair system of distribution, allowing scarce goods to go not to the people who need them most but to those with a lot of money (especially given that most of that money today is intangible, existing only in the memories of machines)?

So far I have argued that we cannot meet the goals of sustainable living on a small planet until we can unpack the concept of consumption, challenge the dogma of eternal growth, and question the doctrine that free markets are the only efficient way to apportion resources and goods. We can see early movement toward more sustainable forms of consumer culture in the rapidly expanding "degrowth" movement (see Whitehead 2013), the Slow Money organization (Tasch 2010), and in proposals for rationing energy and other key goods (Cox 2013). What other critiques can anthropology offer?

Metaphor Theory

One of the basic distinctions anthropologists learn to make is that between emic and etic concepts, although the boundary between them has been subjected to a withering intellectual critique. The ambiguity of the terrain between emic and etic, folk and analytical concepts, is well described by Latour's term *factish* (1999), a state of limbo where science meets politics, where the very tools social scientists use carry their own assumptions and cultural meaning.

The metaphor theory developed by cognitive linguists can help us to understand *consumption* as a term that is both scientific and culturally bound (Lakoff and Johnson 1980; Lakoff 1987, 2008; Kronenfeld 2008). Metaphor theory argues that people mostly think prelogically and metaphorically and that intellectual comprehension is grounded in bodily experience (Lakoff and Johnson 1999). More specifically, in each individual culture, particular metaphors structure the understanding of worldly phenomena to the point where people rationalize or do not even perceive objects, thoughts, and actions that do not fit their metaphorically structured world. Different metaphors of the same phenomenon can compete in the same culture, and people often use more than one set of folk categories, just as they can learn to speak and think in different languages.

Metaphor theory says that humans work with categories and meanings that form fuzzy sets, with *prototypes* at their center, consisting of an image of an action or object that has all the essential qualities of the category. At the center of the category of dog lies a generalized image, so all other members of the set are more or less doglike, and the transition to other categories such as coyotes or foxes is gradual rather than categorical. Most tellingly for the purpose of thinking about consumption, the members of a category are not necessarily related to one another at all—they are bound together by their common relationship to the prototype. This is very different from classical Platonic theory, which sees categories as discrete, clearly bounded parts of the natural order.

The meaningful emic world consists of categories of things and actions that are bound together in ways that seem completely natural and obvious to members of that culture and are rarely questioned. The concepts are made tangible through a metaphorical linkage to objects with which people have direct physical experience, so they feel the rightness of their perception as much as they think it. At the same time, to a person from another culture, or in an etic analytical standpoint, that same set looks arbitrary, imprecise and inaccurate.[2]

Ethnography is the practice that allows anthropologists to learn another culture and that group's emic categories. But if we are interested in applied social science, change, and transformation, we cannot stop there. We also need to be able to translate and evaluate, and we need an etic language or we will lose the advantage of comparison and distance that provide the necessary parallax for applied social science.

The irony is that if we are going to translate research findings effectively into useful policy, we have to follow another process of translation. Latour has been perhaps the most effective ethnographer of technology policy, showing how thoroughly categories of "policy options," legitimate actors and agents, and expected physical outcomes are determined by the folk categories of the technoscientific world (1996, 2013). Rather than being objective descriptions of reality, they are cultural constructs. There have also been effective studies of the cultural specificity of mechanical technicians (Dorsey 1995), medical administrators (Bowker and Star 1999), and genetic laboratory scientists (Rabinow 1997). This research tells us that it is impossible to communicate social research findings directly into effective policy recommendations. Instead we need to make a serious effort at an ethnography of the communities that we hope to address, to understand the metaphors that structure the possibilities they can imagine. We might think, for example, about how economists think about consumer desires as a natural force like gravity. In their minds, it might take the application of energy to move consumption in another direction.

George Lakoff, with his work on framing, has played a major part in bringing the tools of cognitive linguistics to the political world. He argues that the way a public policy debate is framed, by the vocabulary and metaphors used to lay out the terms and stakes, tends to predetermine the outcome. Once the issue of terrorism and violence is successfully defined as a War on Terror, the only question is which side you support. For this reason, if social scientists simply accept the metaphors created and used by expert communities, public opinion, news media, and politicians, they will be unable to find effective ways to affect and change the world. Following this logic, we can ask how the problem of overconsumption is being framed by both scientists and citizens.

Consumption

What happens if we assume that *consumption* is a folk category with a specific history rather than a rigorously defined analytical term?[3] The category of consumption includes very disparate activities, some of which are environmentally destructive, while others are benign or even necessary parts of the cycles of nature. The way we use the word *consumption* is essentially metaphorical; it is a fuzzy set without clear boundaries, so the members of the set are not clearly related to one another. The term *consumption* allows us to map abstract concepts and categories onto common concrete and physical experiences. Consumption is also what Lakoff calls a "graded category" in that some things strike us as better examples of consumption. One scholar trying to define consumption says, "plants consume carbon, animals consume plants" (Borgmann 2000: 418), but the transformation of carbon through photosynthesis seems much less like consumption than cows eating grass. Why?

The basic metaphor in many Indo-European languages for *consumption* has a prototype of "fire." We therefore expect consumption to involve destruction, like a fire burning, which liberates needed energy and produces worthless and dirty waste. Some kinds of everyday consumption fit this metaphor very well, particularly activities like purchasing, using, and throwing away consumer products that have short use-lives, such as toilet paper. Something is "used up," and in the process we realize some tangible benefit and have to deal with resulting waste. But what about the consumption of services? Or objects that are curated for long periods, such as furniture, houses, and landscaping, gaining value over time and leaving no waste? The burning metaphor also fails when waste from one process becomes input to another, such as composting. When substances mix and have complex lives, we cannot easily determine what is really being used up. The power of the consumption-as-fire metaphor is that it leads us to assume that consumption is a linear process, during which a valuable substance produces a tangible benefit to the consumer, leaving behind some form of waste.

In physical reality there is a huge difference between a natural fire, which renews a grassland in the spring, and a destructive forest fire set by land developers who want to turn rainforest into pasture for cattle. Both are "consumption," but one is part of a natural cycle of renewal, and the other creates a pasture that may take a thousand years to regenerate into forest. Eating an apple grown in your back yard and listening to an MP3 music file are both forms of consumption, but it is objectively difficult to identify what the two actions have in common, so it is hard to see how any single measure could be used to compare them.

A second important metaphorical construction of consumption is "eating," which can be seen as a subset of "fire," in the sense that the body "burns" food to produce energy. The immediate bodily nature of the process of eating makes it a powerful and rich source of metaphors for consumption. The stages of eating—hunger, finding food, purchasing, preparing, eating, digesting, and excreting—all have metaphorical analogues. Desire for consumer goods becomes a kind of hunger, a physical need that, unsatisfied, leads to increased hunger and starvation; it can also be indulged as a serious pleasure, a frivolous entertainment, or uncontrolled gluttony with dire physical and social consequences.

The metaphorical equivalence of shopping to hunting and gathering draws our attention to the skill, taste, and choice of the shopper as the prototypical activities of consumption. Yet factually, most of our purchasing behavior has nothing to do with choice—it results from locked-in long-term decisions and takes the form of dribbles of money spent incrementally on utilities and credit, often by institutions instead of individuals. The powerful metaphor, nevertheless, constantly persuades us to think of consumption as a kind of eating that can be controlled, like appetite, whereby disorders are revealed—anorexia or obesity.

From the perspective of policy, the metaphorical construction of consumption leads to a failure of focus on the one hand and moralism on the other. The lack of focus results from the fuzzy, graded nature of the category, so people find it difficult to distinguish the kinds of activities that have serious environmental consequences (for example, driving a car) from those that are relatively benign (for instance, collecting antique cars). A failure to see consumption more clearly can keep people from focusing on other kinds of activities that use huge amounts of resources and have terrible environmental consequences but that do not easily fit into the emic category of consumption. Sport, political rallies, and investing are examples of activities that can consume many resources, but they do not match any of the stages of the consumption- as-eating metaphor and therefore receive much less scrutiny than something as obvious as shopping.

The moral tone of eating is based on metaphorical constructions of the problems of consumption in terms of weakness and willpower, good and bad decisions, even when the actual use of resources is completely beyond individual control. There is no way for me to make a choice between electricity generated by local coal or foreign natural gas. So the simple message to "consume less" may make metaphorical sense, but like dieting it can have perverse effects that increase consumption and cause rebound.

We eat for nutrition, as a social act that draws communities together and for our own pleasure. In every community there is an intense moral scrutiny of equity and balance in every stage of eating, from production through waste disposal (for example, Francione and Charlton 2013). The metaphor of consumption as eating makes it possible to divide overconsumption and underconsumption neatly. Underconsumption can be explained only by a failure of resources, ability, or group social conscience, and gluttony as a failure of the will, of morality, or social constraint. This is why the mixture of pleasure and political awareness in the Slow Food movement is so dissonant and controversial (Pietrykowski 2004).

The metaphor of consumption as eating leads us to envision environmental problems as the result of "using up" resources. But rather than using up a fixed stock of the earth's nonrenewable natural valuables, we are most in danger of overexploiting renewable resources such as timber and fish. The most immediate ecological dangers from pollution, extinction, and climate changes are due as much to waste, unregulated emission, the inherent growth mechanisms of capitalism, and political corruption as they are to the gluttonous use of resources.

Standard of Living

The eating metaphor also leads us into dead-end conundrums, eternal problems of moral philosophy that have no universal resolution. One perennial distraction is the attempt to find a dividing line between needs and wants. The question of "how much is enough" has no objective solution, despite efforts to find one by dividing up the world's resources into per-capita pie slices or universal metrics of well-being (Durning 1992; Kasser 2007; Jackson 2009). Historical evidence shows instead that what is defined as a basic need and what as a luxury always change (Horowitz 1988).[4] Sociologists provide good evidence that well-being is also relative, so that groups measure themselves in comparison with others, rather than against an absolute scale.

For these reasons, the standard of living is better understood as a metaphor for normality and community consensus, embodied in images of level surfaces at different heights—a stairway or a movable platform such as an elevator. The platform can move up and down, so height is the key attribute that allows comparison. The standard of living metaphor has another important property—like a jack or an elevator, it moves more easily in one direction than another (Shove 2003). If the standard of living stands still (held down) or gets lower (falls back), pressure increases, so we can expect a bounce or jump when pressure is released. If you read business journals you will find a lot of metaphors, some liquid (bubbles, flows) and others solid (cracked, shattered, collapsed).

Following a metaphor of progress as achieving greater height, poverty is like a dark smelly basement, rock bottom. It is represented as a hand--to-mouth existence of destitution and hunger. Wealth is at the top, a state where anything is available in abundance. In between are a series of steps or levels, each of which looks down on those below. Moving downward is "falling" and upward is "climbing," and both are forms of mobility. Confined by this linear metaphor, lower levels of wealth and fewer possessions will always be associated with poverty.

Studies of countries undergoing rapid inflation, economic crisis, and rising unemployment suggest that people cling to their ideas of middle-class standards of living, that loss of income can be traumatic, and that periods of recession create pressure to compensate during later periods of expansion and "boom" (Newman 1989; O'Dougherty 2002). The trauma of lost standards is described in such terms as *burnout* and *collapse*, which require "picking yourself up" and getting "back on track."

Recognizing the metaphorical nature of the standard of living helps us to recognize some of the peculiar properties that make the topic such a distraction from realizing substantial change in environmentally significant behavior. The linear quality of the metaphor will always draw us into debate about what is the best metric—be it GDP, HDI, SWB—and the relationship between this best metric and other variables. Telling people that money and material goods will not make them happy does not work well (established religions have been sending this message for thousands of years) (Miller 2004). There may even be a "moral rebound effect" through which reiterating the message creates guilt, which in turn drives the continuing bulimic cycle of binge and purge so characteristic of contemporary consumer culture (Nichter and Nichter 1991; Wilk 2014). Restraint creates the need for release.

Instead using a linear metric, we should recognize that quality of life is complex and multidimensional. Most of us can easily think of things that make us happy and unhappy, satisfied and unsatisfied, which are not in any way mutually exclusive or reducible to a single scale of well-being. Anthropologists find that people who live hand to mouth with few possessions often think of themselves as superior to people who are slaves to their possessions, locked into boring routines and endless work (Day, Papataxiarchis, and Stewart 1999).

Open-ended forms of questioning reveal that while the interviewers know and use a linear notion of wealth and poverty based on possessions and wealth, most Americans and Europeans also have few illusions about their ability to achieve life satisfaction by earning more. The metaphor is powerful, but in practice when people buy things, use energy, and accumulate possessions they are fulfilling social obligations; the demands of work, education, and careers; and gendered role expectations, while seeking fun and distraction in a process best likened to creative improvisation (for example, Holt 1995; Miller 2009; Holt and Thompson 2004; and my own research). Normal people are just as likely to use metaphors such as "rat race" and "treadmill" for the reality of middle-class existence, particularly in times when the benefits of economic growth go mostly to a small minority of the rich.

It is surprising to find out how little anthropological research is done on daily life and the challenges of consumer society. We need to know a lot more about the metaphors people use to understand their position in society, their movement through life, and the rewards and pitfalls of daily existence. This knowledge will give us a basis for reframing the issue of wealth and poverty more effectively in terms that do not require forms of consumption that use large amounts of energy and materials. Only the power of metaphor can make us ignore our own perception and experience of life to the extent that we are willing to accept a concept like standard of living as a natural and objective social fact that can be measured on a linear scale. This metaphor also leads us astray by helping us think that degrees of consumption (levels of wealth) are equivalent to degrees of environmental harm (levels of destruction). In reality, some forms of extravagant displays of great wealth—a Rolex watch, an electric car—are environmentally benign, and some of the kinds of consumption characteristic of the poor—driving an old clunker, using illegal waste dumps, buying disposables in small packages—have far greater proportional impact. So far the discussion of sustainable consumption has tended to shy away from this complex and difficult set of issues.

Freedom

Perhaps the most difficult problem facing us in imaging effective policy responses to climate change and resource depletion is the way in which possible solutions are often constrained by a frame that opposes government regulation with consumer freedom. George Lakoff has done us the favor of dissecting the metaphors of North American conceptions of freedom, including the whole strange, loose agglomeration of different things such as property rights, political speech, wealth, justice, gun ownership and shopping, which are brought together not by any kind of natural order or logic but by their common role in constructing the metaphor of Freedom in its modern and particularly North American cultural form (2006). He defines at the core something he calls "simple freedom," which is based on an embodied experience, nonmetaphorical and uncontested:

> Freedom is being able to do what you want to do, that is, being able to choose a goal, have access to that goal, pursue that goal without anyone purposely preventing you. . . . Political freedom is about the state and how well a state can maximize freedom for all its citizens. . . . A free society is one in which such "basic freedoms" are guaranteed by the state. (2006: 25–26)

Freedom of this sort is visceral, because it is metaphorically connected to the experience of being restrained, confined, or threatened. Freedom is experienced bodily by escaping confinement or reaching a destination, acquiring a desired object, or performing a desired action, all of which involve moving legs and hands.[5] The metaphors of simple freedom connect desire—the physical experience of wanting something—with autonomous action to satisfy that desire. This definition of freedom connects directly with the metaphor of consumption as eating, in which we experience desire (hunger), set out to satisfy that desire by finding and eating a desired food, and then experience satiation and happiness, the reward of free action.

Lakoff goes on to dissect the North American folk-model of how freedom works, through the action of a homunculus-like agent residing in our heads called the *will*. Americans believe that this individual will motivates action, which can be strong or weak, firm or easily melted, and that true freedom comes when the will prevails over internal weakness, emotions and passions, and external obstacles and temptations. Freedom is therefore a kind of self-mastery. From there Lakoff goes on to show how this simple embodied form of freedom underpins American folk theories of rights, justice, property, security, law, and the role of the state.

In this free world, there are only two possible situations in which the free will can legitimately be thwarted or limited—by nature and through competition. Nature imposes limits—we may disagree about what is natural, but everyone agrees that when an earthquake strikes, or we are struck down by an injury, this is not an abridgement of freedom (Lakoff 2006: 53). We can certainly contest what is determined by human nature—whether or not homosexuality is caused by a gene or is the result of choice—but once the limitation falls under the laws of nature, it is acceptable. The other legitimate way freedom is limited is through competition for scarce resources. Freedom to compete is accepted, as long as the rules of competition are seen as "fair." As Lakoff says: "If you are free to enter the competition, there is no abridgement of freedom. If you lose or are eliminated on the basis of rules, there is no abridgement of freedom" (2006: 57). These principles are demonstrated on reality television many times a day.

This notion of fairness under competitive rules leads into contested territory, because it turns out that there are a wide variety of different principles that can be used to decide what is "fair."[6] This fact is crucially important for thinking about consumption, because it underlies ideas about equity of distribution and rights to use and own goods, and it may underlie our persistent failure to understand the values and processes used by people when making choices in the marketplace.[7]

Lakoff defines the following kinds of fairness, some of which are mutually compatible and others of which cannot coexist. They should be familiar to all of us from childhood arguments and family negotiations:

Equality of distribution (one child, one cookie)

Equality of opportunity (one person, one raffle ticket)

Procedural distribution (playing by the rules determines what you get)

Equal distribution of power (one person, one vote)

Equal distribution of responsibility (we share the burden equally)

Scalar distribution of responsibility (greater abilities, greater responsibilities)

Scalar distribution of rewards (the more you work, the more you get)

Rights-based fairness (you get what you have a right to)

Need-based fairness (you get what you need)

Contractual distribution (you get what you agree to) (2006: 50–51).

It is not hard to see that many of the debates about responsibility for causing and fixing climate change are founded in conflicting definitions of fairness. More fundamentally, framing issues in this way means accepting that emissions are the product of free choices among autonomous individual entities with person-like qualities and who are acting according to a set of rules. Nations in this metaphorical construction are analogous to families, and their goal is to reach settlements on the behalf of their members that will be "fair." Among themselves, nations should each have a "fair share" of resources and emissions, acting as proxy people for accounting and negotiating purposes.

The metaphor of nations as individuals competing with one another obscures a key reality of the distribution of wealth, power, and environmental impact on the planet; every country has its own rich classes who have adopted a high-throughput lifestyle that uses large amounts of energy and materials. These cosmopolitan groups have a lot more economic, cultural, and political interests in common with each other, across national boundaries, than they have with the destitute and working poor of their own countries. The problem of regulation may be more manageable if we think about identifiable groups that are heavy users of goods, energy, and services, rather than nations or individuals (for example, Chakravarty et al. 2009).

The other consequence of using the folk definition of freedom as a valid analytical construct is that it leads into a sterile debate that opposes freedom to regulation, as if there were a single linear scale with slavery at one end and absolute freedom at the other. As we see from Lakoff's analysis, freedom is actually a radial category, and at its center there is a bodily sense of unfettered ability to reach a goal or obtain a desired substance. The logic that binds law, rights, political action, and the marketplace is metaphorical, and there are many possible alternatives.

The metaphor of the nation-as-family, according to Lakoff, has two versions in North America—one in which the parents should be nurturant and caring and the other in which parents should be stern disciplinarians. In one model the children (citizens) are inherently reasonable, and the state should give them the incentives and education they need to act responsibly. Then the children will build a commonwealth concerned with equal opportunities and caring for the environment. In the other model, children are inherently unruly and immoral—even dangerous; a stern father has to teach them moral rules and respect for authority. The role of the state is to provide a structured set of rules that allow fair competition and then to impose discipline on those who transgress (Lakoff 1996).

These two cultural models of family authority are rife with contradiction. For example, although the nurturant parent model places a high value on the democratic power of an educated citizenry, real systemic change always begins at the top—it is the responsibility of the state. Conversely, the strict-father model is based on an idea that people require strict rules to get them to behave morally, but it denies that the state should have this power. These inconsistencies emphasize Lakoff's point that these are cultural models held together by metaphor and embodied experience, not logical models that produce systemic explanations or reasonable policy, or analytical models that explain how the United States government actually works.

I would point out that in both the nurturant and strict parent models, freedom is always in danger. In the liberal nurturant parent model, the state may be neglectful and not give its citizens enough information to make informed choices, or it may not protect the vulnerable members of the community. In the conservative stern-father morality, children should make their own choices when they grow up, and deal with the consequences themselves. The state takes away freedom if it interferes and smothers initiative, makes unfair (or overly complex) rules, or fails to act according to strict moral principles. In both constructions the state can help only by providing information and education or by creating a level playing field that ensures fair rules of competition.

If we accept these folk-models, we lock ourselves into a logic that poses regulation and freedom as opposing principles, implying that one will always grow at the expense of the other. If we cannot force people to behave better through the discipline of higher taxes or regulations (justified as fair), we try to appeal to their conscience and morality or teach them a better set of values than materialism. The nurturant parent expects that properly educated citizens will follow Maslow's Hierarchy of Needs (1943) and Galbraith's Affluent Society (1962), realizing that material wealth is superficial and doesn't make them happy, then becoming more community-minded and spiritual (Etzioni 2009; see critique in Trigg 2004 and Slater 1997: 129).

Moral Calculus

In this era of "green capitalism" we have seen an outpouring of hopeful, cynical, polemic, and analytical writings about the need to bring moral and environmental issues more directly into the marketplace. There is a long history of political and moral consumerism going well back into the 18th century, when boycotts of "slavery sugar" were common and North American colonists burned English furniture and took pledges to buy only local produce. A close look at the history of advertising and marketing shows that morality has always been a part of the appeal of consumer goods and a normal part of everyday business. Indeed, we might ask what would lead anyone to imagine a marketplace in which commodities are completely anonymous and have no history, meaning, or moral connections with society and nature (Wilk and Cliggett 2006).[8]

Nevertheless, the dominant academic models of shopping and purchasing are still variations of rational choice and/or game theory, usually modified by constraints on information and peoples' cognitive abilities (for example, Parnell and Larsen 2005). In contrast, ethnographic work on daily practices of shopping reveal a world that is intimately moral, revolving around value judgments and balances between the interests of the self, immediate others, and broader communities (for instance, Miller 1998; Östberg 2003; also see Wilson and Dowlatabadi 2007). Once again, it is easy to get distracted by the folk-models' contrasts between individualistic and group-oriented models, selfishness, and altruism.

I think it is more productive to follow up on the fact that people do not live up to their commitments, consistently pursuing neither their own interests nor those of others. People claim they are concerned about animal welfare but then eat hamburgers. They buy one pound of expensive fair-trade coffee but then drink a hundred cups of uncertified Brazilian blend in the office cafeteria. They say they are doing only what is best for their family, but then they spend the college fund on a wide-screen TV so they can watch a football game.

This inconsistency leads some scholars to deep cynicism about the depth of individual commitment or to the belief that it is impossible to make consistent informed choices in a world glutted with information and full of difficult contradictions between different kinds of "good" and "bad" products (for instance, Belk, Devinney, and Eckhardt 2005). And even positive choices in the marketplace do not necessarily have consistent environmental benefits. As Connoloy says: "Consumers, even when they are environmentally concerned, are still consuming, only they consume perceived green products and recycle more. The actual level of consumption is not identified as a problem" (2003: 288).

It is possible that some of these behavioral inconsistencies are just the results of hypocrisy, poor decision making, or lack of knowledge. Perhaps our expectations for ordinary people in a complex market environment during an economic recession are unreasonable, and once people feel prosperous again they will return to moral marketing with labels of organic, no-sweat, and fair trade (although as Connoloy argues, this does not really address the underlying problem). I think it more likely, however, that there is a persistent problem with moral consumerism based on a metaphorical structure that draws directly on the consumption-as-eating core metaphor.

How do we avoid getting too fat or too thin? To some extent we can depend on our natural appetite to keep us from starving, although there are disorders that cause people to starve themselves. But we are also constantly exposed to incentives to give in to temptation, which can lead us to overeat. The metaphorical extension of starvation is poverty, and gluttony is overspending. The way to avoid peril at both extremes is to find balance, by controlling impulses and temptations, on the one hand, and making sure we eat the right things and get proper nutrition (even if it tastes bad), on the other. Periodically, we are likely to end up bingeing, which can be corrected only by going on a diet and engaging in exercise and other activities that remove or balance the "sin." In other words, we are using a system of moral accounting to guide us in addressing problems of self-control, and temptation within a world of good and evil forms of consumption (see Thaler 1990; Shefrin and Thaler 2004).

The visual metaphor is probably something like a see-saw, which balances when equal weights are put on both sides, as seen in the image of the scales of justice. My impression is that people seek balance in different ways and along different time scales. Achieving balance can be an improvisational process with a running total, which is never added up, or people may use a budget model with planning and periodic totals, followed by penalties or rewards. Following either accounting method, we can easily use a virtuous act, purchase, or performance—such as joining a CSA or subscribing with a "green" power company that charges higher rates—to justify sins and guilty pleasures, including a shopping trip to Wal-Mart or a rainforest hamburger.

This is not to say that the balances people reach through the see-saw metaphor have anything at all to do with environmental impacts as defined by scientists. Nor is there any evidence that people keep separate mental accounts for goods labeled or identified as green, fair trade, ethical, healthy, or even cheap and low in calories. We do not know the degree of *transitivity* between different kinds of sin and virtue—it could vary from person to person, or region to region, but we should not expect the partitions to match any etic analytical categories. We should also recognize that people often seek to achieve balance not by adjusting their behavior but by adjusting their assessment of the moral valence of the things they buy, use, eat, and own. When the price of organic fruit suddenly goes up, we might decide that organic food is really not much better for us, so instead of a form of virtue, organic shopping becomes a luxury that can be counted as a sinful indulgence. Note the paradoxical effect of the see-saw metaphor. When we are convinced that organic food is really good and then buy it, we also consume more sinful food; then when we consume less virtue, we also can consume less sin. The net result is that green shoppers can end up buying more of everything, especially if they live in a household whose members disagree on what constitutes good and bad purchases.

It is very possible that the balance metaphor has wider generality for other kinds of environmentally relevant behavior. We know that Americans like their news to be "fair and balanced," so that when they are told the north polar ice is disappearing and permafrost is melting, they expect that there is another side to the story, that the news cannot really be so bad. To some extent the very process of governance and political decision making in the United States is envisioned through the metaphor of two parties that are in balance; whatever one says, the other will say the opposite. The truth (the state of balance) will always lie in between, so any strong opinion or new proposal is automatically labeled extreme, way out on the edge, threatening to throw the whole machine out of kilter. For the same reason, as a folk theory of personality, the balance metaphor would lead us to think that other people, groups, or products that may appear to be virtuous must have another perhaps hidden side that is sinful and corrupt. This way of thinking is very different from the cynicism that is often blamed for Americans' failure to respond to green messages— rather than a bad attitude, environmental inaction is instead a consequence of a fully formed and highly consistent folk theory based on a coherent metaphorical structure.

Invitation

Rather than closing with a general conclusion, I would instead like to issue an invitation to take emic models of consumption and the environment more seriously. As I have described them, the fire and eating metaphors of consumption, the elevator model of standard of living, the parental metaphors of freedom and government, and the balancing model of choice all offer openings for us to reframe environmental issues in ways that give us some realistic expectations of changing behavior. At the same time, I have also shown just how deeply the purchase, use, and disposal of material goods has become embedded in North American culture. Consumer freedom is a core value that is very hard to question. Consumption metaphors show us how North Americans think about their bodies, public and private morality, families, and government.

Lakoff argues that U.S. political parties and partisan conflict frame issues using metaphor, and these metaphors can have a tremendous influence on how politicians and the public perceive an issue—think, for example, of the controversy over immigrants who are depicted either as poor refugees seeking safety or as toxic and dangerous aliens intent on damaging the nation. Because of his understanding of metaphor, Lakoff has been a consultant for political parties and others seeking to sway public perception. In his most recent works, he addresses the language of climate change and questions the metaphors that frame the debate (2014). But to a cultural anthropologist Lakoff's work lacks ethnographic depth. One immediately important way for anthropologists to address the problems of climate change is to get out in the field and find out what kinds of metaphors people really use in daily speech about weather and climate. Effective strategy and action depend on an ethnographically sound understanding of the language and folk-models that guide scientists, politicians, and the public. We still have a long way to go in understanding how consumer culture grows and expands, and so far we have not found a convincing alternative that might be more sustainable.

Notes

1. This chapter is a heavily revised version of a paper originally published in 2010 as "Consumption Embedded in Culture and Language: Implications for Finding Sustainability" in a special issue of *Sustainability: Science, Practice and Policy* devoted to sustainable consumption (6[2]: 1–11.) I would like to thank Maurie Cohen and the other members of the Sustainable Consumption Research and Action Network (SCORAI), as well as Hal Wilhite and the University of Oslo Center for Environmental Change and Sustainable Energy for my continuing education on sustainability.
2. Unfortunately, metaphor theory as originally defined has tended toward a static analysis, just like previous forms of structuralism. We are still working our way toward an understanding of how systems of metaphor constantly change, and here I depend on the concept of reframing as formulated in Lakoff's recent applied work through his Rockridge Institute, which is devoted to progressive political change.
3. I discuss the metaphorical nature of consumption at greater length in a previous paper (2004), which I have drawn from here. I have also written previously about why metaphorical thinking leads easily to moral judgments about over- and under-consumption (2001).
4. These definitions change within a lifetime as well—the basic needs of college students are hardly the ones these people will experience later in life.
5. In the United States early experiences of this nature are often part of games, fights, competition, and sports, leading to people's lifelong tendency to use sports metaphors in thinking about freedom.
6. Lakoff originally worked out the different folk rules of fairness in his book *Moral Politics* (2002), which is the source behind his more recent political work.
7. There is now a substantial literature on the failure of consumers to live up to their stated green morality—for example, Belk, Devinney, and Elkhardt (2005), Östberg (2003), and the collections edited by Boström and Klintman (2008) and Bevir and Trentmann (2007).
8. Max Weber thought that modern capitalism rested firmly on a moral foundation. In this tradition there is a whole literature in contemporary economic anthropology that makes this point—value-free markets in which goods are truly anonymous are exceptional. Research in consumer research, marketing, and advertising has shown how the consumer marketplace in North America and Europe has been deeply embedded in moral discourse about gender, family, cleanliness, patriotism, progress, modernity, and other themes (for example, McGovern 2006; Stanley 2008).

References

Belk, R. W., Devinney, T., and Eckhardt, G. 2005 Consumer ethics across cultures. *Consumption, Markets and Culture* 8(3): 275–89.

Benson, M., and O'Reilly, K. 2012. *Lifestyle Migration: Expectations, Aspirations, and Experiences*. Burlington, VT: Ashgate Publishing.

Bevir, M., and Trentmann, F. 2007. *Governance, Consumers, and Citizens*. London: Palgrave Macmillan.

Borgmann, A. 2000. The moral complexion of consumption. *Journal of Consumer Research* 26(4): 418–22.

Boström, M., and Klintman, M. 2008. *Eco-Standards, Product Labelling, and Green Consumerism*. London: Palgrave Macmillan.

Bowker, G. C. and Star, S. 1999. *Sorting Things Out*. Cambridge, MA: MIT Press.

Chakravarty, S., Chikkatur, A., de Coninck, H., Pacala, S., Socolow, R., and Tavoni, M. 2009. Sharing global CO_2 emission reductions among one billion high emitters. *Proceedings of the National Academy of Sciences* 106: 11884–88.

Cohen, L. 2003. *A Consumer's Republic: The Politics of Mass Consumption in Postwar America*. New York: Alfred A. Knopf.

Connoloy, J., and Prothero, A. 2003. Sustainable consumption: Consumption, consumers, and the commodity discourse. *Consumption, Markets and Culture* 6(4): 275–98.

Cox, S. 2013. *Any Way You Slice It: The Past, Present, and Future of Rationing*. New York: New Press.

Day, S., Papataxiarchis, E., and Stewart, M. 1999. *Lilies of the Field: Marginal people who live for the moment*. Boulder, CO: Westview Press.

Dorsey, D. 1995. *The Force*. Ballantine Books.

Durning, A. T. 1992. *How Much Is Enough? The Consumer Society and the Future of the Earth*. New York: W.W. Norton.

Etzioni, A. 2009. Spent: America after consumerism. *The New Republic*, June 17, 2009, www.tnr.com/politics/story.html?id=80661c9c-9c63-4c9e-a293-6888fc845351&p=1, accessed August 8, 2009.

Francione, G., and Charlton, A. 2013. *Eat Like You Care: An Examination of the Morality of Eating Animals*. Louisville, KY: Exempla Press.

Frey, B. S., and Stutzer, A. 2010. *Happiness and Economics: How the Economy and Institutions Affect Human Well-Being*. Princeton, NJ: Princeton University Press.

Galbraith, J. K. 1962 [1958]. *The Affluent Society*. Harmondsworth: Penguin.

Holt, D. B. 1995. How consumers consume: A typology of consumption practices. *Journal of Consumer Research* 22(6): 1–16.

Holt, D. B., and Thompson, C. J. 2004. Man-of-action heroes: The pursuit of heroic masculinity in everyday consumption. *Journal of Consumer Research* 31(9): 425–40.

Horowitz, D. 1988. *The Morality of Spending*. Baltimore: Johns Hopkins University Press.

Jackson, T. 2009. *Prosperity without Growth? The Transition to a Sustainable Economy*. Commission on Sustainable Consumption.

Kasser, T. 2002. *The High Price of Materialism*. Cambridge, MA: MIT Press.

Kempton, W., and Montgomery, L. 1982. Folk quantification of energy. *Energy—The International Journal* 710: 817–28.

Kronenfeld, D. B. 2008. *Culture, Society, and Cognition: Collective Goals, Values, Action, and Knowledge*. Berlin: Mouton de Gruyter.

Lakoff, G. 1987. *Women, Fire, and Dangerous Things: What Categories Reveal about the Mind*. Chicago: University of Chicago Press.

———. 1996. *Moral Politics*. Chicago: University of Chicago Press.

———. 2006. *Whose Freedom? The Battle over America's Most Important Idea*. New York: Farrar, Straus, and Giroux.

———. 2008. *The Political Mind: A Cognitive Scientist's Guide to Your Brain and Its Politics*. London: Penguin Books.

———. 2014. *The ALL NEW Don't Think of an Elephant! Know Your Values and Frame the Debate*. Burlington, VT: Chelsea Green Publishing.

Lakoff, G., and Johnson, M. 1980. *Metaphors We Live By*. Chicago: University of Chicago Press.

Lakoff, G., and Johnson, M. 1999. *Philosophy in the Flesh*. New York: Harpercollins.

Latour, B. 1996. *Aramis, or the Love of Technology*. Cambridge, MA: Harvard University Press.

———. 1999. *Pandora's Hope: Essays on the Reality of Science Studies*. Cambridge, MA: Harvard University Press.

———. 2013. *An Inquiry into Modes of Existence: An Anthropology of the Moderns*, C. Porter, Trans. Cambridge, MA: Harvard University Press.

Markus, H., and Schwartz, B. 2010. Does choice mean freedom and well-being? *Journal of Consumer Research* 37(2): 344–55.

Maslow, A. H. 1943. A theory of human motivation. *Psychological Review* 50: 370–96.

McGivern, C. 2006. *Sold American: Consumption and Citizenship, 1890–1945*. Chapel Hill: University of North Carolina Press.
Miller, D. 1998. *A Theory of Shopping* London: Polity.
Miller, D. 2009. *The Comfort of Things*. London: Polity.
Miller, V. 2004. *Consuming Religion: Religious Beliefs and Practices in Consumer Culture*. London: Continuum Press.
Newman, K. 1989. *Falling from Grace: The Experience of Downward Mobility in the American Middle Class*. New York: Vintage Books.
Nichter, M., and Nichter, M. 1991. Hype and weight. *Medical Anthropology* 13: 249–84.
O'Dougherty, M. 2002. *Consumption Intensified: The Politics of Middle-Class Daily Life in Brazil* Durham NC: Duke University Press.
Orwell, G. 2000. *George Orwell: In Front of Your Nose, 1946–1950*, in S. Orwell and I. Angus (Eds.). Boston: Nonpareil Books.
Östberg, J. 2003. *What's Eating the Eater? Perspectives on the Everyday Anxiety of Food Consumption in Late Modernity*. Institute of Economic Research, Lund Business Press.
Owen, D. 2012. *The Conundrum*. New York: Riverhead Books.
Parnell, R., and Larsen, O. P. 2005. Informing the development of domestic energy efficiency initiatives: An everyday householder-centered framework. *Environment and Behavior* 37(6): 787–807.
Pietrykowski, B. 2004. You are what you eat: The social economy of the Slow Food movement. *Review of Social Economy* 62(3): 307–21.
Rabinow, P. 1997. *Making PCR: A Story of Biotechnology*. Chicago: University of Chicago Press.
Shefrin H. M., and Thaler, R. 2004. Mental accounting, saving, and self-control, in C. Camerer, G. Loewenstein, and M. Rabin (Eds.). 2004. *Advances in Behavioral Economics*. Princeton, NJ: Princeton University Press.
Shove, E. 2003. *Comfort, Cleanliness, and Convenience: The Social Organization of Normality*. Oxford: Berg.
Slater, D. 1997. *Consumer Culture and Modernity*. Cambridge: Polity.
Stanley, A. 2008. *Modernizing Tradition: Gender and Consumerism in Interwar France and Germany*. Baton Rouge: Louisiana State University Press.
Tasch, W. 2010. *Inquiries into the Nature of Slow Money: Investing as if Food, Farms, and Fertility Mattered*. Burlington, VT: Chelsea Green Publishing.
Thaler R. 1990. Anomalies: Saving, fungibility, and mental accounts. *Journal of Economic Perspective* 4: 193–205.
Trentmann, F. 2009. *Free Trade Nation: Commerce, Consumption, and Civil Society in Modern Britain*. Oxford: Oxford University Press.
Trigg, A. B. 2004. Deriving the Engel Curve: Pierre Bourdieu and the social critique of Maslow's Hierarchy of Needs. *Review of Social Economy* 62(3): 393–406.
Whitehead, M. 2013. Degrowth or regrowth? *Environmental Values* 22(2): 141–45.
Wilk, R. 2001. Consuming morality. *Journal of Consumer Culture* 1(2): 245–60.
———. 2004. Morals and metaphors: The meaning of consumption, in K. Ekström and H. Brembeck (Eds.), *Elusive Consumption*. Oxford: Berg, pp. 11–26.
———. 2014. *Poverty and excess in binge economics. Economic Anthropology* 1(1): 66–79.
Wilk, R., and Cliggett, L. 2006. *Economies and Cultures* (2nd ed.). Boulder, CO: Westview Press.
Wilson, C., and Dowlatabadi, H. 2007. Models of decision making and residential energy use. *Annual Review of Environment and Resources* 32: 169–203.

Chapter 20

"Climate Skepticism" inside the Beltway and across the Bay[1]

Shirley J. Fiske

In the first edition of *Anthropology and Climate Change* I wrote about climate change policy from "inside the Beltway"—an insiders' view from within the politically charged arena of the United States' capital. I wrote the chapter while I was working as a senior legislative aide in the U.S. Senate (2000–2009), drafting legislation and staffing hearings on climate change and cap and trade bills for a senator who was a senior member of the influential Senate Committee on Energy and Natural Resources. At the time all signs pointed to a national carbon emissions policy.

However, the cap and trade legislation did not materialize, even when proponents controlled both houses of the U.S. Congress and the presidency in the aftermath of the 2008 presidential election (2008–2010). By 2015 Republicans had taken control of Congress, and climate change and cap and trade bills disappeared from the legislative agenda, possibly for the foreseeable future.

In this chapter I update what happened to national climate policy and then assess where the nation stands now, detailing how support for a national carbon policy dissipated and defaulted to states and local policies and responses. To these ends I first describe climate skepticism as an important political force at the national level that catalyzed opposition to climate legislation; then I turn climate skepticism on end to question what it means at the local level—whether it captures the ethnographic reality on the ground—since the term is widely used to describe denial, doubt, and skepticism about global warming in demographic pockets of the American landscape. When interpreted through cultural models and consensus, climate skepticism is clearly not an adequate category to explain people's beliefs, which may stem from a different epistemology of nature and causation. I examine climate skepticism at these two levels—namely, the politicizing and polarizing nomenclature in policy dialogue and in cultural models and belief systems among communities dealing with climate change on the ground in the United States

---

Whatever Happened to National Climate Policy?

In the early years of the millennium, climate change policy was bipartisan. In 2003 Republican Senator John McCain (R-AZ) introduced a bill with then-Democratic Senator Joseph Lieberman (D-CT), the classic Climate Stewardship Act, that became the primary vehicle in the Senate and the primary model of cap and trade legislation for reducing carbon emissions through a national policy. Additionally, in 2005, with bipartisan support, the Senate went on record that global warming is real and mandatory limits are needed. This "sense of the Senate" resolution advocated for a national policy of "mandatory market-based limits and incentives" on greenhouse gases in a manner that "will not harm the U.S. economy and that will encourage other nations who are trading partners and key contributors to do the same."[2]

Following the presidential election of 2008 it seemed likely that a national climate change policy would be enacted, buoyed by the new president's expressed priorities and control of both houses of Congress. What happened?

Several things conspired against mandatory national carbon emission limits. An ongoing battle in Congress and the public media over the credibility of climate science was being waged by climate skeptics, starting well before Barack Obama was elected. The bipartisanship of the early cap and trade bills began to fall apart with the increasing politicization of the concept, promoted by Republican leaders, including Senator James Inhofe (R-OK), who, even to this day, continues to argue that "Global warming is the biggest hoax ever perpetrated on mankind." His line of reasoning was abetted by a well-funded campaign of contrarian interpretations supported by conservative think tanks and institutes (McCright and Dunlap 2000; Dunlap and Jacques 2013) that funded research to delegitimize climate science (Oreskes and Conway 2010; Dunlap and Jacques 2013).

Simultaneously, a well-orchestrated and well-funded public relations campaign by the fossil fuel industry and utility groups with economic stakes in the debate (for example, American Petroleum Institute, American Electric Power) conflated carbon reduction policies with tax increases and higher energy costs for consumers. This campaign recruited scientists who voiced their skeptical views of global warming and also clung to their deep-seated disciplinary values in climate science (Lahsen 2005, 2008, 2013). The high point of the climate denial campaign came in 2009, with the events leading up to "Climate Gate." Immediately following, public belief in global warming dipped several percentage points (Cohen and Agiesta 2009; Pew Research Center 2010).[3]

On top of this battle of climate views, the proximate factor in the defeat of cap and trade was the protracted battle over the Affordable Care Act for national health care, which ensued immediately after the president was inaugurated. The epic political debate exhausted Congress, and there was no Congressional or public appetite for another divisive fight. Although

House Majority Leader Nancy Pelosi was able to pass a climate bill in the House in 2009, the Senate could not agree on a bill, and the lack of action was a final blow to its momentum. Climate skepticism took its toll, as polls reported increasing partisan polarization (Pew Research Center 2013). What was previously a climate and earth science issue of accumulation of greenhouse gases (GHGs) became politically polarizing, clouded by pronouncements of doubt and denial. To be sure, climate skepticism was not the only reason for the demise of cap and trade—related factors were the 2008 housing mortgage boom bust, the Wall Street implosion and resulting recession, and, ultimately, the weight of the legislation's own complexity.

In the subsequent mid-term elections of 2010, the House majority changed hands, and a critical block of Republican Tea Party members were elected who opposed any policy that threatened to raise taxes or enlarge the scope of government. It became clear that there was a lack of political will, and the issue was too divisive for either the American public or Congress to debate and adopt. Indeed, the annual poll by the Pew Research Center for People and the Press shows that "dealing with global warming" has been a low public priority consistently from 2007–2014, ranking 29 out of 30 as a "top priority for the president and Congress" (Pew Research Center 2014). As Table 20.1 shows, global warming is consistently low priority for national legislative action (2007–2014). In the list of 20 policy priorities it ranks 19th, below "improving bridges, roads, public transit" and above "dealing with global trade issues."

The Administration turned to executive and administrative actions to move climate change policies into law. Once the Supreme Court upheld the Environmental Protection Agency's (EPA's) authority to regulate carbon dioxide as a pollutant under the Clean Air Act, the Obama Administration pursued carbon emissions regulations on new power plant emissions and coal-fired electricity generators. In 2013, the Administration proposed regulations on existing coal-fired and natural gas electric power plants (Eilperin 2013). Most recently, in 2014, the president has focused federal efforts more explicitly on the local—specifically, on community—resilience. President Obama established a Task Force on Climate Preparedness and Resilience, directing it to advise the administration on how to respond to the needs of communities nationwide dealing with the effects of climate

Table 20.1 Percentage of people responding that "Dealing with global warming" should be a top priority for the President and Congress, 2007–2014 (source: Pew Research Center 2014a)

Year							
2007	*2008*	*2009*	*2010*	*2011*	*2012*	*2013*	*2014*
38%	35%	30%	28%	26%	25%	28%	29%

change. The creation of the task force is symbolic of the president's pivot to the local; appointed members are from tribal, county, and municipal organizations and are directed to use their own experience of preparedness and resilience in their communities.[4] The announcement coincided with the release of the National Climate Assessment, which became public in May 2014. This report to Congress highlights local, regional, and sectoral impacts of climate change.

The most recent election, in 2014, portends continuing partisan battles over national climate actions over the past eight years. In the Senate, one of the most vocal opponents—and a climate denialist—became chair of the influential Committee on Environment and Public Works in January 2015. The committee is likely to reverse national policies achieved administratively through the EPA (regulation of carbon dioxide emissions from power plants) and discontinue funding for federal initiatives. These developments ensure that Congress and the public will continue to see climate skepticism playing a role front and center on the American political stage.

Climate Change and Skepticism

As climate change policy defaults to community-level adaptations and decisions, state politicians, management agencies, counties, and municipalities are attempting to prepare for the challenges that climate change will bring and to mobilize support for mitigation and adaptation policies. The Georgetown Climate Center[5] reports that 15 states have adaptation policies and plans, primarily on the West Coast and Atlantic coasts. Municipalities in high-risk zones, including New York City and Norfolk, Virginia, are undertaking climate change planning efforts, especially in the aftermath of Hurricane Sandy (2010). In the case of Norfolk, the city has daily tidal flooding in its downtown area owing to rising sea level, which creates a sense of urgency for climate adaptation efforts. Given the growing number of adaptation efforts in vulnerable counties and coastlines, we must more than ever work to understand local constituencies and communities' concerns and views of climate change.

In the mid-Atlantic, climate change will bring more coastal storms, flooding, and imperilment of local infrastructure, including public roads and state parks, sewage treatment systems and plants, and private homes and property. As community and state leaders assess options (for example, beach nourishment, building sea walls, property buyouts, retreat from the coast) decisions will be undertaken in the context of public opinion, including beliefs about changing climate and the appropriate role of governments. In Maryland a recent survey revealed that 73% of residents surveyed would "like local and state governments to take actions to protect their communities against climate harms" (Akerlof and Maibach 2014), an apparent endorsement for the governor to pursue adaptation policies.

Climate skeptics, representing about one-third of Americans, according to national polls, are a vexing challenge to decision makers implementing adaptation policies and a phenomenon of interest to politicians, media, journalists, and social scientists. Climate skeptics are the disengaged, the doubtful, and the dismissive when it comes to global warming, and they are the least concerned and least motivated to do anything about it (Leiserowitz et al. 2010).

Anthony Leiserowitz and colleagues led the way with finely tuned interpretations of national surveys. They identified "six Americas" based on latent class analysis and discriminant functions around 36 variables representing four distinct constructs of global warming beliefs, issue involvement, policy preferences, and behavior. Their analysis shows six "interpretive communities" that fall on a continuum from alarmed and concerned to disengaged, doubtful, and entirely dismissive of climate change (Leiserowitz, Maibach, and Light 2009; Leiserowitz et al. 2010).

In 2014, 61% of the public believed there was solid evidence that the average temperature on earth has been getting warmer over the past few decades; but more than one-third of the public (35%) still say there is no solid evidence that the Earth is warming (Pew Research Center 2014).[6] We know that people's beliefs, attitudes, and values about climate change affect how they respond to the science and policies about it. We also know that these differences are ultimately cultural differences and can help to explain how subgroups of populations have divergent understandings of climate change.

Parsing Climate Skepticism on the Eastern Shore: A Cultural Perspective

Climate skepticism seems to be a uniquely American belief system, with its own exceptional constellation of elements in resistance to pronouncements of scientists, the use of science to set environmental policy, fear of government intrusion, and support for entrenched interests of fossil fuel industries that support a dominant lifestyle. Climate skepticism has become politicized and is closely associated with political party affiliation and conservatism and with growing schisms between political parties (Figure 20.1).

Climate skepticism is a substantial enough cultural phenomenon to merit special attention by social scientists, culminating in a special issue of *American Behavioral Scientist* that examined it comprehensively from many of the perspectives mentioned earlier in the chapter, including its genesis and meaning, its distribution across the populace, its impact as a political force, and the role of conservative think tanks and industry in promoting it (Dunlap 2013). The term *climate skeptic* has multiple semantic components and dimensions—dubious, doubtful, dismissive, and outright denial. The moniker covers broad territory, and, like most labels, the term

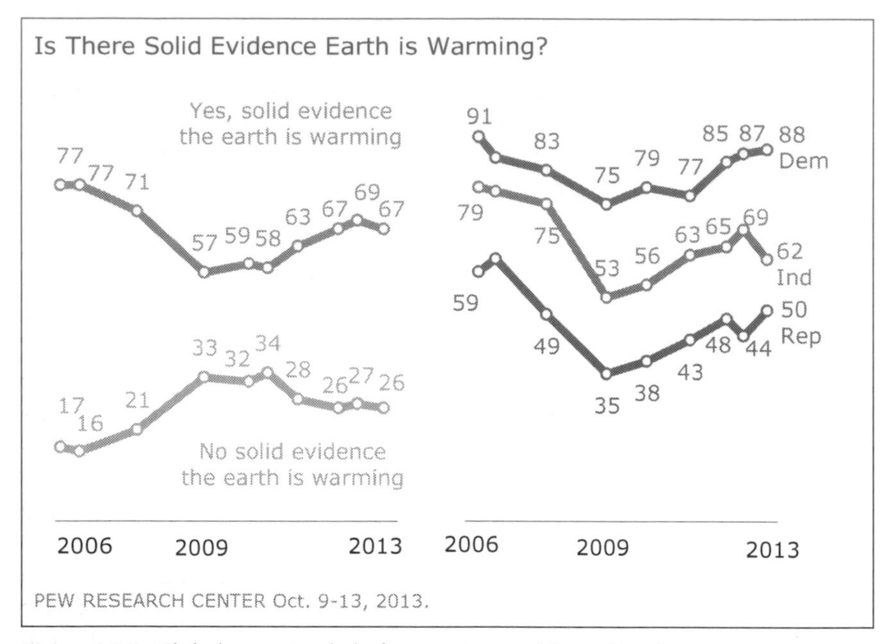

Figure 20.1 Global warming beliefs over time and by political party

itself masks important variability among subgroups. As Dunlap suggests, the concept is perhaps best considered to be a continuum of beliefs from climate skepticism to denial, rather than two distinct ways of thinking (Dunlap 2013). While one might assume that conservative businesspeople and shop owners may be "climate skeptics," I argue that we must understand people's underlying belief systems before we use the label inclusively.

What anthropologists add to the conversation on climate skepticism is a synthesis of several perspectives: (1) attention to *context*—that is, the embeddedness of communities' and groups' deeply held beliefs (the role and animism in nature, society, humans) in historical circumstance; (2) generation of place-based local knowledge (epistemological factors) about natural causation and human agency; and (3) structural features of the political ecological landscape (Marino and Ribot 2012). With respect to the last point, Bartels and colleagues have shown, for example, that farmers' apparent conservative choices with respect to climate adaptation are not rooted in denial but in the way the local political economy constrains their choices of alternative crops—via subsidies and local processors' capacities in their region (Bartels et al. 2013).

In addition, anthropologists bring methodological and theoretical attention to how people frame their beliefs and utilize implicit propositions (schema) to make sense of the world. The study of cultural models and cultural consensus in belief systems is complemented by the study of the structural and historical context of communities, adding valuable interpretation to the models (Holland and Quinn 1987; Strauss and Quinn 1997; Quinn 2005; Weller 2007). American ranchers and farmers in the American Southwest have an ontological view of "living with the climate" that is contrasted with the dominant view of "controlling the climate" espoused in the dominant culture through science and resource management (Brugger and Crimmins 2013), as shown in the discourse and interviews of people as they talk about the environmental changes around them (Vasquez-León 2009; Brugger, Crimmins, and Owens 2011). The belief in alternative epistemologies that *do not utilize* the core Western scientific beliefs of anthropogenic global warming is well documented by anthropologists in the non-Western world from Tanzania to Papua New Guinea (for example, see Crate 2008; Henshaw 2009; Lipset 2011; Shaffer and Naiene 2011; Sheridan 2012; Marino and Lazrus 2014), but it is only beginning to be addressed among belief systems of core American subgroups.

In the sections that follow I draw on data from a recently completed study on climate change and cultural models and consensus on Maryland's Eastern Shore[7] that led me to question how social science (and the public) uses the term *climate skeptic*. The theoretical and methodological approach of the study was explicitly cultural. We sought to identify frameworks of meaning and shared sets of implicit knowledge that enable people to communicate and understand one another's perspectives (Kempton et al.1992; D'Andrade 1995; Kempton et al. 1995; Strauss and Quinn 1997; Garro 2000; Maloney and Paolisso 2006; Johnson and Griffith 2010; Paolisso, Weeks, and Packard 2013). Cultural models and cultural consensus are methodologies that discern the filters and implicit knowledge that people use to understand something—in this case, environmental and climate changes that they have observed. These frames also motivate people's responsiveness, indifference, or resistance to a myriad of actions, programs, and policies, from long-range planning efforts to choices about stabilizing the shoreline.

Climate Change Models

The intent of our research project was to learn through ethnographic research, conducting interviews and surveys and performing textual analysis, what residents thought about local environmental change and how they talk, think, and frame climate change. We identified three subgroups of residents central to the economic and cultural fabric and heritage of the counties on the Eastern Shore—commercial fishermen (watermen), farmers (family owned and operated farms), and recent "amenity migrants" to the Eastern Shore, the last group largely retirees from the "other side" of the Chesapeake, primarily the Washington-Baltimore metropolitan area, who seek refuge in the idyllic rural countryside.

I discuss what we discovered from farmers who live on, own, and operate farms in the center of Maryland's agricultural heartland, Dorchester County. Farmers are a visible and economically important group in the United States who face risks and challenges to their livelihood in the context of climate change. Although farmers and agro-forestry communities internationally have been a focus for anthropologists for many years, a growing number of anthropologists are now paying attention to the challenges climate change has for U.S. farmers, often focusing on their beliefs, knowledge, and concerns (Vasquez-León 2009; Crane et al. 2010; Bartels et al. 2013; Brugger and Crimmins 2013;).

We collected ethnographic information through semistructured, in-depth interviews with 15 farmers and through a survey (186 respondents, 44 of whom were farmers) to determine the extent of shared knowledge within and across subgroups. Cultural consensus analysis indicated that farmers utilize a highly shared system of knowledge to generate their understandings of climate change, using two cultural models to do so.

We identified a core cultural model based in farmers' lived experience and the experience of generations of family members (uncles, fathers, wives, and grandfathers) who have farmed on the Eastern Shore; it is a powerful epistemological model that includes an understanding of causation and nature, the place of humans and society in nature, and a guide for human agency.

Along a continuum from "dismissal" to "alarmist," a majority of farmers believed that climate change may be happening but that humans were not causing most of it. About one-third are dubious that it's happening. Only a few farmers rejected (dismissed) climate change phenomena categorically. On the other end of the "Six Americas" continuum (Leiserowitz et al. 2010), a small number of farmers were quite concerned about climate change and believed climate change and sea-level rise were occurring. Figure 20.2 shows how farmers are distributed along the continuum.

We discovered a second model—also widely shared, part of the same overall model for climate change but a contrasting set of epistemological assumptions—one that "sees" climate change as analogous to the well-known and problematic environmental changes in the Chesapeake, where human mediation and agency are primary for the solution of those problems (Fiske, Paolisso, and Clendaniel 2014).

The First Cultural Model

The first cultural model is *climate change as natural change*. In this model farmers interpret the changes they notice (for example, warmer winters, more intermittent and unpredictable thunderstorms in the summer, periods of summer drought, "rising tides") as part of the patterns and cycles of nature; the patterns will cycle back and /or change yet again. These are all natural processes in the earth's aging and evolving; they are not caused by

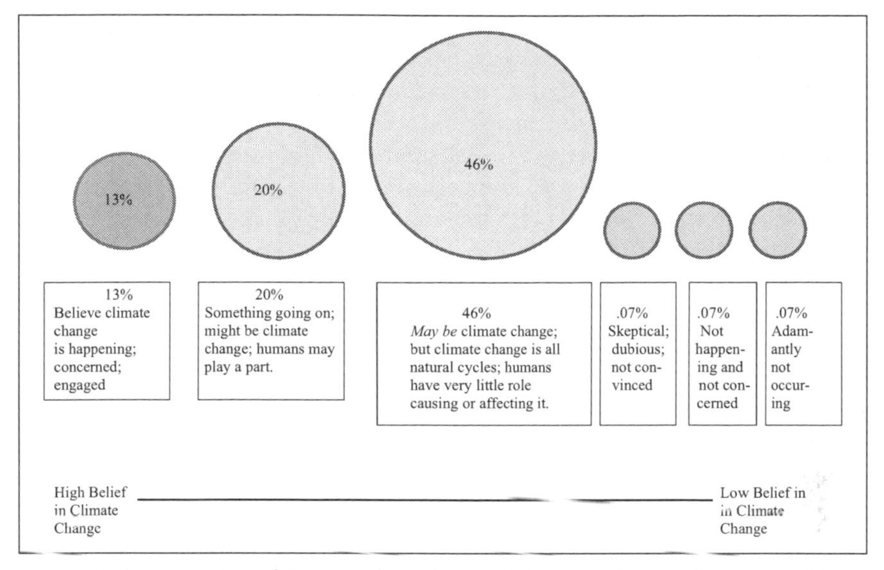

Figure 20.2 Proportion of Eastern Shore farmers' views on climate change ($n = 15$) (graphic design adapted from Leiserowitz, Maibach, and Light 2009)

humans. Farmers expressed doubt about climate change, but they are not all dismissive or denialists. They are very observant about any changes around them and acknowledge many specific changes in the environment on the Eastern Shore; but they do not buy the *canonical* theory of climate change, particularly the human-caused or anthropogenic elements and causation. This means that farmers generally question the idea of the buildup of greenhouse gases (GHGs) from humans' use of fossil fuels, causing rising temperatures, glacial melting, and sea-level rise (SLR). Their experience, and that of their relatives, tells them that weather goes in patterns and cycles and will vacillate. They all shared stories of weather cycles and patterns from grandfathers, uncles, and fathers, all of whom lived on the Eastern Shore and most of whom farmed for a living—for a minimum of two and maximum of six generations.

This understanding of how nature works—"nature runs in patterns" and "cycles"—is a key proposition, or schema, of their traditional cultural model and is widely shared. Further, their cultural model tells them that climate change, while perhaps newly identified by scientists, has been with us "forever," from the beginning of time—and consequently did not begin with the Industrial Revolution. In other words, while climate change may be happening, and has always happened, it is not primarily caused by humans. Generally speaking, it is a natural change driven by Earth's evolving biogeophysical dynamics, or by larger forces, or both, and humans have done little to cause it and can do little to change it.

In addition, perhaps unique to the Chesapeake, the geological phenomenon of subsidence supports farmers' cultural models of rising tides as natural change. The land masses around the Chesapeake, particularly the Eastern Shore, have been sinking relative to the Chesapeake Bay waters since the Pleistocene glaciers retreated (Boesch et al. 2013). This is well known among Eastern Shore residents living in coastal areas, and it explains in part a nearly universal observation by farmers—that the tides are rising. Farmers told us that "the tides are rising" and "the creeks are higher," attributing these changes to subsidence ("we're sinking") rather than climate change phenomena such as sea-level rise. This situation informs the recent survey of Maryland residents that showed more than one-third say they don't know what is causing sea- level rise; a quarter (23%) say it is caused equally by natural and human influences; and another 17% say that natural causes are the primary cause, leaving 18% who assert that it is rising primarily because of human causes (Akerlof and Maibach 2014).

When contemporary theories of climate change, with its long-term, unidirectional changes in sea level and in atmospheric composition, challenge farmers' well-validated cultural models, the long-held traditional cultural models are likely to trump the emerging cultural model of climate change as human-induced change.

The Second Cultural Model

We found that that farmers used the second cultural model, which we call *climate change as environmental change,* to interpret recent changes. This model reinterprets climate change as having the same processes and outcomes as recent environmental issues in the Chesapeake Bay region—including the decline of oyster populations and blue crab, the growth of hypoxic areas in the summer, outbreaks of harmful algae ("blooms"), loss of sea-grass beds, and increasing shoreline erosion and turbidity/siltation. All these phenomena are anthropogenically driven and take human actions to reverse them. This model frames climate change as an "environmental problem" in the same category with Chesapeake Bay cleanup, requiring human agency to manage or control them, through moratoria on fishing, refuges for oysters, set-backs and buffer zones along the shoreline, and regulations on fertilizers and chicken waste entering the tributaries of the Bay. Farmers implicitly make a distinction between what they see as the order of *natural* change and what they see as human-caused *environmental* changes (which are always problems).

Farmers' cultural models tell them that if climate change becomes identified by politicians and policy makers as an environmental (human-induced) problem, that will lead to human intervention and ultimately regulations being put on farmers. This framing is similar to that of farmers in California who are much more fearful of the unknowns of climate change policy than they are of the physical attributes of climate change (Niles, Lubell, and Van Haden 2013).

Farmers in Maryland have little confidence that any climate policy will benefit them; they told us that "most politicians don't have a clue what we do out here every day," meaning that laws and regulations did not bear any semblance to farmers' reality in their fields (for example, how much fertilizer was used, the nature of run-off from fields). For farmers, the risks of climate change lie in the opaque and ominous arena of political action and regulation. In short, farmers are more concerned about regulations and costs that may be imposed on them than about the physical realities of climate change in the mid-Atlantic and Chesapeake Bay.

The certitude on the part of Maryland farmers stems in part from the regulatory context of this generation of farmers in Maryland. The last 40 years has brought increasing regulations from the Clean Water Act (CWA) and pursuant state regulations and county requirements that are meant to reduce nutrients, pesticides, insecticides in the Bay's waters, as well as the passage of the Coastal Zone Management Act in 1972 and subsequent state laws that required changes in farming operations (buffer zones, set-back areas along tributaries and the Bay). The most recent CWA regulations, spurred on by President Obama's Executive Order in 2013, are currently being implemented by counties in each state, requiring reductions by farmers (each farm must have a nutrient management plan) and other stakeholders in nutrient loads entering the Bay.

The salience of farmers' experience with regulations and the perception of "climate policy risk" mirrors the findings in the Yolo County region of the Sacramento Valley. Niles and colleagues reported: "We found that *the past matters*. Farmers' past perceptions of different environmental policies had a larger impact on their climate change beliefs and policy behaviors than their actual experience with climate change" (Niles 2013; italics inserted by S. Fiske). "Climate policy is the highest priority risk perceived by farmers," (Niles, Lubell, and Van Haden 2013). Similarly, Dorchester County farmers are not too concerned about future flooding but are concerned about potential regulations put on them in response to climate change.

This parallel background, living on and adjusting to the land over generations, may be why farmers are not concerned in general with "adaptation." They do not feel vulnerable to climate changes such as higher temperatures, more sporadic rainfall, droughtlike conditions, and more frequent floods as the tides rise higher.[8] The majority of farmers told us they were not concerned about adaptation to climate change, because they have been adapting to changes in climate and weather all their lives and have been doing it successfully. "It's what we do." "We don't talk about it, we just live with it." In addition, most farmers had options to deal with changes—they can switch to more drought-tolerant varieties of corn and grain; they now have more efficient fertilization with GPS precision; they can invest in irrigation systems; and conservation programs exist at both the state and federal levels that provide incentives for taking farmland out of production as marshes move inland. These options are tools that allow farmers to adapt over time to environmental changes or the expected effects of climate change and that make them feel less vulnerable.

This attitude is consistent with farmers and ranchers in the American Southwest, who see themselves as "living with climate" throughout their lives, and it contrasts sharply with the dominant approach to climate in the 20th century, which has been "overcoming the climate" (Brugger and Crimmins 2013). Brugger and Crimmins contrast these two belief systems as having different ontologies, epistemologies, logic, practice, and values. The "living with climate" belief system includes an ontological view that nature and society are mutually connected and that peoples' values include an attachment to place that means adjusting human activities to fit changes in the climate (2013). These elements of the belief system also characterize Eastern Shore farmers' cultural model of "climate change as *natural* change"—primarily the connection to place and the relationship of humans to nature being a mutual adaptation. Eastern Shore farmers' second, and competing, cultural model of climate change frames the environment and nature as distinctly different entities, with the logic being to manage or engineer the environment to "fix" problems and accommodate human activities.

Conclusions

Much attention has been paid to climate skepticism as an American political force inside the Beltway—by journalists, the media, and social scientists. Climate skepticism, as a uniquely American belief system, has been an important instrument in buttonholing climate change legislation at the national level. Even after the 2014 elections, and during the presidential campaign leading up to the election in 2016, with new conservative leadership in Congress we continue to see attempts to roll back the administrative and judicial gains at the national level regarding coal-fired emissions; but it is anticipated that state-level and municipal-level adaptation and climate change policies will continue.[11]

Beyond being a political force, skepticism is a broad classification of "thinking about climate change" that consistently characterizes about one-third of the American public. This chapter makes the point that the classification, as generally used, is broad enough that we need to examine it critically. By deconstructing it through cultural models we see ontological and epistemological systems in operation among farmers on the Eastern Shore of Maryland—well-organized, shared systems of belief based on substantially different experience, beliefs, and values from the dominant scientific paradigm about climate change. Eastern Shore farmers tend to believe climate change is natural change, with cycles and patterns and very little human causation. They do not feel vulnerable to climate change but believe they have constantly been able to adapt and will continue to do so. Similarly, western ranchers and farmers in the United States believe that they are "living with the climate" and assume that there is a mutuality between nature, society, and adaptation (Brugger and Crimmins 2013: 1831).

On-the-ground understandings of people's beliefs about climate change such as these contrast with the dominant paradigm of Western science, whereby there is dualism and separation of humans and natural spheres; and these understandings are different from "climate denial" fueled by industry associations, corporations, and their political sympathizers (Dunlap 2013; Singer 2015). In a climate change and culture review chapter, Adger and colleagues point out that policies and programs for adaptation and resilience will be more accepted and effective if they approach it with a cultural view of climate change (Adger et al. 2013).

Across the Bay, the majority of farmers expressed doubt about the climate change theory and its anthropogenic nature, and they expressed concern about what might happen if climate change is moved into the political and policy domain of environmental problems. They were keenly aware of the many environmental changes going on around them on their farms and in the Chesapeake Bay region. They told us that "the tides are higher," winters are warmer, more time elapses between storms in the summer, and crops are faster to mature, and they detailed the changes in wildlife on the Eastern Shore.[9] Nonetheless, they are willing to address environmental changes that are tangible problems and issues in Dorchester County. Niles and colleagues found, similarly, that despite past policy experiences, farmers are willing to consider governmental programs structured as incentive programs for climate change adaptation and mitigation (2013: 1757).

Understanding farmers' cultural models helps point to ways of communicating with people who might be labeled "climate skeptics" and points to ways to design policies that may not be rejected. Farmers' ways of thinking about climate change cultural models are built and shared from lifetime experience situated in making a living raising crops and livestock, paying close attention to weather and climate changes. Their cultural models come from farming under increasing regulations, an arena where farmers feel they have already done a great deal to improve water quality and be good stewards of the land (Paolisso, Weeks, and Packard 2013). The models did not stem from the campaigns of right-wing think tanks, media personalities, or from active doubts about the veracity of data. In a similar vein, on the other side of the North American continent, the Yolo County study concluded that by considering farmers' perceptions of climate change and climate policy, policy makers have an opportunity to reduce the amount of time before programs become accepted by farmers, when their views of past policies affect how they respond to future environmental issues (Niles 2013).

On the Eastern Shore, farmers' views have their basis in situated experience, place-based knowledge and values, and a cultural model of nature and society in which humans adapt to the natural cycles that occur rather than resisting and trying to change natural cycles on global scales. This sense of humans' place in nature and climate change may be a commonality among farmers and ranchers elsewhere in the United States, with generations of working the land. Where these contextual conditions are present, it is likely that additional research and studies that try to "prove" the validity of climate change, or argue that we must change our behavior because of the effects of increasing GHGs, will likely not change people's cultural models. However, talking about current problems held in common may be a way to open the conversation. Discussions and policies that help people to deal with real issues at hand—the "rising tides," erosion along the shorelines, and loss of cropland to marsh conversion—might be productive trajectories.[10] For policy makers to work effectively with various publics, they must understand and acknowledge the cultural factors in climate change and meet people "where they are" in their assessment of climate change, rather than imposing a top-down, one-size fits all approach to adaptation and climate change.

Notes

1. "Inside the Beltway" is an U.S. political phrase that refers to the interstate freeway (beltway) that encircles Washington, D.C., and the activities that are important within the political enterprise—the federal government, its contractors, lobbyists, Congress, and the media that cover them. It can be used to denote something that is of very limited importance to the general public, as well as something that is important insider information. "Across the Bay" refers to a wholly different geographic and cultural lifestyle on the eastern side of the Chesapeake Bay—the somewhat mythical image of tranquil waterfront living, retirement homes, and slower, more honest rural lifestyle—in contrast to the highly charged and political existence on the other side of the Chesapeake Bay "inside the Beltway."
2. Senate Amendment 866 to H.R. 6 passed as a resolution by voice vote, after a motion to table (kill) the amendment failed. It started by acknowledging the buildup of greenhouse gases and their effect on climate change, quite remarkable considering it was during the George W. Bush administration, which had tried to sweep climate change under the table. While a Congressional resolution conveys the "sense of the Senate" and is not a legally binding instrument, it does set a precedent of Senate support for regulation of GHGs. Seeing the voted bipartisan agreement on this resolution, which was taken up immediately after the Senate voted on the Energy Policy Act of 2005 (from which it had been stripped in conference), was part of my reason for optimism with respect to a national climate change policy.
3. The so-called Climate Gate included the well-funded climate skepticism campaign, conflict of interest charges that were leveled at the IPCC Chairman at the time, and a Republican think-tank that hacked emails of climate scientists in the peer-review process, which were then selected, exaggerated, and distorted by climate denial groups (see Leiserowitz et al. 2013).
4. The president announced a new $1 billion climate change resilience fund during a visit to California in spring 2014, prompted in part by the extreme drought in the state and its impact on farmers; the fund appears to cobble together resources from a number of federal agencies and private foundations and focuses on community resilience. It follows on the heels of the passing of a Farm Bill that creates several regional "climate hubs" aimed at helping farmers and rural communities. The president announced that he would ask Congress for $1 billion in new funding for a "climate resiliency" program to help communities invest in research, development, and new infrastructure to prepare for climate disasters.

5. www.georgetownclimate.org/adaptation/state-and-local-plans, accessed November 2014.
6. The Pew report identified "steadfast conservatives" and "business conservatives" as the core segments of the public who are the denialists (Pew Research Center 2014b: 69).
7. The author gratefully acknowledges funding by the National Science Foundation (NSF) Anthropology Award # BCS-1027140, 2011-2014, Cultural and Consensus Models of Climate Change Adaptation for Chesapeake Bay, Co-PIs, Michael Paolisso and Shirley J. Fiske. The study investigated climate and environmental change perceptions among three subgroups of residents on the Eastern Shore—farmers, commercial watermen, and new residents who have moved to the Eastern Shore.
8. Not surprisingly, the Maryland public thinks of agriculture as a sector very vulnerable to climate change; 56% of the Maryland public said that agriculture was most likely to be at risk in the next several years, being at the top of their list, along with people's health (55%) (Akerlof and Maibach 2014).
9. Many farmers supplement their income by allowing seasonal hunting (duck, geese, pheasants, deer, muskrat) or alternative recreational use on their lands.
10. A majority of state residents support policies that protect shorelines and low-lying lands from sea-level rise, such as regulations, zoning laws and set back distances for buildings (67%); long-range planning (66%); tax incentives for property owners to take protective actions (55%); and government funds to buy natural areas as buffers against rising waters and storms (55%) (Akerlof and Maibach 2014)—all reasonable approaches tailored to regional and sectoral needs and ones that farmers might consider.
11. State-level policies such as in California and Maryland, and municipalities on coasts such as New York City, Norfolk, VA, and Charleston, SC, are examples of legislative and planning commitments in the face of climate change that are likely to continue.

References

Adger, W. N., Barnett, J., Brown, K., Marshall, N., and O'Brien, K. 2013. Cultural dimensions of climate change impacts and adaptation. *Nature Climate Change* 3(2): 112–17.

Akerlof, K., and Maibach, E. W. 2014. *Adapting to Climate Change & Sea Level Rise: A Maryland Statewide Survey, Fall 2014*. Fairfax, VA: Center for Climate Change Communication, George Mason University.

Bartels, W.-L., Furman, C. A., Diehl, D. C., Royce, F. S., Dourte, D. R., Stein, B., Zierdan, D., Irani, T. A., Fraisse, C., and Jones, J. W. 2013. Warming up to Climate Change: A Participatory approach to engaging with agricultural stakeholders in the southeast US. *Journal of Regional Environmental Change* 13(1): 45–55.

Boesch, D. F., Atkinson, L. P., Boicourt, W. C., Boon, J. D., Cahoon, D. R., Dalrymple, T. E., Horton, B. P., Johnson, Z. P., Kopp, R. E., Moss, R. H., Parris, A., and Sommerfield, C. K. 2013. *Updating Maryland's Sea-Level Rise Projections. Special Report of the Scientific and Technical Working Group to the Maryland Climate Change Commission.* Cambridge, MD: University of Maryland Center for Environmental Science, https://www.pwrc.usgs.gov/SeaLevelRiseProjections.pdf.

Brugger, J., and Crimmins, M. A. 2013. Living with climate change: The art of adaptation in the rural American Southwest. *Global Environmental Change* 23: 1830–40.

Brugger, J., Crimmins, M. A., and Owens, G. 2011. Finding a place for climate science in the rural west. *Rural Connections* 5(2): 5–10.

Cohen, J., and Agiesta, J. 2009. On environment, Obama and scientists take hit in poll. *Washington Post*, December 18.

Crane, T. A., Rocoli, C., Paz, J., Breuer, N., and Broad, K. 2010. Forecast skill and farmers' skills: Seasonal climate forecasts and agricultural risk management in the southeastern United States. *Weather, Climate, and Society* 2: 44–59.

Crate, S. A. 2008. Gone the bull of winter? Grappling with the cultural implications of Anthropology's role(s) in global climate change. *Current Anthropology* 49(4): 569–95.

D'Andrade, R. 1995. *The Development of Cognitive Anthropology.* Cambridge: Cambridge University Press.

Dunlap, R. E. 2013. Climate change skepticism and denial: An introduction. *American Behavioral Scientist* 57(6): 691–98.

Dunlap, R. E., and Jacques, P. J. 2013. Climate change denial books and conservative think tanks exploring the connection. *American Behavioral Scientist* 57(6): 699–731.

Eilperin, J. 2013. Obama unveils ambitious agenda to combat climate change, bypassing Congress. *Washington Post.* June 25, www.washingtonpost.com, accessed September 9, 2014.

Fiske, S. J., Paolisso, M. J., and Clendaniel, K. 2014. *Climate Change among Farmers on the Eastern Shore of Maryland.* Paper presented in session Changing Climate and Community Resiliencies in the Chesapeake Bay region. American Anthropological Assocation meetings, Washington, D.C., December 2014.

Garro, L. 2000. Remembering what one knows and the construction of the past: A comparison of cultural consensus theory and cultural schema theory. *Ethos* 28(3): 275–319.

Henshaw, A. 2009. Sea Ice: The sociocultural dimensions of a melting environment in the Arctic, in S. A. Crate and M. Nuttall (Eds.), *Anthropology and Climate Change.* Walnut Creek, CA: Left Coast Press, Inc., pp. 153–65.

Holland, D., and Quinn, N. 1987. *Cultural Models in Language and Thought.* Cambridge: Cambridge University Press.

Johnson, J. C., and Griffith, D. C. 2010. Finding common ground in the commons: Intracultural variation in users' conceptions of coastal fisheries issues. *Society and Natural Resources* (23): 837–55.

Kempton, W., Boster, J. S., and Hartley, J. A. 1995. *Environmental Values in American Culture.* Cambridge, MA: The MIT Press.

Kempton, W., Feuermann, D., McGarity, and A. E. 1992. "I always turn it on super": User decisions about when and how to operate room air conditioners. *Energy and Buildings* 18(3-4): 177–91.

Lahsen, M. 2005. Technocracy, democracy, and U.S. climate politics: The need for demarcations. *Science, Technology, and Human Values* 30(1): 137–69.

———. 2008. Experiences of modernity in the greenhouse: A cultural analysis of a physicist "trio" supporting the conservative backlash against global warming. *Global Environmental Change* 18(1): 204–19.

———. 2013. Anatomy of dissent: A cultural analysis of climate skepticism. *American Behavioral Scientist* 57(6): 732–53.

Leiserowitz, A., Maibach, E., and Light, A. 2009. *Global Warming's Six Americas | Center for American Progress.* Washington, D.C.: Center for American Progress, http://www.americanprogress.org/issues/green/report/2009/05/19/6042/global-warmings-six-americas/, accessed July 9, 2013.

Leiserowitz, A., Maibach, E., Roser-Renouf, C., and Smith, N. 2010. *Global Warming's Six Americas, June 2010.* Yale University and George Mason University. New Haven, CT: Yale Project on Climate Change.

Leiserowitz, A., Maibach, E., Roser-Renouf, C., Smith, N., and Dawson, E. 2013. Climategate, public opinion, and the loss of trust. *American Behavioral Scientist* 57(6): 818–37.

Lipset, D. 2011. The tides: Masculinity and climate change in coastal Papua New Guinea. *Journal of the Royal Anthropological Institute* 17(1): 20–43.
Maloney, S. R., and Paolisso, M. 2006. The "art of farming": Exploring the link between farm culture and Maryland's nutrient management policies. *Culture & Agriculture* 28(2): 80–96.
Marino, E., and Lazrus, H. 2014. *Time and Flexibility: A Cross-Cultural Comparison of Climate Change Adaptation, Disaster Preparedness, and Bureaucratic Constraint in Alaska and Tuvalu.* Paper presented at the Society for Applied Anthropology annual meeting, Albuquerque, NM.
Marino, E., and Ribot, J. 2012. Special issue introduction—Adding insult to injury: Climate change and the inequities of climate intervention. *Global Environmental Change* 22(2): 323–28.
McCright, A. M., and Dunlap, R. E. 2000. Challenging global warming as a social problem: An analysis of the conservative movement's counter-claims. *Social Problems* 47: 499–522.
Niles, M. T. 2013. Policies worry farmers more than climate change, says new study. *UCDavis News and Information*, September 10, http://news.ucdavis.edu/search/printable_news.lasso?id=10714&table=news.
Niles, M. T., Lubell, M., and Van Haden, R. 2013. Perceptions and responses to climate policy risks among California farmers. *Global Environmental Change* 23(6): 1752–60.
Oreskes, Naomi, and Erik M. Conway. 2010. Merchants of Doubt. Bloomsbury Publishing USA.
Paolisso, Michael, Weeks, Priscilla, and Packard, Jane. 2013. A Cultural Model of Farmer Land Conservation. Human Organization 72(1): 12–22.
Pew Research Center. 2010. *Opinions about Global Warming: 2006–2010*, http://www.people-press.org/2010/10/27/little-change-in-opinions-about-global-warming/.
———. 2013. *GOP Deeply Divided over Climate Change*. Report on Survey Conducted October 9–13.
———. 2014a. Thirteen Years of the Public's Top Priorities. Washington, D.C., http://www.people-press.org/interactives/top-priorities/, accessed September 9, 2014.
———. 2014b. *Beyond Red vs. Blue: The Political Typology—Fragmented Center Poses Election Challenges for Both Parties*, http://www.people-press.org/2014/06/26/section-7-global-warming-environment-and-energy/, accessed September 27, 2014.
Quinn, N. 2005. *Finding Culture in Talk: A Collection of Methods.* Basingstoke: Palgrave Macmillan.
Shaffer, L. J., and Naiene, L. 2011. Why analyze mental models of local climate change? A case from southern Mozambique. *Weather, Climate, and Society* 3(4): 223–37.
Sheridan, M. J. 2012. Global warming and global war: Tanzanian farmers' discourse on climate and political disorder. *Journal of Eastern African Studies* 6(2): 230–45.
Singer, M. 2015. Climate change denial: The organized creation and emotional embrace of unsupported science claims. *Anthropology News* 56(1-2): 19.
Strauss, C., and Quinn, N. 1997. *A Cognitive Theory of Cultural Meaning:* Cambridge: Cambridge University Press.
Vasquez-León, M. 2009. Hispanic farmers and farmworkers: Social networks, institutional exclusion, and climate vulnerability in southeastern Arizona. *American Anthropologist* 111(3): 289–301.
Weller, S. C. 2007. Cultural consensus theory: Applications and frequently asked questions. *Field Methods* 19(4): 339–68.

Chapter 21

When Adaptation Is Not Enough: Between the "Now and Then" of Community-Led Resettlement

Kristina Peterson and Julie Koppel Maldonado

In the seven years since the chapter appeared in this book's first edition, the Grand Bayou Atakapa-Ishak Tribe in Louisiana has experienced many physical changes brought by continual land loss, industrial and Western development, major storms, a tsunami-type wave, and an oil disaster. The community has also experienced a growth in participation with agencies, coastal traditional environmental knowledge (TEK) research, collaboration with other long-established and tribal communities, and the formation of organizations that bring voice and knowledge to political and legal decision making. The community and others are aware that the rapid onset of current disasters and changes in the physical surroundings means that TEK, although essential in relationship to resources, is no longer enough to navigate these changes. Creation of new knowledges and the adaptation of TEK with other coastal communities is becoming both a political and survival mechanism. The families and the individuals that give of themselves to develop knowledge and adaptation skills would much rather be with their community and families and enjoy life as they knew it before human-induced devastation occurred.

Grand Bayou, as well as other coastal communities in the Gulf of Mexico region, is committed to adaptation in place as long as it is humanly possible—but few options exist that are available for people whose subsistence and identity is place based on estuaries. To move would be to separate community from a place of familiarity because of the lack of land and its lessening affordability as people move "up-bayou" from the lands that are drowning. People who do move have to go into other locations and cultures not compatible to subsistence and tribal practices. Historically, the tribes and the historied communities had flexibility for moving during traditionally special seasons of hunting and fishing. But these practices became more difficult as Western development competed for land and the oil and gas industries interfered with fishing resources. As the coast familiar to inhabitants of Grand

Bayou becomes like that of other coastlines—gentrified and industrialized—people are experiencing angst over their future. Some communities are seeking to move farther inland, because their families are being separated from their subsistence life; but without a state or federal resettlement program in place, moving becomes the financial responsibility of the subsistence community. Other communities, similarly to the Grand Bayou, want to stay in place in order to continue to address the issues of climate change and to maintain their cultural integrity and preserve traditional ways. This chapter explores the difficulty of being in a situation that has few acceptable alternatives; it examines how life might be between the time of staying and the time of eventually leaving.

Background

The rapid rate of environmental change in coastal Louisiana, especially of land loss, is outpacing the traditional means of adaptation of many communities who have lived in the region for centuries. In response to the devastating land loss, Louisiana has developed a Comprehensive Master Plan for a Sustainable Coast (CPRA 2012), the state's directive for restoration projects in a quickly disappearing coastal region. However, the CPRA pays little attention to the well-being of some communities that are most at risk of catastrophic land loss in coastal Louisiana's bayou region (Dalbom, Hemmerling, and Lewis 2014). In areas excluded from the state-led restoration activities or levee systems, adaptation measures are mostly left to the imagination of the people living there.

The high level of vulnerability communities have to erosion, land loss, storms, and toxins, has, for the most part, resulted from extractive-industry-driven economic and political entities. The settlements along the bayous and in water-placed locations were previously safely inland, protected by cheniers (ridges formed by sediment deposits), tree ridges, miles of marsh, and the communities' own mutual aid and reciprocity. However, previously sustainable, resilient communities have been placed in harm's way through the loss of land between them and the Gulf of Mexico owing to the industrial buildup of the region and the growth machine of oil, gas, and other industries (Figure 21.1).

Now the once largely self-sufficient communities are squeezed onto less land with fewer resources, such as farming, hunting, and fishing, for their subsistence. The communities are forced to decide whether to stay in place to protect what is left of their historical ancestral lands, lifeworlds, and livelihoods; to relocate as individuals; or to resettle as a community farther north of proposed levee protection. The issue is complicated by race and class differences, rising costs of real estate and rural gentrification (Solet 2006; see Figure 21.2), soaring rates of homeowner and flood insurance, and the lack of governance systems with a legal or policy structure to address resettlement for communities as a whole.

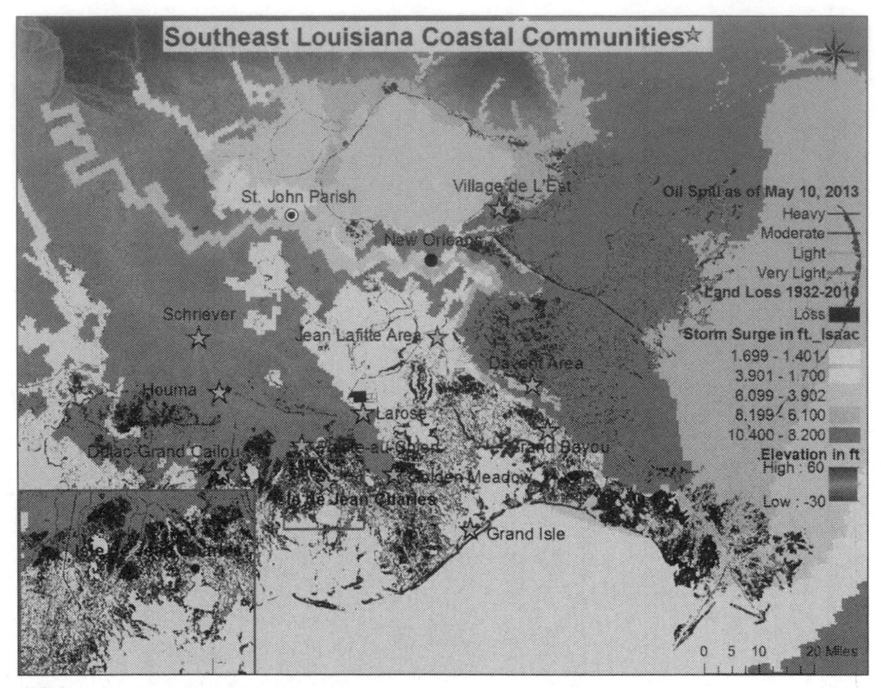

Figure 21.1 Map of southeast Louisiana coastal communities[1] (created by M. Reddy Avula, University of New Orleans, 2015)

Figure 21.2 Example of rural gentrification (photo by Kristina Peterson)

Using ethnographic and participatory action research (PAR) conducted with communities in the region, this chapter considers (1) the economic and political forces that have caused the environmental degradation and communities' socially constructed vulnerability, (2) the consequences of the layers of disasters, (3) the adaptation strategies communities are employing, and (4) the actions needed to support communities if the only option left to maintain community and culture is relocation.

1. Economic and Political Forces

Over 2 million people call coastal Louisiana home, and their lives are intricately intertwined with the region's ecosystems (CPRA 2012). These elaborate networks of habitats and landforms are connected in the world's seventh-largest delta and lie at the end of the Mississippi River, the fourth-largest drainage basin in the world (CLEAR 2006; Campanella 2010; Couvillion et al. 2011). Multiple forces have transformed this complex web of land, water, and communities: oil and gas corporations dredging canals to lay thousands of miles of pipelines; the construction of levees, and dams; cypress logging; large-scale agricultural development; anthropogenic-caused climate change; hurricanes; and oil disasters have led to drastic coastal erosion, saltwater intrusion, subsidence, land loss, and environmental degradation, meaning that many coastal residents have no choice but to relocate (Barry 1997; Laska et al. 2005; Austin 2006; Button and Peterson 2009; Freudenberg et al. 2009). Over 2,000 square miles of delta land have become water (Marshall 2014), altering the biological and ecosystem characteristics of the areas and perpetuating social and cultural consequences, many of which are likely irreversible. For example, Leeville, a once thriving community, recently gave up its effort to save its cemetery to the rising waters (Wilson 2014; see Figure 21.3).

Additionally, communities have experienced rapidly increasing rates of diabetes, cancer, and high blood pressure. Decades-long industrial contamination has led to areas of the region being dubbed "cancer ally" (Bullard 1993; see Figure 21.4). Encroaching toxic industries and chemicals from dispersants, oil disasters, and poststorm debris contaminate the communities' air, soil, and water, creating severe adverse health and livelihood effects, along with a forced change in diet (Laska and Pine 2007; Coastal Louisiana Tribal Communities 2012; Maldonado et al. 2013).

It is the sum of these interacting forces that has increased residents' vulnerability to displacement. It is not the single hurricane, oil spill, or event of sea -level rise alone that forces people to relocate but rather the combination and layering of disasters and vulnerabilities that displace people (Laska et al. 2015). With each additional hurricane, oil disaster, and/or continuing degradation to the water and other resources important to residents' livelihoods, more people are forced to relocate. This necessity, in turn, threatens the cultural sovereignty and social fabric

Figure 21.3 Highway 1 near Leeville, Louisiana (photo by Kristina Peterson)

Figure 21.4 Image of cancer alley petrochemical plant, Louisiana (photo by Kristina Peterson)

of communities, many of which have spent centuries dwelling in-place. Coastal Louisiana had been a region where Native American, Acadian, Vietnamese, Caribbean, Isleños, and other population groups sought refuge from marginalizing political and social forces. Even though oil was first drilled in 1868 the vast amount of drilling did not start until 1940, and the introduction of offshore platforms in the late 1940s, transforming the coastal area into an energy sacrifice zone—"a place where human lives are valued less than the natural resources that can be extracted from the region" (Buckley and Allen 2011: 171; also Colten 2012). The region can be considered among "the new geographies of domination" (Reid and Taylor 2010: 11) in which increasingly at-risk populations become even further marginalized (Peterson 2005).

2. Consequences of Layered Disasters

Coastal Louisiana's residents often experience "toxic uncertainty" (Auyero and Swistun 2009: 66) arising from the multitude of contamination events of unknown toxic chemicals and pollutants over the last 80 years from oil and gas extraction and the use of toxins to clean up oil spills. Such contamination was compounded by the 2010 British Petroleum (BP) Deepwater Horizon oil disaster, during which nearly 5 million barrels of oil spilled into the Gulf of Mexico (National Commission on the BP Deepwater Horizon Oil Spill and Offshore Drilling 2011), and approximately 1.84 million gallons of toxic dispersant were sprayed in the Gulf waters by boats and airplanes to sink the oil (U.S. Coast Guard 2011). With each storm that hits, residents are left in fear of oil and toxic chemicals washing ashore (Figure 21.5). Three and a half years after the Deepwater Horizon's Macondo well explosion, scientists found a layer of oil the size of the state of Rhode Island surrounding the site of the disaster (Valentine 2014).

Many residents are also worried about the future health effects as the consequences of toxic chemicals and contamination become more prevalent and understood. The Louisiana Bucket Brigade has been monitoring air

Figure 21.5 Cleanup workers place a boom around Isle de Jean Charles to try to prevent oil from the BP Deepwater Horizon Oil Disaster coming ashore (photo by Julie Maldonado, 2010)

quality in the region for many years and has charted an increase of reported spills and gas released. In 2009 there were 3,636 reported spills/releases, and that number increased to 4,888 in 2010. All these releases have a serious impact on human and environmental health (Lasley 2010). For example, the BP disaster in 2010 affected approximately 1,100 miles of coastal wetlands, and sediment erosion increased where the oil damaged vegetation and root systems (National Academy of Sciences 2013). The diminished health of people and the marshes have affected the overall well-being of the subsistent coastal communities, adding an additional layer of stressors.

Historically, coastal Louisiana communities were subsistence-based, predominantly engaged in fishing, trapping, farming, and hunting. However, since the 1960s the effects of oil- and gas-related development, changing waterways, and, increasingly, climate change have resulted in a loss of those historic livelihoods. The farming, trapping, and hunting lands are inundated by saltwater, eroded, or completely gone, and the seafood has been affected by oil disasters, saltwater intrusion, toxic chemicals, and the neoliberal policies driving the global seafood industry (Harrison 2012; Maldonado 2014).

Additionally, the majority of jobs available are with the oil and gas industries and supportive industries—a situation that has moral implications for inhabitants, who have to then work for the very industries that are responsible for the environmental degradation of their homeland and for their original livelihood loss. Working "on the clock" restricts the flexible time available for community residents to be able to participate in time-honored practices of reciprocity and mutual aid. Retaining livelihoods and community rituals is increasingly difficult, and, when it is unsuccessful, losing shared resources changes community dynamics—people no longer gather and share as they did, and the everyday traditions and cultural practices that tied families and communities together gradually diminish (Coastal Louisiana Tribal Communities 2012).

The loss of place, or placelessness, that results from loss of cultural space and identity (Cernea 1999), affects residents' cultural identity and practices for current and future generations (Fullilove 2005). Such loss of place, subsistence, and traditional knowledge is often manifested through feelings of *solastalgia*, which is "the distress that is produced by environmental change impacting people while they are directly connected to their home environment" (Albrecht et al. 2007). Watching your lifeworld disappear and being ignored in the fight to save it causes solastalgia and is anger-producing:

> It is like having a person who needs emergency care, and you can see a hospital on the horizon but nobody wants to help get you there. Well that is what it is like for our wetlands; people want the oil and gas, but they don't care about what it is doing to destroy the land and the people. We have a right to be here. (P. Bergeron, Bayou Blue, Louisiana 2006)

In activities since Hurricane Katrina, residents report a number of damaging social and cultural effects, including the loss of a shared family livelihood; diminished sharing and resource exchange; loss of traditional medicine; loss of traditions and cultural practices; and loss of social networks, social cohesion, reciprocity, and mutual aid. As the land and waterscape where residents carried out their livelihoods and cultural practices rapidly changes (Figure 21.6), many people experience a sense of dislocation without leaving their community, because of a shift in their sense of place and lifeworlds as their connection to the water and land diminishes. Fishing, farming, hunting, and trapping on the same lands and in the same waters for generations connected people, but the loss of subsistence together with severe water and landscape changes have led to residents expressing feelings of dislocation and uncertainty that pervade their everyday lives (Maldonado 2014). When asked what they would want if they had a magic wand, tribal leaders from several coastal communities said, "to restore their lands" (Peterson 2014).

3. Adaptations and Collaborations

While experiencing loss, communities are often challenged to reinvigorate their cultural identities and to recover their historical roots as a method to move forward. Thus many communities and leaders have taken action to mitigate such social and cultural losses, including some historied and tribal communities in coastal Louisiana who have renewed their cultural

Figure 21.6 Ghost forest, coastal Louisiana (photo by Kristina Peterson)

traditions and maintained their food and community traditions (Laska et al. 2010; Maldonado et al. 2015). For example, the fast-disappearing area of lower St. Bernard Parish was home to the historical communities of the Isleños, settlers from the Canary Islands. As the Gulf waters took their lands, the communities worked diligently to preserve their culture through festivals, museum exhibits, and the collection of oral histories (West 2009).

Many residents have pursued adaptation strategies, such as elevating houses above the floodwaters, owing to increased flooding and a lack of protection from diminished wetlands and levees (Laska 2012; see Figure 21.7). However, elevating houses puts residents, especially those with special needs, at risk; unless means of entry is universal for all houses, the continuity of community relationships is broken and contributes to the loss of reciprocity (Laska et al. 2010, Laska 2012; Maldonado et al. 2015). Some residents are implementing alternative farming and protection methods, such as using raised-bed gardens and Hesco baskets to plant trees, medicinal plants, and vegetables (Heifer Project 2007; Coastal Louisiana Tribal Communities 2012; see Figure 21.8). The communities are cultivating and saving seeds that are salt-tolerant and participating in ethnobotany and phenology work as a way to secure against the loss of critical plants.

Figure 21.7 House elevated above the floodwaters from Hurricane Isaac, Isle de Jean Charles (photo by Julie Maldonado, 2012)

Figure 21.8 Dirt and debris used for raised-bed gardens in Grand Bayou, Louisiana (photo courtesy of Kristina Peterson, 2011)

Another adaptation strategy includes forming collaborations and partnerships across communities and with a variety of agencies. In 2009, at the request of Fr. Nguyen, then priest of Mary Queen of Vietnam Church, Kristina Peterson coordinated with her graduate students and Shirley Laska (University of New Orleans Center for Hazards Assessment, Response, and Technology) a gathering of coastal communities called the Cross Coastal Forum (CCF). This gathering was the result of people's desire to know other coastal communities with diverse ethnic backgrounds and from multiple bayous. This event, held in the East New Orleans Vietnamese community, brought 15 communities together for the first time for a meal and to share common concerns over conversations and coastal maps. The gathering helped to establish a level of trust that facilitated communication essential for building coalitions following the Deepwater Horizon oil spill and the ensuing storms and hurricanes in the region. Building trust through collaborations became critical for the work that was done by the Gulf Organized Fisheries in Solidarity and Hope (GoFish) and a group working to form a Regional Citizens Advisory Council like the one formed in Prince William Sound following the *Exxon Valdez* oil disaster in March 1989. (Some community members that attended the CCF traveled together to Alaska to learn from the Prince William Sound Regional Citizens Advisory Council.) Such gatherings and knowledge sharing helped to form bonds of friendship, establish partnerships, and provide mutual aid, especially for the coastal populations most at risk.

Other coalitions that were instrumental in getting people together included the Gulf Coast Fund, which supported gatherings based on issues and coalition-building. These collaborations continued through the support of the Lowlander Center and a Presbyterian congregation, with representatives from Louisiana and Alaska at-risk communities coming together to share stories, knowledge, and lessons learned about the damage caused by extractive industries and human-induced environmental changes (Peterson 2011; Maldonado et al. 2013).

The emerging friendships bonded by mutual issues of land loss, splintering of communities, and the layering of disasters have fostered many additional forms of collaboration. For example, through the support of the U.S. Department of Agriculture's Natural Resources Conservation Service (USDA-NRCS), Wisconsin Tribal Conservation Advisory Council, and the Lowlander Center, four Native American communities in coastal Louisiana created the First People's Conservation Council to address natural resource issues occurring in and around tribal and coastal lands (Jessie Noyes 2013; Maldonado et al. 2015; see Figure 21.9).

Residents' multigenerational knowledge of the water and landscape and weather patterns can help to identify adaptation strategies to maintain

Figure 21.9 First Peoples' Conservation Council leaders are introduced to a local USDA-NRCS plant materials center to work on vegetation restoration (photo by Julie Maldonado, 2012)

cultural sovereignty. For example, the Native American tribes who have spent centuries dwelling in coastal Louisiana living off the land and water "are seeing the changes that are happening, whereas much of the rest of Western society is not paying close attention" (Petersen 2014: 4). In fact, "very few [Western] people have lived in a place long enough within a community to know what's normal and what isn't normal, what's changed and what hasn't changed" (Wildcat 2014: 4). By living in flexible patterns that changed with the seasons and physical locations the tribes that predated the Western colonization of the region utilized the bayous and land in ways that assured the restoration of animals and fish. These patterns continued until as recently as the 1960s, reflected in bayou students' report cards, which indicated time away from school as "gone trapping" (Philippe 2004), which was true for not only the tribes but also for many other long-dwelling, subsistence-based communities in coastal Louisiana. However, as the environment rapidly changes and degrades, people's traditional knowledge about the environment is not always applicable. The socially constructed vulnerabilities and environmental changes require decision making under increasingly uncertain and complex conditions (Austin 2004; also Watts 1983).

4. Actions: Facing Displacement and Relocation

Community leaders and chiefs from the at-risk coastal communities face the challenge of doing what they can, with limited time and resources to maintain their communities in place as long as possible. At the same time they recognize that community-led relocation plans take many years to develop and carry out, so they must also plan toward relocation. But if community leaders decide that the community is too much at risk to continue to adapt in-place and that they must pursue relocation, the community could be cut off from funding to support their existing infrastructure. For example, the villages of Newtok and Shishmaref, Alaska, voted to relocate both in 1994 and in 2001 but have not been able to move (see the chapter by Marino and Schweitzer in this volume). And they were cut off from state and federal funding for much needed infrastructure once they declared the resolve to relocate (Tom 2010; Rising Voices 2014).

Louisiana's Comprehensive Master Plan for a Sustainable Coast had the option of voluntary relocation for individual households (CPRA 2012), but relocation as a community was not included in the plan (Dalbom 2014). The state's offering of support only for individual relocation will further scatter communities, tear apart social fabric, and can lead to the loss of cultural practices that connect people. Furthermore, as the result of benefit-cost analysis, some communities or parts of communities were left out of the plan. For example, Isle de Jean Charles, a tribe in southeast Louisiana that has experienced near complete community land loss, was included in the original plans for the U.S. Army Corps of Engineers-led Morganza to the Gulf [of Mexico] Hurricane Protection System, a project included in the Master Plan for a Sustainable Coast, but the community was excluded from the plans several years later because the U.S. Army Corps of Engineers determined that it was not cost efficient to include it. Such analysis and determination of who is included in hurricane protection systems ignore significant social and cultural losses, as well as the equitable distribution of costs and benefits (Cernea 2000; Sand 2012; Maldonado et al. 2013).

Some leaders from coastal communities, such as the Isle de Jean Charles Tribe, have been actively trying to relocate their communities for nearly two decades. With so many people having relocated individually, tribal leaders fear cultural genocide—their communities will cease to exist if residents are geographically separated. While the Isle de Jean Charles Tribe is working to bring people back together through a sustainable resettlement plan, they are also working to mitigate further degradation of their current and ancestral lands (Laska et al. 2014). Community leaders are challenged in pursuing community-led relocation, owing to the lack of both a federal government agency mandated to manage communities' relocation efforts and funds/institutional support to move a community (Bronen 2011; Maldonado et al. 2013; Bennett et al. 2014). These layers of disasters and socially constructed vulnerabilities (Wisner et al. 2003) are decreasing many communities' options to relocate, and fewer and fewer options exist for in-situ adaptation (Laska and Peterson 2013; Maldonado 2014). However, several communities in Alaska and Louisiana may become models for climate adaptation resettlements if funded through the HUD NDRC (Housing and Urban Development National Disaster Reduction Competition). These proposed pilots might be able to help develop models needed for federal policy change (Peterson 2015 HUD NDRC Proposal).

Conclusion

Ethnographic and PAR work with communities in the Southeast Louisiana coastal region has demonstrated that it is not a single, isolated event that leads to devastation but rather layers of disasters and socially constructed vulnerability, on top of a long-standing foundation of extractive-industry-driven economic and political forces, that more often place communities in harm's way. However, despite constraints and oppression, community leaders and members can take actions to maintain their communities and cultures.

A socially just adaptation process needs to be forged that includes "indigenuity" (Wildcat 2009, 2013), equitable knowledge sharing, and traditional knowledge in decision making in a way that respects and honors the knowledge being shared. Adaptation needs to include a people-centered framework in which human and environmental rights are upheld. Relocation as a form of adaptation cannot become an excuse for government and private interests to either ignore or force communities out of their homes (Maldonado 2014). Community resettlement, as opposed to individual relocation, is preferred by many of the traditional, tribal, and historied communities along the coast (Peterson 2015 HUD NDRC Proposal).

Anthropologists play a crucial role in the establishment of resettlement and adaptation measures. Often not included in the agency and academic mix—except for archeological concerns for historic sites—issues pertinent to culture and cultural survival need to be included in deliberation as well as community/tribal concerns, to assure that the people are not overlooked.

A White House report pointed to the important social and environmental elements that cannot be given a monetary value and are at significant threat of loss if policy actions to mitigate climate change are delayed (White House Council of Economic Advisers 2014). Direct action needs to be taken, exemplified by the recommendation by Rising Voices, a community of engaged Indigenous and non-Indigenous leaders, environmental experts, students, and scientific processionals, to the Presidential Task Force on Climate Preparedness and Resilience, which states: "Convene a Climate Migration Task Force. A Federal Task Force on Climate Migration would address the identified need to establish a legal mechanism, institutional framework, and financial support to directly support marginalized communities (Indigenous and non-Indigenous) who are facing displacement due to climate change impacts and who desire to migrate safely and with dignity" (Rising Voices 2014).

Dislocation and the loss of cultural practices, traditions, and ways of life are occurring throughout coastal Louisiana. And although coastal Louisiana faces the highest rate of relative sea-level rise (the combined phenomena of land subsidence and sea level rising) worldwide (Marshall 2013; Osborn 2013), its communities are not alone in being at risk of displacement. Climate change effects such as sea-level rise, occurring in conjunction with other human-induced environmental, economic, and sociocultural changes, are placing communities in coastal and low-lying locations around the world at risk of inundation and displacement.

As we envision a future that does not sacrifice communities but that respects people's traditional forms of knowledge, "it's important to apply the insights and knowledge native people hold today when we think about planning or adapting" (Wildcat 2014: 3). Making this cultural shift is one significant step in addressing the issues that communities now face when they consider whether they can continue to adapt in-place or when they may have to relocate—and what happens between now and then. Until the Western world changes its anthropocentric framing of human relationships with the environment, adaptation to the climate crisis will need serious attention and enaction, which should come in the form of embracing diverse forms of knowledge.

Note

1. References for Figure 21.1, Map of Southeast Louisiana Coastal Communities:
   Maximum Oiling Status during the Gulf Water Horizon Oil Spill, accessed January 14, 2015, https://usresponserestoration.files.wordpress.com/2012/11/erma-gulf-response-oiling-composite-sar-index_large.png.
   Land Loss of Coastal Louisiana from 1932–2010, accessed January 15, 2015, http://pubs.usgs.gov/sim/3164/downloads/.
   Storm Surge during Hurricane Isaac, accessed January 15, 2015, www.nhc.noaa.gov/gis/archive_psurge_results.php?id=al09&year=2012&name= Hurricane%20 ISAAC.
   Louisiana Digital Elevation Dataset, January 14, 2015, http://catalog.data.gov/dataset/louisiana-digital-elevation-dataset-from-ldeq-source-data-utm-zone-15-nad83-losco-2004-24k-2004#.

References

Albrecht, G., Sartore, G.-M., Connor, L., Higginbotham, N., Freeman, S., Kelly, B., Stain, H., Tonna, A., and Pollard, G. 2007. Solastalgia: The distress caused by environmental change. *The Royal Australian and New Zealand College of Psychiatrists* 15: S95–98.

Austin, D. 2004. Partnerships, not projects! Improving the environment through collaborative research and action. *Human Organization* 63(4): 419–30.

———. 2006. Cultural exploitation, land loss, and hurricanes: A recipe for disaster. *American Anthropologist* 108(4): 671–91.

Auyero, J., and Swistun, D. A. 2009. *Flammable: Environmental Suffering in an Argentine Shantytown*. Oxford: Oxford University Press.

Barry, J. M. 1997. *Rising Tide: The Great Mississippi Flood of 1927 and How It Changed America*. New York: Simon and Schuster.

Bennett, T. M. B., Maynard, N. G., Cochran, P., Gough, R., Lynn, K., Maldonado, J., Voggesser, G., Wotkyns, S., and Cozzetto, K. 2014. Indigenous peoples, lands, and resources, in J. M. Melillo, T. Richmond, and G. W. Yohe (Eds.), *Climate Change Impacts in the United States: The Third National Climate Assessment*. Washington, D.C.: U.S. Global Change Research Program, pp. 297–317.

Bronen, R. 2011. Climate-induced community relocations: Creating an adaptive governance framework based in human rights doctrine. *New York University Review of Law and Social Change* 35: 356–406.

Buckley, G. L., and Allen, L. 2011. Stories about mountaintop removal in the Appalachian coalfields, in M. Morrone, G. L. Buckley, and J. Purdy (Eds.), *Mountains of Injustice: Social and Environmental Justice in Appalachia*. Athens: Ohio University Press, pp. 161–80.

Bullard, R. (Ed.). 1993. *Confronting Environmental Racism: Voices From the Grassroots*. Boston: South End Press.

Button, G. V., and Peterson, K. 2009. Participatory action research: Community partnership with social and physical scientists, in S. A. Crate and M. Nuttall (Eds.), *Anthropology and Climate Change: From Encounters to Actions*. Walnut Creek, CA: Left Coast Press, Inc., pp. 327–40.

Campanella, R. 2010. *Delta Urbanism: New Orleans*. Chicago: American Planning Association.

Cernea, M. M. 1999. Why economic analysis is essential to resettlement: A sociologist's vbiew, in M. M. Cernea (Ed.), *The Economics of Involuntary Resettlement: Questions and Challenges*. Washington, D.C.: The World Bank, pp. 5–49.

———. 2000. Risks, safeguards, and reconstruction: A model for population displacement and resettlement. *Economic and Political Weekly* 35(41): 3659–78.

CLEAR (Coastal Louisiana Ecosystem Assessment and Restoration). 2006 *Reducing Flood Damage in Coastal Louisiana: Communities, Culture and Commerce*, www.clear.lsu.edu, accessed November 18, 2014.

Coastal Louisiana Tribal Communities. 2012. *Stories of Change: Coastal Louisiana Tribal Communities' Experiences of a Transforming Environment (Grand Bayou, Grand Caillou/ Dulac, Isle de Jean Charles, Pointe-au-Chien)*. Workshop report input into the National Climate Assessment. Pointe-aux-Chenes, LA: January 22–27.

Colten, C. E. 2012. An incomplete solution: Oil and water in Louisiana. *The Journal of American History* 99(1): 91–99.

Couvillion, B. A., Barras, J. A., Steyer, G. D., Sleavin, W., Fischer, M., Beck, H., Trahan, N., Griffin, B., and Heckman, D. 2011. *Land Area Change in Coastal Louisiana from 1932 to 2010: U.S. Geological Survey Scientific Investigations Map 3164*, http://pubs.usgs.gov/sim/3164/downloads/SIM3164_Pamphlet.pdf, accessed July 27, 2014.

CPRA (Coastal Protection and Restoration Authority of Louisiana). 2012. *Louisiana's Comprehensive Master Plan for a Sustainable Coast*. Baton Rouge, LA: CPRA, www.lacpra.org/assets/docs/2012%20Master%20Plan/Final%20Plan/2012%20Coastal%20Master%20Plan.pdf, accessed December 4, 2014.

Dalbom, C., Hemmerling, S. A., and Lewis, J. A. 2014. *Community Resettlement Prospects in Southeast Louisiana: A Multidisciplinary Exploration of Legal, Cultural, and Demographic Aspects of Moving Individuals and Communities*. New Orleans, LA: Tulane Institute on Water Resources Law and Policy.

Fullilove, M. Thompson. 2005. *Root Shock: How Tearing Up City Neighborhoods Hurts America and What We Can Do about It*. New York: One World Press.

Freudenberg, W. R., Gramling, R., Laska, S., and Erikson, K. T. 2009. *Catastrophe in the Making: The Engineering of Katrina and the Disasters of Tomorrow*. Washington, D.C.: Island Press.

Harrison, J. A. 2012. *Buoyancy on the Bayou: Shrimpers Face the Rising Tide of Globalization*. Ithaca, NY: Cornell University Press.

Heifer Project. 2007. *World Ark Magazine*. Little Rock, AR: Heifer International.

Jessie Noyes Foundation. 2013. *Newsletter*, Spring. New York.

Laska, S. 2012. Dimensions of resiliency: Essential, exceptional, and scale. *International Journal of Critical Infrastructure* 6(3): 246–76.

Laska, S., Laska, T., Gough, B., Martin, J., Naquin, A., and Peterson, K. 2014. Proposal for Isle de Jean Charles relocation planning. *Indigenous Roots for Sustainable Futures: Proactive Solutions for a Time of Change*.

Laska, S., and Peterson, K. 2013. *Between Now and Then: Tackling the Conundrum of Climate Change*. CHART Publications, Paper 32: 5–8, http://scholarworks.uno.edu/chart_pubs/32.

Laska, S., Peterson, K., Alcina, M. E., West, J., Volion, A., Tranchina, B., and Krajeski, R. 2010. *Enhancing Gulf of Mexico Coastal Communities' Resiliency through Participatory Community Engagement*. CHART Publications, Paper 21, http://scholarworks.uno.edu/chart_pubs/21.

Laska, S., Peterson, K., Rodrigue, C., Cosse, T., Philippe, R., Burchett, O., and Krajeski, R. 2015. "Layering" of natural and human-caused disasters in the context of anticipated climate change disasters: The coastal Louisiana experience, in M. Companion (Ed.), *Disasters' Impact on Livelihood and Cultural Survival: Losses, Opportunities, and Mitigation*. Boca Raton, FL: CRC Press, pp. 225–38.

Laska, S., and Pine, P. 2007. *Collaborative Research—Small Grant for Exploratory Research (SGER): An Interdisciplinary Approach to Coastal Vulnerability and Community Sustainability*, Shirley Laska, Awardee: University of New Orleans Award Number: 0408496.

Laska, S., Wooddell, G., Hagelman, R., Grambling, R., and Teets Farris, M. 2005. At risk: The human, community, and infrastructure resources of coastal Louisiana. *Journal of Coastal Research* 44: 90–111.

Lasley, C. B. 2010. *Gas and Oil Release Map*. Poster made for the Louisiana Bucket Brigade.
Maldonado, J. K. 2014. *Facing the Rising Tide: Co-Occurring Disasters, Displacement, and Adaptation in Coastal Louisiana's Tribal Communities*. Ph.D. Dissertation, Department of Anthropology, American University.
Maldonado, J. K., Naquin, A. P., Dardar, T., Parfait-Dardar, S., and Bagwell, K. 2015. Above the rising tide: Coastal Louisiana's tribal communities apply local strategies and knowledge to adapt to rapid environmental change, in M. Companion (Ed.), *Disasters' Impact on Livelihood and Cultural Survival: Losses, Opportunities, and Mitigation*. Boca Raton, FL: CRC Press, pp. 239–53.
Maldonado, J. K., Shearer, C, Bronen, R., Peterson, K., and Lazrus, H. 2013. The impact of climate change on tribal communities in the U.S.: Displacement, relocation, and human rights. *Climatic Change* 120(3): 601–14. Reprinted in J. K. Maldonado, R. Pandya, and B. Colombi (Eds.), 2014, *Climate Change and Indigenous Peoples in the United States: Impacts, Experiences, and Actions*. Cham, Switzerland: Springer International Publishing.
Marshall, B. 2013. New research: Louisiana coast faces highest rate of sea-level rise worldwide. *The Lens,* February 21, http://thelensnola.org/2013/02/21/new-research-louisiana-coast-faceshighest-rate-of-sea-level-rise-on-the-planet/.
———. 2014. Losing Louisiana, August 28, *Pro-Publica*; *The Lens*.
National Academy of Sciences. 2013. Assessing impacts of the Deepwater Horizon Oil spill in the Gulf of Mexico. *Science Daily*, July 10, www.sciencedaily.com/releases/2013/07/130710122004.htm.
National Commission on the BP Deepwater Horizon Oil Spill and Offshore Drilling. 2011. *Deep Water: The Gulf Oil Disaster and the Future of Offshore Drilling. Report to the President,* www.gpo.gov/fdsys/pkg/GPO-OILCOMMISSION/pdf/GPOOILCOMMISSION.pdf, accessed July 22, 2014.
Osborn, T. 2013. Keynote Comments: *Critical Needs for Community Resilience*. Presentation at the Building Resilience Workshop IV: Adapting to Uncertainty Implementing Resilience in Times of Change, March 2–9, New Orleans, LA.
Philippe, R. 2004. Oral history project, Grand Bayou, Louisiana.
Petersen, S. 2014. Lessons in resilience: How indigenous tribes are helping lead the way on climate change. *NOAA Coastal Services*, April/May/June 2014 17(2).
Peterson, K. J. 2005. *Participatory Action Research: The Case of Recovery in Grand Bayou, Louisiana*. Presentation at the Natural Hazards Research and Applications Workshop, July 10–13, Broomfield, CO.
———. 2011. *Transforming Researchers and Practitioners: The Unanticipated Consequences (Significance) of Participatory Action Research (PAR)*. Ph.D. dissertation, Department of Urban Studies, University of New Orleans.
———. 2014. Community conversation with Point au Chien Tribe and the OGHS staff from Presbyterian Church (USA), Tribal Center, October 28, 2014.
———. 2015. HUD-NDRC Phase II proposal, https://www.hudexchange.info/news/ndrc-webinar-series/.
Reid, H., and Taylor, B. 2010. *Recovering the Commons: Democracy, Place, and Global Justice*. Urbana: University of Illinois Press.
Rising Voices. 2014. *Adaptation to Climate Change and Variability: Bringing Together Science and Indigenous Ways of Knowing to Create Positive Solutions*. Workshop Report, National Center for Atmospheric Research, Boulder, CO, June 30–July 2, www.mmm.ucar.edu/projects/RisingVoices, accessed January 5, 2015.
Sand, M. G. 2012. *Transforming Sustainability through Adaptive Co-Management: A Critique of Louisiana's Coastal Master Plan*. Master's Thesis, Department of Urban and Regional Planning, University of New Orleans.

Solet, K. 2006. *Thirty Years of Change: How Subdivisions on Stilts Have Altered a Southeast Louisiana Parish's Coast, Landscape, and People.* Master's thesis, Department of Urban Studies, University of New Orleans.

Tom, S. 2010. Center for Natural Resources and Economic Policy Conference, May 2010, New Orleans.

U.S. Army Corps of Engineers, Louisiana Coastal Protection and Restoration Authority Board, and Terrebonne Levee and Conservation District. 2013. *Summary of the Morganza to the Gulf of Mexico, Louisiana Final Post Authorization Change Report,* www.mvn.usace.army.mil/Portals/56/docs/PD/Projects/MTG/M2GPACReportMay2013.pdf, accessed July 27, 2014.

U.S. Coast Guard. 2011. *BP Deepwater Horizon Oil Spill: Incident Specific Preparedness Review,* www.uscg.mil/foia/docs/DWH/BPDWH.pdf, accessed June 4, 2014.

U.S. Geological Survey. 2004. *Southeast Louisiana Land Loss: Historical and Projected Land Loss in the Deltaic Plain.* Lafayette, LA: U.S. Geological Survey, National Wetlands Research Center, www.nwrc.usgs.gov/factshts/2005-3101/2005-3101-figure2.htm accessed July 18, 2014.

Valentine, D. 2014. *Proceedings of the National Academy of Sciences*, October 27.

Watts, M. 1983. The political economy of climatic hazards: A village perspective on drought and peasant economy in a semi-arid region of West Africa. *Cahiers d'Études Africaines* 23(89/90): 37–72.

West, J. J. 2009. *Negotiating Heritage: Heritage Organizations amongst the Isleños of St. Bernard Parish, Louisiana, and the Use of Heritage Identity to Overcome the Isleño/Tornero Distinction.* Master's Thesis, University of New Orleans, http://scholarworks.uno.edu/td/960/.

White House Council of Economic Advisers. 2014. *The Cost of Delaying Action to Stem Climate Change,* www.whitehouse.gov/sites/default/files/docs/the_cost_of_delaying_action_to_stem_climate_change.pdf,,accessed December 17, 2014.

Wildcat, Daniel R. 2009. *Red Alert!: Saving the Planet with Indigenous Knowledge.* Golden, CO: Fulcrum Publishing.

Wildcat, D. R. 2013. Introduction: Climate change and indigenous peoples of the USA. *Climatic Change* 120(3): 509–15.

—— . 2014. *NOAA Coastal Services* 17(2).

Wilson, X. 2014. Group gives up on saving Leeville Cemetery from the water. *The Houma Courier*, October 28.

Wisner, B., Blaikie, P., Cannon, T., Davis, I. 2003. *At Risk: Natural Hazards, People's Vulnerability, and Disasters.* London: Routledge.

Chapter 22

Narwhal Hunters, Seismic Surveys, and the Middle Ice: Monitoring Environmental Change in Greenland's Melville Bay

Mark Nuttall

Extreme weather events have become increasingly common for the hunters who travel, move, and camp around the coasts, islands, and peninsulas of Melville Bay in northwest Greenland, whether on the sea ice by dog sledge in winter and spring or by small open boats and kayaks in summer and autumn. The hunters remark how shifting currents, more powerful waves, stronger and sharper-edged winds, and seemingly unyielding fierce storms are becoming a part of daily life, and they experience far more difficult ice conditions than they did even just several years ago. They notice that the many glaciers descending from the inland ice to the coast are receding at a fast rate and calving smaller icebergs, that polar bears are found in different places than previously cited, and that marine mammals are either moving into areas that they did not previously frequent or are exhibiting unusual behavior. Some species, such as orca whales and Greenland right whales, are increasingly observed in places where they have not been seen in living memory.

These changes in the environment and in the movement of animals, while considered interesting, astonishing, and sometimes alarming, are the subjects of much household conversation in the communities of Savissivik and Kullorsuaq and in the Melville Bay hunting camps where residents from these northern villages meet. They are also consistent with what people say about weather and animals elsewhere in northwest Greenland (for example, Hastrup 2009 and this volume; Nuttall 2009). However, contrary to popular understandings of climate change effects in the Arctic, such changes do not necessarily constrain or affect people's lives or affect them in ways that are considered to bring dramatic shifts or impending catastrophe. People remark that, although travel on noticeably thinner sea ice in the winter or navigating across open water or through iceberg-choked stretches of sea in ferocious summer storms gives cause for concern and sometimes makes

Anthropology and Climate Change: From Actions to Transformations (2nd ed.) by Susan A. Crate and Mark Nuttall (Eds.)

them anxious, they have to find ways of approaching and dealing with such conditions safely as they become regular occurrences. Furthermore, when animals such as narwhals move around the coast in ways that are considered unusual compared to how they usually migrate, people remark that this phenomenon could be the result of something that is happening "out there in the water," or it could be entirely consistent with how narwhals behave, given that they are notoriously skittish. Hunters say nothing can be taken for granted, not least animals.

What *is* troubling for hunters is the imposition of quotas and regulations that restrict their abilities to pursue certain marine mammals (such as narwhals and belugas, or to fish Greenland halibut) and the presence of seismic survey vessels and people who work in the oil and minerals industries arriving in the north by ship or helicopter. Despite Greenland government requirements for companies to inform communities about their plans, there is increasing public disquiet in Greenland concerning a lack of adequate community consultation, with many residents close to planned mines and potential oil fields saying that they do not feel knowledgeable about intended exploratory activities (Nuttall 2013). Greenland's far north is not exempt from interest in probing beneath the subsurface. Recent geological exploration and drilling for rock and mineral samples at sites on the Melville Bay coast, close to areas used by Savissivimmiut ("people of Savissivik"), have taken place, people complain, without mining companies visiting the community to explain their projects and what their activities will mean, yet Savissivimmiut have nonetheless since heard about the surveys and are worried about the prospects for development and the consequences for the environment and their livelihoods. At the same time, environmental groups wishing to protect the north Greenland ice in the name of biodiversity conservation are attempting to assert influence over those concerned with environmental governance, but they are seemingly privileging a view of the region that does not necessarily acknowledge the particularities of place or the actualities of living there. In short, climate change and increasingly uncertain and difficult weather conditions are not always talked about as a major problem. Hunters and fishers are far more worried about the decisions made by the Greenland government related to environmental management and quota systems, the restrictions imposed on the sale of things that are produced from hunting and fishing, and the possible impacts of extractive industries.

In this chapter I discuss the importance of Melville Bay and the sea ice to the communities in the region and consider aspects of an interdisciplinary community-based project to understand and monitor environmental change, with a focus on the possible effects of seismic activities on narwhals and narwhal hunting. This project, which I have been carrying out since summer 2014, began as a collaborative venture with marine biology researchers from the Greenland Climate Research Centre at the Greenland

Institute of Natural Resources (GINR) in Nuuk, and with hunters from Kullorsuaq and Savissivik. It aims to synthesize acoustic data and catch/interview material, as well as to develop a methodology for community-based participatory research, to investigate if a detected change in the distribution of narwhal catches is related to the sound intensity from seismic surveys in Baffin Bay and Melville Bay in the protected wildlife reserve close to the central part of the coast. The project has a central anthropological component—we seek to understand local observations and experiences of environmental change in Melville Bay, including changing sea-ice conditions, changes in travel routes and animal movement, and adaptive strategies and anticipatory knowledge; yet the ambition is for a transformative process that improves dialogue between marine biologists and hunters and incorporates local knowledge into decision-making and management processes, something that, compared with other places in the northern circumpolar world such as Alaska and the Canadian North, is currently minimal in Greenland even if such a process happens at all. This work is preliminary to what we intend to be a multiyear project that will contribute to the monitoring of environment change and the understanding of narwhal movement and migration routes by integrating scientific knowledge and the local knowledge of hunters from Melville Bay communities.

Melville Bay and the Middle Ice

Melville Bay (*Qimusseriarsuaq* in Greenlandic: "the great dog sledging place") is in the northeast corner of Baffin Bay on northwest Greenland's continental shelf. While it is often a geographical description used to refer to a vast area of northwest Greenland from the Svartenhuk peninsula in the southern part of the former Upernavik municipality to the Cape York area in what was previously known as the Thule district, or the old Avanersuaq municipality, its southernmost boundary is usually accepted to be in the northern part of the Upernavik archipelago (an area that is now part of the municipality of Qaasuitsup), close to the villages of Nuussuaq and Kullorsuaq. Settlement patterns in Melville Bay reveal how the area has been used extensively by Greenlandic Inuit societies for several thousand years, as well as being an important migration route between northern Greenland and the west coast. In this part of northern coastal Greenland, from the southern reaches of the Upernavik area and north to Cape York, people have long depended on hunting marine mammals, including seals, walrus, narwhals, beluga, fin and minke whales, and polar bears and also fishing for fjord cod, Greenland halibut, salmon, and Arctic char. Land animals such as reindeer and Arctic foxes have also been of some importance, and, in some areas, musk-ox.

Local livelihoods are often affected by outsider views of Melville Bay as an exceptional or extraordinary place—these views have an historical provenance and influence contemporary decision making about environmental and wildlife management systems. Nineteenth- and twentieth-century images of Arctic sea ice derived and emerged from accounts of direct and immediate encounters with it, as well as with icebergs and the glacial ice that hindered and defeated explorers on discovery expeditions to the High Arctic, frustrated those who sought a route to the North Pole through Smith Sound, and confounded whalers whose ships were often trapped in the pack ice. Nineteenth-century whalers approached Melville Bay with fearful anticipation. It was a place they had to cross to reach the North Water polynya and the whaling grounds of Canada's Lancaster Sound—despite the dangers, the route was chosen because a dock could be cut in the fast land ice adhering to the coast should the heavy pack ice endanger a ship. In this part of the Arctic, whalers referred to the sea ice as the middle ice. In the 1860s Isaac Israel Hayes pointed to a difference between the cartographic representation of the shores of Melville Bay as a simple curved line of the Greenland coast and the area as experienced by those who sailed into and through it. The "Melville Bay of the geographer comprehends much less of the mariner" he wrote and defined it as "the expansion of Baffin Bay which begins at the south with the 'middle ice' and terminates at the north with the 'North Water'" (Hayes 1867: 58). The middle ice was also known as "the pack,"

> made up of drifting ice-floes, varying in extent from feet to miles, and in thickness from inches to fathoms. These masses are sometimes pressed close together, having but little or no open space between them; and sometimes they are widely separated, depending on the conditions of the wind and tide. They are always more or less in motion, drifting to the north, south, east, or west, with the winds and currents. The penetration of this barrier is usually an undertaking of weeks or months, and is ordinarily attended with much risk. (Ibid.)

In 1873 Albert Hastings Markham sailed northward through Davis Strait and Baffin Bay on the Dundee steam whaling ship *Arctic*. Leaving Upernavik in early June, the crew encountered an area of "large open water, entirely free from ice, which astonished all on board, promising a fair and easy passage through Melville Bay" (Markham 1874: 104). The optimism was short-lived. Passing Devil's Thumb at 75°N (where the hunting settlement of Kullorsuaq was established in 1928), Markham remarked that "our troubles are about to commence, for stretching out from the nearest point of the shore to the northward and westward, as far as the eye can reach, is our great enemy, the dreaded floe ice of Melville Bay" (ibid.: 105). *Arctic*, being a steam whaler, found it easier to progress through the ice than many of its sail ship predecessors had done in earlier decades. The crews of whaling ships from England and Scotland knew Melville Bay as the bergy hole and the whaleship graveyard. It could take several weeks to get through the ice, but many ships never made it across the bay—they struck icebergs or were crushed by what the whalers called "thick-ribbed ice."

In 1875, when Markham returned north to Upernavik and Melville Bay as second-in-command under George Nares on HMS *Alert* during the British Arctic Expedition, in an attempt to reach the North Pole by way of Smith Sound, it remained a dangerous voyage. Whaling crews and discovery expedition members still prepared themselves for the possible abandonment of their ships as they ventured into the middle ice. In *The Great Frozen Sea* Markham wrote that "many a well-equipped ship has been caught in the fatal embrace of this bay. What tales of woe and disaster could its icy waters unfold." Once it had been crossed, and with the North Water in sight, he wrote with relief that "Melville Bay, with all its terrors, was behind us; a beautifully smooth unruffled sea devoid of ice was before us." Some of the best descriptions of the Melville Bay ice come from the writings of whalers and from explorers such as Markham and Hayes, and they offer insight into weather and environmental conditions, especially useful given the absence from the literature of indigenous accounts from that time. (Notwithstanding difficulties of communication, when Greenlandic knowledge of ice is noted it is often with reference to information needed for navigation by the whaling ships and discovery expeditions.) Some of Hayes's earlier work, such as *The Open Polar Sea*, gives a sense not only of the ice itself but also of the trials and tribulations of those caught up in it. However, writing about a relatively carefree voyage in the summer of 1871 aboard the *Panther* Hayes reported that the ice did not cause any problems. Nor did it for Allen Young—in *The Cruise of the Pandora* (1876), he wrote that "we could hardly believe we were in the dreaded Melville Bay." In April 1894 Eivind Astrüp, a member of Robert Peary's 1893–1894 expedition to northwest Greenland set out east of Cape York by dog sledge with two Inughuit hunters "to get a close view of the unexplored shores of Melville Bay" (Astrüp 1894: 159). He found the coastline "continually broken by large and active glaciers" (ibid.: 164), and, following a day of hazy travel, the fog cleared away to reveal "a grand and impressive scene": "High, dark mountains, gigantic glaciers, and lofty bluish-tinted snow peaks, all illuminated by the brilliant rays of the sun, lay scattered along the horizon in wild disorder and formed the attractive picture of Melville Bay" (ibid.: 165).

In such accounts of Melville Bay, little attention, if any, was given to recording indigenous understandings of the environment or climatic conditions. Just as Hayes pointed out that the geographer's understanding of Melville Bay diverged from the mariner's, both understandings also diverge considerably from those who inhabit it and travel in and across it. Astrüp does make brief mention of overland and cross-glacial routes his companions would take in winter and spring if open water was ever encountered in certain places, but this is a passing reference to the movement of the people who were indigenous to the region through an environment that was rendered wild, empty, and unexplored by the likes of Peary and many others. Melville Bay was perceived as a barrier between the west coast of Greenland and what was known as the Polar North, or the Thule region. The Polar Inuit—the Inughuit—around Cape York and areas farther north around the coast, referred to by John Ross in 1818 as Arctic Highlanders, were considered isolated and untouched by civilization and Danish colonization. In *The People of the Polar North*, Knud Rasmussen (1908) imagined Thule as a place apart, a polar region separated by Melville Bay from the Danish trading posts farther south on the west coast. (Into the 20th century the activities of the Royal Greenland Trade Company extended as far north as the Upernavik district, allowing Rasmussen and Peter Freuchen to establish their own trading post at Thule in 1910.) And in the latter half of the 19th century and into the early 20th century, Upernavik was written about by whalers and explorers as the last outpost of civilization before heading into the icy wastes of Melville Bay and beyond.

This brief reference to explorers' and whalers' descriptions of Melville Bay is important, because their accounts provide significant information on historic ice conditions (and in a related project I am probing deeper into 19th- and early 20th-century narratives, diaries, and logs and their descriptions of the north Greenland environment). But it is also necessary, because today the area is still considered something exceptional, viewed as a sensitive ecosystem, a last ice area under threat from climate change and also part of a Greenlandic resource frontier depicted on maps as a collection of license blocks available for oil exploration and potential development and in high-resolution seismic images of hydrocarbon migration paths and shallow gas accumulation (Nuttall 2012). North Atlantic Mining Associates (NAMA) and Red Rock Resources, U.K.-registered companies working in partnership to determine the potential for iron ore mining in the area immediately to the northwest of Savissivik, describe the area (and Greenland more generally) on their respective websites as a "last mining frontier."

People in Savissivik and Kullorsuaq are not unaware of these historical and contemporary representations and how they influence their lives—stories of European and American explorers and whalers continue to be told in these northern villages, and these people's spectral presence is felt in the environment through the non-indigenous names still imprinted on maps and charts and used to refer to glaciers, islands, and stretches of coast. Awareness of how visitors, sojourners, and others have approached Melville Bay, and have usually disregarded indigenous knowledge and experience, comes to the fore in community discussions about how the area should be used and managed and how conservation measures need to recognize indigenous rights. While explorers or their sponsors are memorialized through names such as Peary Glacier, Rink Glacier, and Lauge Koch Coast, and while such historic traces of exploration continue to haunt the global imagination about Melville Bay and other Arctic places such as

Canada's Northwest Passage, indigenous place names refer to the things that one can expect to find, or that one used to find, places that are good to hunt seals, narwhals, reindeer or polar bears, or geese, for example. Local residents do not talk about Melville Bay as a whaleship graveyard, a barrier between imagined northern places, or as a last ice area. It is a place of travel, activity, social association and close social relatedness, a place of routes, trails, passages, a place to hunt and fish, a place connecting people to their ancestral memoryscapes, contemporary settings, and anticipated futures. Yet, local people feel this indigenous use and knowledge of Melville Bay is overlooked and ignored in discussions and decisions about environmental protection and resource development and in the formulation and implementation of wildlife management.

The Experience of Living in a Shifting World

Scientific studies of dramatic changes in Greenland's environment, in the sea ice and inland ice (for example, Gillet-Chaulet et al. 2012) are often borne out by indigenous experience of climate change. In Melville Bay communities, when hunters talk about animals, for example, they often frame their discussions with an account of the conditions of weather, sea, and ice that they experienced on the hunt, or they begin their narrative with an account of the appearance, behavior, and movement of seals, polar bears, narwhals, walrus, and other animals as they encountered, watched, and pursued them. Overall, their comments confirm that extreme weather events have become increasingly common. For instance, as one hunter put it to me in July 2014: "The biggest change of all is the weather. It is more rainy and windy in the summer." This situation, he said, contrasts with previous summers, when hunters experienced sunny, clearer weather with very little wind.

In northern Greenland, sea ice (*siku* in kalaallisut/Greenlandic; *hiku* in the Inuttun spoken in Savissivik and communities farther north) is central to people's lives for several months of the year. Each physical shifting state of the sea and the ice opens up all kinds of possibilities for hunting, fishing, and traveling, and sea ice connects communities in a way significantly different from that of the open water. Yet considerable changes to the sea ice are being observed and experienced. The sea ice now tends to form later and break up earlier than many people have known it to do in their lifetimes. By way of example, the period of travel by dog sledge on good, solid sea ice is now around two or three months during winter and spring. Hunters concur that around both Savissivik and Kullorsuaq the ice is best in March and April (January and February were also months when hunters could expect good sea ice until a decade or so ago), which reduces significantly the amount of time hunters are able to hunt and fish. At the same time, around Kullorsuaq, hunters have, in recent years, been hunting by boat during some periods when there is open water during winter and spring. In summer 2014 one hunter from Savissivik told me that polar bears (*nannut*; sing. *nanoq*) used to be hunted far out on the sea ice, in places where hunters would lose sight of land, and around areas where icebergs were locked in the ice. These are places called *qorfiit*, "where the polar bears hunt." He remarked: "Now it's only a story to tell. We can't hunt them there anymore." I had seen these places myself when I was first in Savissivik in the spring of 1988, and in March 2015 I traveled with him by dog sledge to see the receding boundary between ice and open water. There is an increasing level of risk to travel far out on the sea ice—it can no longer be relied on to be solid—while the ice edge shifts and open water is now closer than before. Another reason for increasing risk is that the ice is constantly moved around by wind and sea currents. Another hunter from Savissivik told me that, while he was out hunting for polar bears with his father-in-law some 25 km south of the community during spring 2013, the ice suddenly broke up in front of them, an occurrence that was rare even a decade ago. Now that *qorfiit* are all but gone, polar bears are being observed closer to Savissivik, and hunting them takes place nearer the community. As one hunter put it: "Now we are living with *nannut*."

Hunters report that the sea ice is also of a different texture and consistency. As I described in my chapter in the first volume of *Anthropology and Climate Change*, the instability of sea ice has been making ice-edge hunting in northwest Greenland more difficult and dangerous. For example, some hunters in communities north of Upernavik have remarked that the experience of traveling to the edge of the ice in the last 10 to 15 years makes them more anxious than ever before—they say that the ice is "more slippery" than they have known it to be and that they feel more secure traveling by dog sledge on the solid ice that is attached to the shore, thereby limiting how far on the ice they can travel (Nuttall 2009). Changes in snow cover are also causing difficulty in accessing hunting and fishing areas by dog sledge or snowmobile, making local adjustments in winter travel and in hunting and fishing strategies necessary. Hunters and fishers are utilizing new adaptive strategies that include traveling to new fishing grounds, preparing for an increased reliance on boat transportation during the winter and when there are increasingly ice-free waters, and seeking alternative sources of income (which are somewhat limited).

In summer and early autumn 2014 I traveled with hunters from Savissivik and Kullorsuaq and marked on maps and charts the extent of glacial melt and retreat in Melville Bay. My companions pointed out just how far some glaciers had receded, according to their experience and that in community memory. I asked them if there were Greenlandic names for the glaciers. They responded that there were too many to name, and so calling every glacier *sermeq* was easier. Some glaciers are used occasionally as crossing places when the sea ice does not make for safe dog sledging;

otherwise people do not venture into Melville Bay to ascend or to travel on them unless they provide sledging routes to hunting or fishing grounds. However, my companions remarked that just as these bodies of dense ice are disappearing, so too are the names given to them. As one said: "Peary Glacier is disappearing, and so is the name. New land is being revealed, and we can name it in a Greenlandic way." Thus climate change not only erodes features of the environment; it also erases commemorations of particular views of Arctic landscapes and icescapes. Some people may lament the retreat and possible eventual disappearance of the Peary Glacier as the loss of polar heritage inscribed in a mass of ice formed from the accumulation of snow, but from a local Greenlandic point of view the movement of a glacier, its surging and receding, and either the covering or undraping of land are entirely consistent with the world as undergoing a continual process of becoming (Nuttall 2009). Receding glaciers reveal water that can be fished in summer or traveled on by dog sledge in winter when it turns to ice, and land that can be walked, traversed, and camped upon.

The Regulated Narwhal Hunt

Narwhals are distributed across a vast stretch of the North Atlantic Arctic, where they feed on halibut and squid. They dive deep for their prey and are attracted to ice-covered areas—along with their restricted core areas of distribution, this fact makes them sensitive to climate change (Laidre et al. 2008). The greatest abundance is in the eastern Canadian Arctic and west and northwest Greenland, with populations in East Greenland, and in the Greenland Sea and northern Barents Sea. Using aerial surveys and satellite telemetry, marine biologists estimate that there are upward of 16,000 narwhals in Baffin Bay; their summer range includes most of the waters of the Canadian Arctic Archipelago and northwest Greenland. The main summering areas in Greenlandic waters are in Melville Bay and Inglefield Bredning (Born 1986; Born et al. 1994; Heide-Jørgensen 1994; Heide-Jørgensen et al. 2010). Part of Melville Bay is a marine wildlife sanctuary, and a narwhal protection zone has been designated within it; the sanctuary was established in 1980, but hunting with motorized boats had been prohibited since 1946, which represented an early conservation measure—and, although there are restrictions on hunting, occupational hunters are permitted to hunt in a borderline area of the sanctuary. Quotas for narwhal hunting have been in place since they were introduced by the Greenland government in 2005 following scientific advice that the hunt was unsustainable, a view that hunters continue to challenge. The quota and the hunts are monitored by the municipal authorities and the Greenland government's hunting, fishing, and agriculture department through a license and catch report system. The Canada-Greenland Joint Commission on Conservation and Management of Narwhal and Beluga (JCNB) provides management advice, with scientific advice provided by a JCNB working group and the North Atlantic Marine Mammal Commission's (NAMMCO) scientific working group. Greenland government quotas are then based on the JCNB recommendations, and Greenland's hunting council and the municipal authorities distribute the quotas. Hunters apply for a license from the local authorities and must report the catch after each narwhal is taken. While residents of Greenland can sell and purchase products made from narwhal tusk and teeth, from jewelry to an entire tusk, since 2006 there has been a ban on visitors and tourists exporting them, under the Convention on International Trade in Endangered Species (CITES).

Like other marine mammal hunting activities in this part of Greenland, the regulated narwhal hunt in Melville Bay has both economic and cultural importance. It is an essential part of the annual seasonal hunting and fishing round, contributing to the maintenance of the social and economic fabric of family and community life. Animals are vital for the communities in the Melville Bay area, and hunting is a fundamental basis for how people think about sustainable livelihoods and resilience. There is currently a quota of 81 narwhals per year for the Upernavik area and Melville Bay, and many are hunted during the open water season from August to September. Local hunting regulations do not permit the use of motorboats to kill narwhals (they can be used for transport to and from hunting areas), so kayaks are used to track, follow, and harpoon the narwhals from.

A hunter who catches a narwhal will use much of the animal for himself and his household, but money can be earned from the sale of *mattak*, the skin and thin layer of blubber that is prized for its vitamin C and as a delicacy. *Mattak* enters local markets, with hunters selling in local networks or in the town of Upernavik, or to hotels and restaurants along the Greenland coast. Facebook, email, and mobile telephones provide ways of finding distribution channels and markets throughout the country, and there is always an eager market in Nuuk. Yet, although much *mattak* is sold, the needs of the hunter and his family come first. Sharing patterns remain vital in places such as Kullorsuaq and Savissivik, as well as in other communities along the northwest coast. However, hunters point out that the quota system now makes it much more difficult to earn money through the sale of *mattak*. As one hunter remarked to me: "The quota system limits the number of narwhals we can catch per trip and during the year, so it's hard to make money when there are expenses associated with hunting trips, such as fuel for boats." One consequence of the quota system has been a tendency for hunters to sell less from the hunt within wider networks and to keep much of what they catch for their own household consumption and to share within the community. But the quota system often has the opposite effect, too—when more *mattak* and meat are sold, there is less to share within local social networks.

Oil Exploration, Seismic Surveys, and Narwhal Hunting

Climate change is often cited as contributing to the conditions that make many new Arctic resource development projects possible, as accessibility to places where minerals and hydrocarbons are located becomes easier. However, in Greenland, although we should not dismiss climate change—together with the "greening of Greenland" discourse—entirely, the definition, gradual marking out of, and potential industrialization of what is emerging as a resource frontier are more the result of a Greenlandic nation-building and state-formation process underway since Greenland achieved Home Rule within the Kingdom of Denmark in 1979. Identifying climate change as the reason why Greenland is becoming accessible and attractive to multinational corporations is simplistic. The unprecedented growth of international interest in Greenland in the last five to ten years is largely due to an active international marketing campaign and strategic business strategy by the Bureau of Minerals and Petroleum (BMP). Since 2009 Naalakkersuisut, the Greenlandic government, has implemented policies that actively promote exploration and investment in extractive activities. The Self Rule Act of 2009 has given Greenland control of subsurface resources, and developing mines and oil fields is seen as a way of ensuring greater economic and possibly political independence (Nuttall 2012).

The summer and early autumn open-water season is when seismic activities can take place in Baffin Bay and Melville Bay. In northwest Greenland mining companies are engaged in prospecting and developing plans for several projects, but Melville Bay has also been the focus of recent international interest in the prospects for the discovery of oil (Nuttall 2012, 2013). Analysis of magnetic and gravity data allowed geologists to predict the existence of large structures and sedimentary basins in the 1960s (Hood, Sawatsky, and Bower 1967). In the late 1990s these basins were confirmed by the analysis of seismic data, in particular emphasizing that a major sedimentary basin—the Melville Bay Graben, a feature up to 100 km wide—exists in the area, with implications that were considered encouraging for hydrocarbon exploration (Whittaker, Hamann, and Pulvercraft 1997). The first seismic reconnaissance in Melville Bay was carried out in 1992 as part of the KANUMAS project. In 2012 a large seismic exploration campaign was carried out in Baffin Bay. The combined activities, composed of four 2D and 3D seismic surveys and shallow core drillings, represented the most intensive exploration ever carried out in Greenland.

There are several blocks adjacent to Melville Bay (called Tooq, Napu, Anu, Pitu, and Qamut). International companies—Shell, Maersk, Cairn Energy, and ConocoPhillips and their partners—hold active licenses for exploration, including both seismic and potential drilling activities. In 2013 Shell carried out a series of seismic site surveys, some of which overlapped with the narwhal protection zone. After these surveys had finished, hunters from Melville Bay communities reported that narwhal behavior was different and that the hunt had been influenced negatively owing to the seismic activities in the area. Canadian narwhal hunters believe that narwhal migrations have changed since seismic surveys began in Baffin Bay (as was emphasized in a press release by the Inuit Circumpolar Council in September 2013), and in April 2015 the organization representing Clyde River's hunters and trappers went to the Federal Appeals Court in Toronto as part of their legal challenge to overturn the National Energy Board's decision to allow more seismic testing for oil and gas on the Nunavut side of Baffin Bay and Davis Strait. Marine biologists and hunters in Greenland have also begun to express concern over the possible effects of intense seismic survey activities and increased shipping on the future of marine mammals and hunting communities, and they are calling for long-term monitoring programs to be put in place (see, for example, Heide-Jørgensen et al. 2013a). There are concerns that seismic noise affects narwhals, particularly increasing the possibility of ice entrapment (Heide-Jørgensen et al. 2013b). It is likely that exploratory activities will increase and intensify in the coming years.

In 2014 I began an interdisciplinary study with marine biologists—colleagues from the Greenland Climate Research Centre at the Greenland Institute of Natural Resources in Nuuk—to understand some of the consequences of seismic activities on narwhal hunting in Melville Bay and to examine them within the context of a broader community-based project that would monitor and map climate change in relation to other social, economic, political, and environmental changes. The project aims to synthesize acoustic data (gathered using anchored moorings with sound recorders) and catch/interview data to investigate if a detected change in the distribution of narwhal catches is related to the sound intensity in the reserve from seismic surveys in northeast Baffin Bay. Traveling in Melville Bay on GINR's research vessel *Sanna*, accompanied by hunters from Savissivik and Kullorsuaq (the respective populations of which are around 60 and 450), we deployed the sound recorders in July at key narwhal summer feeding sites based on local knowledge (and we retrieved the equipment in September), while in the communities themselves we carried out interviews, held workshops, reported on the scientific data at follow up meetings, and have begun to develop collaborative research questions to guide further research. In our work we draw on hunting reports, engage in dialogue with hunters, make observations of hunts, map the routes traveled and the locations of hunts, and supplement community-based research with monitoring activities in order to investigate possible disturbance effects of oil and gas exploration activities. The project goals are to (1) monitor changes in distribution of catches; (2) monitor changes in search time (effort) for a successful narwhal catch; and (3) understand and learn from hunters' perceptions and knowledge of possible changes in narwhal behavior and hunting success during years both with and without seismic activities.

Furthermore, the project seeks to understand local observations and experiences of environmental change in Melville Bay, changes in travel routes and animal movement, adaptive strategies, and anticipatory knowledge.

We have begun to map the locations of each narwhal hunt and the travel routes of hunters since 2005 and are building a comprehensive knowledge base of local occupancy and use of Melville Bay. Because the narwhal hunt is regulated, hunters are required to complete a catch report for the municipal authorities and Greenland Institute of Natural Resources for every narwhal they have caught. These reports contain rich information on narwhals, including gender, size, hunting method, and location of the catch, and they contribute to and enhance scientific knowledge about the animals—for example, a total of 1,759 catch reports (of one to several narwhals) covering hunts from Siorapaluk to Upernavik Kujalleq between 2005 and 2014 were reported by a total of 356 hunters. We designed the initial phase of the project to scope out a larger research program and with the intention to gather baseline data to inform the development of a set of questions that would guide it. Our research, although preliminary, provides further evidence from local observations that there have been noticeable changes in narwhal behavior in recent years, as well as in the behavior and movement of other marine species. Although it is not possible at this stage to determine the reasons—for instance, climate change, shifting ice, and increased noise in marine environments (including, some hunters also readily admit, noise from the increase in motorized boats owned by people in the communities themselves, as well as from increased shipping activity) may combine to have an impact—in our initial project reporting in 2015 (which went to a seismic impacts working group in Denmark and Greenland and then on to the oil companies), we have emphasized that it is critical that additional work is carried out before more seismic activities proceed. Furthermore, we have recommended that such work be developed and carried out with the participation of northern communities at all stages (Nuttall, Simon, and Zinglersen 2015).

As previous work concludes (Heide-Jørgensen et al. 2013a, 2013b), there remains uncertainty in relation to the effects of seismic noise on narwhal populations, and careful monitoring is needed. However, our research shows that the observations of local people confirm there are changes in narwhal behavior and movement in Melville Bay and along the coast of the Upernavik district. There is general agreement from hunters that narwhals have been restless, anxious, and harder to hunt, and many hunters—particularly those living close to Melville Bay—have said that this situation has been more noticeable in the last few years, especially since 2012. For example, hunters pointed out that narwhals have been observed to now move faster near Tuttulipaluk and Tuttulissuaq—important narwhal feeding places and the areas closest to the seismic survey areas—and that in other parts of Melville Bay narwhals have also changed direction in their migration, or they exhibit changes in their feeding routines. In the town of Upernavik, which is south of the main Melville Bay hunting areas, hunters have also provided us with their accounts of narwhal movement and migration. All agree that the most noticeable change has been that narwhals have been very close to Upernavik in the last few years. One hunter said that narwhals "have been too alarmed to be close to the open sea." Another told us that "narwhals have been very close to the coast in recent years . . . they are obviously afraid of something farther away in the open sea." Another hunter said the most unusual thing he had noticed about narwhals is that they had appeared in the Upernavik area in great numbers during spring 2014. They also pointed out that belugas (of which there seem to be far more than in previous years) have changed their route by moving farther away from the coast when they are going north, although they have not changed their route when they are traveling south. One hunter said that there were fewer narwhals in Melville Bay in 2014, and they have also behaved differently. However, he reinforced the general observation that there are more narwhals in the Upernavik area, and they have been very close to the coast in recent years. Another hunter from Upernavik said: "There have been lots of narwhals very close to Upernavik in the last few years. It means that the narwhals must have changed their routes, even though they have not changed their behavior and there are more of them." Another also said there have been far more narwhals in the Upernavik area, especially in spring 2014, and he believed that there may be a connection between seismic activities and narwhal routes.

Protecting the Northern Ice

Melville Bay and areas farther north are not necessarily written about today as something dreaded or terrifying, as whalers and members of discovery expeditions described them in the 19th century. Nonetheless, the north Greenland ice, the glaciated coastline, and the marine ecosystem continue to be seen as exceptional, ecologically distinct, and worthy of protection. In *Greenland by the Polar Sea* Knud Rasmussen called the glaciers of Melville Bay "extravagant" (Rasmussen 1921: 59) and considered the area a remnant of the ice age—"the journey from Upernivik to Melville Bay represents the last chapters of this history," he wrote. "The land in front of the ice becomes increasingly narrow; every valley is filled with ice. Large glaciers shoot out between and across islands and skerries; near the Devil's Thumb the coast consists as much of ice as of land, and north of this point only occasional small islands and nunataks push up. For miles the coast is one continuous wall of ice" (ibid.: 307).

This description has a contemporary counterpart in how the World Wide Fund for Nature (WWF) outlines its Last Ice Area initiative.[1] As WWF puts it on its website: "This region in the high Arctic of Canada and Greenland was identified using cutting-edge sea ice models, and it's projected to be the last stronghold of summer sea ice as the Earth continues to warm owing to climate change. In the coming years, it will be essential as an enduring home for wildlife and therefore for the communities that depend on those species." Despite WWF's attempts to work with Inuit communities, many of these efforts are confined to eastern Nunavut, and people from Greenlandic communities have expressed concern about the lack of consultation with them and remain suspicious about the conservation organization's real intentions—which many believe to be a concern with protecting polar bears and denying Inuit the right to hunt them. Thus Melville Bay and areas farther north have become contested sites over "saving the Arctic." As Page and colleagues put it, setting aside areas for protection and conservation of wildlife and nature "provide a means of seeing and governing the world that have myriad social effects" (2006: 255). Drawing on Carrier and Miller (1998) they argue that protected areas are a form of virtualism and become a discursive way through which the world is perceived, approached, and remade. Campaigning groups and NGOs located many thousands of miles away from northern Greenland and other parts of the Arctic make widespread use of the imagery of melting ice and starving polar bears as they seek to gain public attention for their activities, appealing in particular to those who may never visit the Arctic but who nonetheless seek to imagine and visualize it, lending support to the organizations and institutions that exert an influence over or have the authority to make decisions about conservation and environmental management (Dodds and Nuttall 2016).

In Greenland's northern communities, hunting and fishing activities satisfy vitally important social, cultural, and nutritional needs of families, households, and communities, as well as economic ones. Foods purchased from the local store supplement diets composed mainly of meat and fish caught locally. Individuals and households lacking the means or ability to hunt find that it is often essential to be able to obtain *kalaalimernit* (Greenlandic foods derived from hunting and fishing) through local distribution channels and sharing networks. In some cases the money earned makes it possible to continue hunting and fishing, rather than contributing to its decline, because cash buys, for example, boats, rifles, and snowmobiles. Cash also meets demands for a rising standard of living: to purchase oil to heat homes, buy consumer goods, or travel beyond one's community. While food procured from hunting and fishing continues to provide people with important nutritional, socioeconomic, and cultural benefits, finding ways to earn money is a major concern for many people in northwest Greenland.

The imposition of quotas and regulations is troubling for hunting households—or perhaps it is more accurate to say that the way quotas and regulations are decided on and who makes the decisions are matters of some controversy and concern. Management systems restrict abilities to go after certain marine mammals such as narwhals or to fish Greenland halibut. Recent research in East Greenland has shown that the introduction of quotas on narwhal hunting in 2009 has had a negative cultural effect (Nielsen and Meilby 2013). Such management policies have been criticized because they are implemented by government authorities and are based on scientific knowledge void of local consultation. Hunters argue that improved dialogue between them, scientists, and policymakers is essential. I have heard hunters talk time and time again about how the hunting regulations make them they feel like criminals, with biologists and bureaucrats defining their activities as an unsustainable harvest of a species at risk and intrusive in an environment threatened by climate change. Citing their awareness of a new narwhal management plan for Nunavut—approved in 2013 by Canada's Department of Fisheries and Oceans and the Nunavut Wildlife Management Board—that has increased the total allowable harvest and, importantly, has drawn considerably on Inuit traditional knowledge, hunters in northwest Greenland have expressed their incredulity that, unlike the situation on the other side of Baffin Bay, their knowledge is ignored and scientific knowledge is privileged in decision-making processes that regulate the hunt and affect local livelihoods.

Conclusions: Toward Community-Based Monitoring

Previous work by narwhal biologists concludes that there remains uncertainty in relation to the effects of seismic noise on narwhal populations, and so careful monitoring is needed (Heide-Jørgensen et al. 2013a, 2013b). However, our research shows that the observations of local people confirm there are changes in narwhal behavior and movement in Melville Bay and farther south along the coast of the Upernavik district. Narwhals are sensitive to noise and especially susceptible to climate change, so although it is difficult to determine a single or major reason for such changes (even if they appear to be more noticeable since the intensive seismic survey campaigns), hunters are calling for improved dialogue between communities, marine biologists, and oil companies and for co-management strategies for environmental protection and narwhal hunting that recognize and incorporate local knowledge.

Beyond the need to understand local knowledge of narwhals and narwhal movement and behavior, a number of questions center specifically on climate change and its consequences for Melville Bay communities. Hunters express their concerns that the effects of climate change on their livelihoods would be exacerbated by seismic activities, as well as by tighter management regulations for narwhal hunting. The impacts of changes in sea ice, environment, and socioeconomic conditions of people's livelihoods in communities in northwest Greenland are not yet fully understood. One challenge is to investigate how changes have affected communities and their adaptive capacities and survival strategies. For example, what kinds of flexibility in technology and social organization do people need to cope with climate change, allowing them to respond both to its associated risks and to seize its opportunities? And how are we to understand cultural and ecological diversity within a context of innovation, flexibility, and resilient coping strategies during periods of extreme change?

However, it is also crucial to understand climate change impacts within the context of other changes and societal and economic transformations in Greenland, including resource development and extractive industries (Nuttall et al. 2005). In this regard, both a long-term monitoring project on impacts of seismic activities on narwhals and discussions on the quota system were highlighted as areas of further work that hunters feel is necessary for the future of sustainable hunting and community economies. Hunters, perhaps not surprisingly, disagree with scientific assessments of narwhal populations and believe there are more narwhals than scientists have counted during aerial census projects, and they point out that there should be greater collaboration between hunters, marine biologists, and the oil companies. As one put it: "The scientists think that narwhals move along particular routes and stay in one place. But the narwhals move all over. We watch them move south, which is what the scientists would think is their migration pattern, but then they come back north again." Such disagreement about whose knowledge counts in the politics of narwhal management was apparent in June 2014, when Finn Karlsen, then minister for fisheries, hunting, and agriculture, approved a proposal put forward by the northern hunting associations that an additional 30 narwhals could be hunted that year on top of the quota of 81. The reason for this change was that the entire quota had been hunted by May of that year, mainly around Upernavik and the southern communities of the district (which, according to statements already mentioned, are witnessing a greater congregation of narwhals nearby), leaving the northern communities of Nuussuaq, Kullorsuaq, and Savissivik with no legal rights to hunt by the time the narwhals had made their migration to Melville Bay later in the summer. Karlsen's decision secured the hunt for those communities but was criticized by biologists at the Greenland Institute of Natural Resources as well as by WWF Denmark for going against scientific advice that the narwhal hunt is not sustainable beyond the current quota.[2]

At community meetings in Kullorsuaq in February 2015, at which we reported on our initial findings and the progress of the project, the village hunting association proposed the need for discussion about developing a co-management structure for Melville Bay that would involve communities as key participants in decision making in relation to seismic activities, possible oil development, conservation of marine resources, and regulated narwhal hunting as well as hunting for other species (they even believe the narwhal protection zone should be enlarged). Their view was that a long-term project to monitor changes in narwhal behavior and movement, one that combines local and scientific knowledge, will contribute to scientific understanding of migratory narwhals and inform social impact assessments and environmental impact assessments. Furthermore, they argued that such a project would have the potential to develop a new methodology for strengthening local community involvement in work that contributes to the environmental review process for oil exploration and possible development, as well as decision-making processes for future governance systems and ecosystem-based management. In this way, say northern hunters and other community members, such a project would, they hope, provide an opportunity for local voices to be heard and for local interests to be recognized.

Acknowledgments

This chapter draws in part on work carried out under the auspices of the Climate and Society research program at the Greenland Climate Research Centre (GCRC), with funding from project GCRC6400, as well as GCRC's Melville Bay Narwhal Project and the EU FP7 ICE-ARC project. Thanks go to Qaerngaaq Nielsen, Olennguaq Kristensen, Martin Jensen, Martin Nielsen, Ole Olsvig, Pipaluk Hammeken, Malene Simon, Lene Kielsen Holm, Carl Isaksen, Peter Hegelund, and Karl Zinglersen, as well as the many hunters and fishers in several northwest Greenland communities who have given generously of their time. My thanks also go to Susan Crate for her comments and suggestions.

Notes

1. www.wwf.ca/conservation/arctic/lia/
2. "*Finn Karlsenip qilalukkat qernertat immeraaassutigai/Finn Karlsen spiller hasard med narvhalerne*" ["Finn Karlsen gambling with narwhals"], *Sermitsiaq* 28, June 11, 2014, pp. 10–11.

References

Astrüp, E. 1894. Reconnaissance of Melville Bay, in R. E. Peary, *Northward over the "Great Ice,"* Volume II. London: Methuen and Co., pp. 157–74.

Born, E. W. 1986. Observations of narwhals (*Monodon monoceros*) in the Thule area (NW Greenland), August 1984. *Reports of the International Whaling Commission* 36: 387–92.

Born, E. W., Heide-Jørgensen, M. P., Larsen, F., and Martin, A. R. 1994. Abundance and stock composition of narwhals (*Monodon monoceros*) in Ingelfield Bredning (NW Greenland). *Meddelelser om Grønland. Bioscience* 39: 51–68.

Carrier, J. G., and Miller, D. 1998. *Virtualism: A new political economy*. Oxford: Berg.

Dodds, K., and Nuttall, M. 2016. *The Scramble for the Poles: The Contemporary Geopolitics of the Arctic and Antarctic*. Cambridge: Polity.

Gillet-Chaulet, F., Gagliardini, O., Seddik, H., Nodet, M., Durand, G., Ritz, C., Zwinger, T., Greve, R. and Vaughan, D. G. 2012. Greenland ice contributions to sea-level rise from a new generation ice-sheet model. *The Cryosphere* 6: 1561–76.

Hastrup, K. 2009. Arctic hunters: Climate variability and social flexibility, in K. Hastrup (Ed.), *The Question of Resilience: Social Responses to Climate Change*. Copenhagen: Royal Danish Academy of Science and Letters.

Hayes, I. I. 1867. *The Open Polar Sea: Narrative of a Voyage of Discovery toward the North Pole*. New York: Hurd and Houghton.

———. 1871. *The Land of Desolation: Being a Personal Narrative of Adventure in Greenland*. London: Sampson, Low, Marston, Low, and Searle.

Heide-Jørgensen, M. P. 1994. Distribution, exploitation, and population status of white whales (*Delphinapterus leucas*) and narwhals (*Monodon monoceros*) in West Greenland. *Meddelelser om Grønland. Bioscience* 39: 135–49.

Heide-Jørgensen, M. P., Guldborg Hansen, R., Fossette, S., Hjort Nielsen, N., Villum Jensen, M., and Hegelund, P. 2013a. *Monitoring Abundance and Hunting of Narwhals in Melville Bay during Seismic Surveys in 2012*. Nuuk: Greenland Institute of Natural Resources.

Heide-Jørgensen, M. P., Guldborg Hansen, R., Westdal, K., Reeves, R., and Mosbech, A. 2013b. Narwhals and seismic exploration: Is seismic noise increasing the risk of ice entrapments? *Biological Conservation* 158: 50–54.

Heide-Jørgensen, M. P., Laidre, K. L., Burt, M. L., Borchers, D. L., Marques, T. A., Hansen, R. G., Rasmussen, M., and Fossette, S. 2010. Abundance of narwhals (*Monodon monoceros*) on the hunting grounds in Greenland. *Journal of Mammalogy* 91(5): 1135–51.

Hood, P. J. P., Sawatsky, P., and Bower, M. E. 1967. Aeromagnetic investigations of Baffin Bay, the North Atlantic Ocean, and the Ottawa area. *Geological Survey of Canada,* Paper 68-1: 79–84.

Laidre, K. L., Stirling, I., Lowry, L. F., Wiig, O., Heide-Jørgensen, M. P., and Ferguson, S. H. 2008. Quantifying the sensitivity of Arctic marine mammals to climate-induced habitat change. *Ecological Applications* 18: S97–125.

Markham, A. H. 1874. *A Whaling Cruise to Baffin's Bay and the Gulf of Bothnia*. London: Sampson, Low, Marston, Low, and Searle.

———. 1878. *The Great Frozen Sea*. London: Daldy, Isbister and Co.

Nielsen, M. R., and Meilby, H. 2013. Quotas on narwhal (*Monodon monoceros*) hunting in East Greenland: Trends in narwhal killed per hunter and potential impacts of regulations on Inuit communities. *Human Ecology* 41(2): 187–203.

Nuttall, M. 2009. Living in a world of movement: Human resilience to environmental instability in Greenland, in S. A. Crate and M. Nuttall (Eds.), *Anthropology and Climate Change: from Encounters to Actions*. Walnut Creek, CA: Left Coast Press, Inc., pp. 292–310.

———. 2012. Imagining and governing the Greenlandic resource frontier. *The Polar Journal* 2(1): 113–24.

———. 2013. Zero-tolerance, uranium, and Greenland's mining future. *The Polar Journal* 3(2): 368–83.

Nuttall, M., Berkes, F., Forbes, B., Kofinas, G., Vlassova, T., Kofinas, G., and Wenzel, G. W. 2005. Hunting, herding, fishing and gathering: Indigenous peoples and renewable resource use in the Arctic, in ACIA *Arctic Climate Impact Assessment: The Scientific Report*. Cambridge: Cambridge University Press, pp. 649–90.

Nuttall, M., Simon, M., and K. Zinglersen. 2015. *Possible Effects of Seismic Activities on the Narwhal Hunt in Melville Bay, Northwest Greenland* Nuuk: Greenland Climate Research Centre, Greenland Institute of Natural Resources. Unpublished report.

Rasmussen, K. 1908. *The People of the Polar North*. London: Kegan Paul, Trench, Trübner and Co. Ltd.

———. 1921. *Greenland by the Polar Sea*. New York: Frederick A. Stokes Company.

West, P., Igoe, J., and Brockington, D. 2006. Parks and people: The social impact of protected areas. *Annual Review of Anthropology* 35: 251–77.

Whittaker, R. C., Hamann, N. E., and Pulvercraft, T. C. R. 1997. A new frontier province offshore Northwest Greenland: Structure, basin development, and petroleum potential of the Melville Bay area. *AAPG Newsletter* 81(6): 978–98.

Young, A. 1876. *Cruise of the "Pandora."* London: William Clowes and Sons.

Chapter 23

Insuring the Rain as Climate Adaptation in an Ethiopian Agricultural Community

Nicole D. Peterson and Daniel Osgood[1]

History of the Project

In 2007, while working with Columbia University's International Research Institute for Climate and Society (IRI) and the Center for Research on Environmental Decisions (CRED), we were asked to join an interdisciplinary team on a project that would create a microinsurance program for farmers in Malawi. The insurance companies and banks offering the insurance were concerned that farmers would misunderstand the concepts involved, and so one of our roles was to show that farmers understood the program. To this end, we developed a communication tool to share the ideas of the program with farmers, many of whom were illiterate. Once on the ground in Malawi, we recognized that some farmers easily understood the core concept of insurance and also the details of microinsurance, with payout depending on the amount of rainfall measured at specific locations. Others required more exposure to the idea through training. Overall, the farmers were quick to learn. One farmer stood up and began using the communication tool to explain it to his colleagues. In a later project in Ethiopia, farmers suggested that insuring their irrigation structure[2] would be a very good thing to do—and they used the English term *insurance*, which they had not heard before, to describe the way the dam's irrigated water already protects them from drought. Although farmers were unfamiliar with several of the central ideas in the contracts, including rain gauges and the contractual division of the growing season, they were eager to learn about them, and many asked for rain gauges for their own fields.

Farmers in Malawi and Ethiopia, among many other countries, have purchased index insurance, the form of microinsurance that responds to an index of rainfall amounts in an area rather than to conditions in individual fields. Offered by NGOs, governments, and private insurers (often jointly), index insurance is intended to be one tool among many to help farmers adapt to new climate patterns resulting from anthropogenic climate change

(Collier, Skees, and Barnett 2009). Because of decreasing rainfall in places such as Ethiopia, groups offering index insurance intend to help reduce the impact of drought on farmers' livelihoods by providing food or money when rainfall has been poor, thereby preventing asset sales, migration, and other strategies that may compromise long-term adaptability.

The intent of these programs is thus to improve the adaptive capacity of farmers in the face of climate change. Our discipline-specific backgrounds have been crucial to the project; Peterson's role as an anthropologist has been to help collect information from the farmers about their insurance preferences, including the desired interannual frequency and yearly timing of premiums and payouts, using interviews, focus groups, surveys, and economic games (for example, Patt et al. 2009). As an independent researcher, Peterson also studied the risks facing farming and coping strategies in Tigray state in northern Ethiopia and how they are distributed within a farming community. Osgood, as an agricultural economist, has been responsible for supporting the design of the contracts, in consultation with farmers, insurers, and others involved with the project. The project itself was developed and implemented by Oxfam America, in partnership with the Relief Society of Tigray, Ethiopian insurance companies, and the reinsurer SwissRe. Originally called HARITA (Horn of Africa Risk Transfer for Adaptation), the program now goes by the name R4 Rural Resilience Initiative and is jointly implemented with the World Food Programme.

We began our work to design index insurance for each location by gathering information about agriculture, first by examining the conditions of farming in Ethiopia and how climate change is affecting agriculture and livelihoods overall. To these ends we asked what the root causes were of the poor state of agriculture in Ethiopia (see also Devereux, Teshome, and Sabates-Wheeler 2005). We next looked at how insurance may interface with the farming/climate change context; for example, we asked questions about how the index insurance would interact with the current coping strategies of agriculturalists and how these strategies might introduce new risks to farming (see Peterson 2012). Understanding some of the basic issues facing agriculture and adaptation in Ethiopia and how policies can or cannot facilitate adaptation should help in planning and policymaking for climate change.

In this chapter we focus on our collaboration with the HARITA project,[3] using the case of an agricultural community in Tigray state, northern Ethiopia, to examine climate change adaptation programs and the problems they intend to address. In the process we illustrate how anthropological investigation can reveal important pieces of the climate change puzzle—in this case, for instance, how the recurrent drought we see in Ethiopia may have less to do with climate change and more to do with long-term trends and socioeconomic issues. Here, combating climate change means looking beyond climate change to the root causes of problems in the area; otherwise, we may be providing only a band-aid and not what farmers need.

Using results from surveys and interviews in Ethiopia in 2009,[4] we examine the kinds of issues raised by farmers and the ways they intersect with climate change and index insurance, including ideas about poverty, risks to farming, and coping strategies. Surveys were administered to 200 randomly selected farmers: 114 from those who purchased insurance and 86 from those who did not. The statistics reported are from analysis of variance (ANOVA) used to examine differences between means of different groups (comparing buyer versus nonbuyer gender, age, and so on). These results are reported with their F-ratio and resulting p value, which indicates the level of significance of the difference. To examine the relationships among variables, the Pearson Correlation Coefficient is reported, and the significance of these correlations is also given as p values.

Twenty-one farmers who had completed the June 2009 survey were invited for follow-up interviews. Farmers were selected based on demographic information to ensure equal representation of farmers from each of four villages, male and female household heads, and participants and nonparticipants in the Ethiopian Productive Safety Net Program (PSNP), which provides food and other resources in exchange for labor on local projects. In addition, four farmers who did not purchase insurance were interviewed, one from each village. Interviews were recorded and coded for themes.

For comparative purposes, we also include information from other sites where we have implemented index insurance. In conclusion, we suggest how this study might contribute to climate change adaptation and to understanding risk more broadly.

Identify Key Issues: Poverty

In interviews and focus groups, we asked farmers about their definitions of poverty—under what conditions is someone relatively deprived?—in order to understand vulnerabilities and risks within the community. In response, almost every person interviewed mentioned the importance of being able (or not) to labor and to access land. Many of the definitions also involved the idea of being able to eat: "those who are poor lack a daily meal because they lack labor or the desire to labor. If they can labor, they can feed themselves, even without assets." In addition, landowning, particularly the ability to work the land oneself, is seen as a crucial element for avoiding poverty. Thus, the two groups considered to be poor are those with no land (for example, young men) and those with no ability to labor (such as older widowed women).

Other factors associated with poverty included age, unequal land distribution, lack of irrigated land, lack of things to sell or exchange, large family size, and, more generally, a lack of sufficient food for the household; some people also mentioned the use of loans for nonproductive items (for instance, food, ceremonies, and the like). The cost of fertilizer was also mentioned by a respondent as a contributing factor to poverty. And, last, wealthy farmers had different ideas about what causes poverty than did poor farmers; for example, wealthier farmers mentioned laziness or bad character as factors more often.

The survey data showed that self-reported higher wealth categories correlated most strongly with nonparticipation in PSNP ($R = -0.467$, $p < 0.01$), more oxen owned ($R = 0.421$, $p < 0.01$), more irrigated land ($R = 0.187$, $p < 0.01$), and more rain-fed land ($R = 0.142$, $p < 0.05$). PSNP eligibility is determined by experiences of food shortages at both the community and family levels (Ministry of Agriculture and Rural Development 2006), matching our expectations that resource access is central to definitions of poverty in the area and that PSNP is targeting more resource-limited farmers. Female-headed households are, on average by all measures, less wealthy than their male counterparts, given significantly lower self-rated wealth (average of 2.08 vs. 2.77, $F = 75.92$, $p < 0.000$), lower numbers of oxen (0.36 vs. 1.46; $F = 48.217$, $p < 0.000$), and higher rates of PSNP participation (76.7% vs. 37.8%; $F = 28.55$, $p < 0.000$).

Our survey data also suggest that poorer farmers (those with fewer oxen and who participate in PSNP) rent out more land to other farmers (an adaptation to having fewer resources, particularly oxen), grow less teff (the staple food crop used to make injira; growing less can be understood as a consequence of having less land), and spend less money on seeds and fertilizer per timad of land (a timad is 0.25 hectare). These results, shown in Table 23.1, suggest that farming differs significantly for poor farmers. Wealthy farmers rent out less land, grow more teff, and spend more on seeds and fertilizer per timad of land. In terms of gender, we see similar patterns. Female-headed households (due to death or divorce) hold less land ($F = 4.714$, $p < 0.031$), grew less teff in the previous year ($F = 46.639$, $p < 0.000$), and expect to plant less in the next season ($F = 59.22$, $p < 0.000$). Female-headed households also differ in that they rent out less rain-fed land than do male-headed households ($F = 37.782$, $p < 0.000$) and share in less rain-fed ($F = 44.497$, $p < 0.000$) and irrigated land ($F = 27.371$, $p < 0.000$), reflecting gendered agricultural roles. Given the similarities of women and poor farmers,[5] one notes that gender is an important dimension of wealth in the community.

The resulting relationships between poverty, assets, and livelihood options do not represent a complete assessment of poverty in the community. However, they do suggest some characteristics related to poverty and vulnerability, including the way that such characteristics as gender, marital status, wealth, and age intersect. In other words, resources and gender are important indicators of wealth and seem to act through access to land and labor; they also determine some aspects of crop production. Given this situation, we aimed to understand the potential value of index insurance for a variety of farmers, including landless men and widowed women.

Table 23.1 Relationships between agriculture, oxen, and PSNP participation in 2009 survey data

*Pearson Correlation Values (Two-Tailed) (*indicates p < 0.05; **indicates p < 0.01)*	*Number of Oxen Owned*	*PSNP Participation*
Timad of rain-fed land you will rent out this season	–.340**	.275**
Timad of irrigated land you will rent out this season	–0.144	.210**
Timad of teff grown last season	.485**	–.357**
Seed cost in birr per timad for teff last season	.438**	–0.107
Fertilizer cost in birr per timad for teff last season	–0.045	–.155*
How many timad will you plant with teff next season?	.501**	–.375**

Risk

In the community in Tigray, Ethiopia, all farmers consider drought their biggest problem. They can list the years in their lifetimes when drought has led to hunger, famine, and even death, most prominently in 1977 and 1984.[6] In 2009 many also indicated that the previous three to five years had been drought years for them, exhausting their ability to cope. For most farmers interviewed, late rains and an early end to rains are the other two largest problems. The first interview question asked participants to remember an event in the previous few years that involved a low yield of teff. Many answered drought in the last year (7 respondents) or previous years (3 respondents); others mentioned hail last year (6 respondents). Two mentioned both drought and hail in the last year. Flooding and water-logging were also mentioned (3 respondents).

Most Ethiopian agriculture is dependent on rainfall, although some farmers have access to irrigation through dams such as the one mentioned at the beginning of the chapter. Those in the community in Tigray with irrigation consider dams a crucial strategy for withstanding drought, allowing farmers to harness consistent water amounts to produce such crops as oranges and peppers. Other farmers would like to access irrigation, either through dams or wells, and they talk about the value of these structures for adapting to drought; but they lack access to them because of land distribution, geography, and absence of funds to construct the channels.

However, drought is not the only risk facing farmers, nor is it often their most pressing issue. Other meteorological problems identified in interviews were hail, water-logging, and flooding. Many of the interviewed farmers also mentioned that a lack of capital or credit for inputs was a large constraint or problem, with fertilizer being the specific expensive input.

And some felt that high food prices in the market were a problem. A few mentioned a lack of livestock or land, and weeds. During focus groups in Ethiopia, while farmers spent some time discussing the years when drought destroyed crops near the end of the growing season, they also emphasized their problems with access to markets and their dependence on their aging irrigation system. (Focus group results differ from interviews, likely because the latter naturally focused on concerns of individuals, not a broader group.) An earlier survey (Teshome et al. 2008) indicated that lack of capital and high prices were a major issue for farmers in Tigray, even more important than drought for over half the farmers surveyed. Many other issues arose in focus groups and interviews, including increasing reliance on and costs of fertilizer, the need to expand the irrigation system to reach more fields, and health problems affecting the communities.

Adaptation

Our interviews and surveys also attempted to identify farmers' strategies for preventive and post hoc coping, including internal household strategies, community strategies, and support offered by external organizations—of which the safety net program (PSNP) was the most frequently listed, albeit not always used as a primary strategy. Most pre- and postevent coping was done through household strategies, such as changing seeds or selling assets (Tables 23.2 and 23.3). In interviews, respondents said that poor farmers cope with poverty by renting out land, taking credit, selling their labor, or renting oxen. Land and water conservation were also seen as important ways to avoid or escape poverty. Two respondents suggested that leadership or advice would help those who were poor by bringing them options for work or improvements in production.

Like poverty and wealth, coping strategies also depended on resources available and other characteristics (gender and age of the farmer). For example, wealthy farmers saw their irrigated land as a form of insurance against drought, like their large numbers of goats and sheep, savings, and available credit. Single female farmers, who were less able to labor on their own land than were male heads of household, or to migrate for wage labor, reported that they coped by leasing their land to landless farmers or others for a third of what it would earn the farmer. (Although they reported doing this less than male farmers did, it was still seen as an important coping strategy.) Asset-building by male farmers was also an important protective strategy, often unavailable to women, landless men, and those with fewer resources in general.

Despite these coping strategies, most farmers argued that, in the past few years, they had found themselves out of options, exhausting their reserves and ability to cope through using their usual strategies. Because of this problem, they were very interested in the index insurance project and very eager to participate without even knowing much about it. Our

Table 23.2 Coping strategies mentioned in response to crop failure during 2009 surveys

Coping Strategy (I = individual, C = community, E = external)	*Type of Coping Strategy*	*Number of Farmers Who Indicated It as Primary*	*Number of Farmers Who Mentioned It*
PSNP	E	104	119
Own funds	I	65	107
Donation	C	53	77
Sell livestock	I	42	157
Livestock share	C	36	55
Additional job	I/C	33	120
Food aid	E	30	75
Borrowing no interest	C	27	95
Loans/borrowing with interest	C/E	15	71
Eat less	I	13	156
Eat less favored food	I	13	147
Grants	E	9	69
Debts (general)	C/E	9	93
Defer expenses	I	8	57
Coop aid	C	8	39
Eat seeds	I	6	68
Resowing	I	5	44
Spend less clothing	I	4	148
Loans (general)	C/E	2	36
Migration	I	2	32
Sell assets	I	2	17
Eat wild food	I	1	31
Spend less medicine	I	0	87
Spend less school	I	0	73
Send children	I	0	23

survey assessed farmers' understanding of the index insurance contracts and concepts and found that the primary reason they did not buy insurance was that they had not yet heard of it and had not had the opportunity to purchase it. Part of the reason they were so eager is that the insurance program was going to be run through Oxfam America, the same international NGO that had helped to construct the dam that allowed for irrigated fields without the intensive labor needed for traditional hand-hewn canals. Many

Table 23.3 Preventative strategies mentioned by respondents

Preventative Strategies	*Type of Strategy (I = individual, C = community, E = external)*	*Number of Farmers Who Indicated It as Primary*	*Number of Farmers Who Mentioned It*
Pray to God	I	82	122
Increase chickpea	I	72	80
Decrease maize	I	50	55
Decrease millet	I	43	48
Change seed varieties	I	30	152
Put money in savings	I	24	90
Plant more irrigated land	I	23	110
Increase pepper	I	22	34
Infrastructure investment	I/C	20	117
Decrease pepper	I	16	22
Decrease banana	I	16	16
Increase land used for other crops	I	15	77
Decrease amount of land used for crops	I	15	76
Increase maize	I	15	17
Livestock sharing	C	14	17
Increase sesame	I	8	9
Decrease orange	I	7	14
Increase banana	I	6	6
Change planting timing	I	5	89
Increase orange	I	5	7
Buy livestock	I	4	8
Get loan for additional seed	E	3	55
Rent out land	I	2	22
Fertilizer use	I	2	16
Decrease other	I	2	3
Get loan for other crops	E	1	50
Share in extra land	I/C	1	21
Increase flax	I	1	1
Other (sell livestock)	I	1	1
Plant more rain-fed land	I		8
Attend training workshops	E		1

farmers said that they would support anything offered by Oxfam America as a way to maintain their connection to the INGO and its resources.

Another reason for their enthusiasm for the index insurance program is its connections to two of the most popular coping strategies, food-for-work participation and taking loans. These connections are now part of the holistic approach within the R4 initiative in addressing risk, which focuses on improved resource management (risk reduction), microcredit (prudent risk taking), insurance (risk transfer), and savings (risk reserves) (Oxfam America 2013). To make the insurance available to a wider clientele—namely, those without cash—the program also was designed to provide insurance in exchange for labor by partnering with the Ethiopian government's Productive Safety Net Program (PSNP). PSNP projects include such environmental conservation activities as terrace-building and tree-planting, which also benefit the community. Farmers report that they are hesitant to take loans because they fear not being able to repay them. In the context of discussing the insurance program, several commented that insurance might give them more confidence to take loans.

Most important, a majority of farmers said that drought was a reason that they purchased insurance, and farmers expect that index insurance will help them to better cope with drought. This expectation indicates that index insurance, even without the connection to Oxfam America, is seen as a valuable coping strategy for drought, particularly for those most vulnerable. Those who purchased insurance were more likely to be female, young, and a PSNP participant and to own less land and grow less teff.[7] PSNP participants bought on average twice what those paying cash purchased, although this relationship may be in part due to a tendency for farmers to undervalue their own labor, especially during the off-season PSNP activities.

HARITA Outcomes

The bulk of the research for this chapter was completed by 2010, but the HARITA project continued until 2012, when, as mentioned earlier, it became the R4 initiative. A recent evaluation suggests that HARITA was successful in many of its aims, seeing an increase in the average savings and number of oxen across the three evaluated districts, including the one examined in this chapter (Madajewicz et al. 2013). The report suggests that although results vary across these districts, in one or more of them farmers with the insurance, particularly female-headed households, are using more compost, fertilizer, and traditional seeds. The interest in the labor-for-insurance option is also outpacing the ability to provide this insurance option, while the insurance program in general is reaching 29% of the farmers in the districts.

Many farmers are excited by the new inputs and techniques offered by the partnership with REST, which provides agricultural extension services. Overall, although "almost all also agree that HARITA is not yet improving livelihoods in a transformative way, . . . it is too early to assess whether the program in its current form can achieve this goal [of improving living standards]" (Madajewicz et al. 2013: 2). From 2009 to 2012, HARITA grew from 200 households in one community to over 18,000 purchasers in 73 communities (ibid.: 8–9).

Intersection of Risks, Vulnerability, and Adaptation

The question remains: is index insurance going to help farmers cope with climate change? Understanding the different ideas of poverty helps to discern vulnerability and how landless men and widowed or divorced women may be more susceptible to hazards, given their lack of land or labor, respectively. Widowed women reported much lower incomes from the land they rent out and more frequent experiences of hunger, compared to men. Female-headed households are also members of fewer community organizations, rent out less rain-fed land, share in less rain-fed and irrigated land, grow less teff from year to year, and will plant less in the upcoming year—thus making gender an important component of wealth in the community and revealing the importance of looking at the intersection of wealth with gender in understanding vulnerability.

However, these gender aspects do not seem to affect the kinds of coping strategies the women employ. Women used a similar number of coping strategies as did their peers (6.3 vs. 6.9 strategies mentioned on average, respectively), albeit perhaps slightly different ones. For example, women place less importance on migration and livestock and more importance on changing seed varieties, increasing the amount of chickpea sown, PSNP, and food aid. Aside from these few differences, they do not differ in their use of other coping strategies or the overall number of either preventive or coping strategies. While insurance is more popular among women, the young, and the poor, perceived importance of coping strategies do not differ significantly among those we surveyed along lines of gender or wealth. This finding suggests that vulnerability and adaptation strategies are not tightly coupled in the community. Similarly, climate risk in the area may focus on drought, but many farmers seem to believe that other risks are equally or more important, primarily those related to the market. Farmers often gauged the insurance program goals and outcomes against the need for better markets and pushed for resolution of this need either in addition to or instead of index insurance for their crops.

The question about insuring the right risk also arose during our investigation. While drought and climate change are important issues in the area for farmers, will it be possible to improve agricultural outcomes by addressing them, or are there more fundamental risks that will continue to cripple Ethiopian farming, even if the effects of climate change can be mitigated (assuming increasing drought does not lead to increasingly unaffordable insurance premiums)? In other words, if current strategies are insufficient, adding another may not really change the situation by much. If understanding vulnerability, risk, and adaptation to climate change is insufficient in this case for conceptualizing agriculture in Tigray, we must examine other possibilities. If climate change is not at the root of these problems, perhaps potential solutions need to look past climate change to address them (O'Brien 2012, 2013). For example, as mentioned, in Tigray farmers almost universally identify drought as their biggest problem, and they can list the extreme droughts that occurred in the previous half-century. Climate change may increase the frequency and intensity of drought in Tigray, but drought itself is an old problem, with specific kinds of risks, experiences, and potential solutions. Since people are adapting to multiple stressors and processes without differentiating their responses, we must then understand adaptation as a broad response, rather than one specifically triggered by climate change or other events (O'Brien 2012).

Similarly, we can examine these problems to identify root causes. Drought in the late Victorian age was a major problem around the world owing to an extremely powerful El Niño weather system. Yet famine occurred only because of the way that food was distributed; for example, food was transported out of drought-stricken areas (Davis 2001). In this case, drought was only one factor affecting health and well-being. Explanations offered for Tigrinya agricultural stagnation include land erosion, desertification, population density, and egalitarian land distribution (Devereux, Teshome, and Sabates-Wheeler 2005). One popular explanation for such stagnation in Ethiopia and other areas is a poverty trap in which farmers lack the kinds of resources that can help them to improve their agriculture profitability (Carter et al. 2004). Interestingly, this explanation is similar to the "too equal" land-distribution explanation and ties into population density with the idea that the small land allocations are insufficient for improved agriculture. All these explanations rely on examining Ethiopia as a closed system—rather than considering international relationships and connections that might be affecting farmers in Tigray and the rest of Ethiopia, as well as the kinds of solutions that they might offer.

In addition, discussions of vulnerability often focus on the ability of a person, household, or community to adapt to change, and these assessments often depend on relative wealth, ability to access resources, or some sense of coping strategies or available options for adapting under different scenarios. In many cases these assessments are linked to older patterns of inequalities, either within the community or within a set of communities, that make one subset less able to adapt to change—for example, the social constraints on female migration, policies on land and labor allocation, and market pricing. Understanding the roots of these inequalities would also suggest ways that vulnerabilities are historical, political, and social in origin and that in many cases predate anthropogenic climate change.

O'Brien (2012, 2013) argues that adaptation and mitigation are not sufficient responses to climate change, in part because the current framing of adaptation "fails to engage with the real 'adaptive challenge' of climate change, i.e., a questioning of the assumptions, beliefs, values, commitments, loyalties, and interests that have created the structures, systems, and behaviours that contribute to anthropogenic climate change, social vulnerability, and other environmental problems in the first place" (2012: 668). Kelman and Gaillard also challenge the dominant framing of climate change as a hazard rather than as a deeply sociopolitical event, which can lead to maladaptation for large events (2010).

While Kelman and Gaillard advocate for understanding climate change as similar to other disaster risks, O'Brien (2012) calls for deliberately transforming our systems and societies (for example, to green economies; see also Gibson Graham 2006). O'Brien (2013) requires greater attention to processes of change—how humans approach change, why change is resisted, and how systemwide changes occur. She then argues that transformative change is needed to move us beyond the kinds of responses that currently dominate—such as those focused on emissions and "attitude, behavior, choice" models—to those that examine deeper power structures by questioning the assumptions on which models and policies are built. To conclude, she recommends a continued focus on systems rather than separate problems, on growth beyond economics, on power relations and interests, and on long-term thinking.

As our work in Ethiopia and other places has moved on to focus on larger systems of longer-term relationships, we find these approaches to climate change useful for thinking through new approaches to climate risks. Our work suggests that climate change, although an important component or aggravator for agricultural risks in the community, might not be the most effective target for addressing agricultural problems. In some cases the answer may be that climate change is the right focus for intervention, particularly in cases of sea-level rise and permafrost melt. However, even in these cases climate change may be exacerbating other kinds of issues of land or resource access. Continuing to ask what creates differential vulnerability to certain risks or hazards will maintain a focus on underlying issues that may be tied to inequitable distribution of resources, economic market risks, or other potential ways for hazard impacts to be magnified for some groups or communities. The 2013 evaluation of the project reveals that farmers and community leaders would like for HARITA to expand its scope and "address the obstacles that impede significant improvements in living standards in rural Tigray. They recommend that HARITA invest in irrigation, since current rainfall may be insufficient to support large increases in agricultural productivity, and that HARITA engage in diversifying rural livelihoods" (Madajewicz et al. 2013: 3). Although HARITA and R4 cannot address all problems facing farmers, staff members see the current program (R4) as a resilience program rather than climate change adaptation, with a suite of risk-management strategies, including index insurance. This kind of holistic approach can examine a broader scope of activities not necessarily limited to agricultural improvements.

The potential for vulnerability to risks to increase for certain groups and communities (and nations) might also lead us to think about risk in a slightly different way. Moving beyond the kinds of hazard-based risks that Kelman and Gaillard critique to those contextualized as outcomes of sociopolitical relationships, we might also think about the ways that risks are distributed in societies. Events like drought and earthquake, of course, may happen to anyone, but the ability to adapt to these events can depend on the ways that farmers in Ethiopia and elsewhere are exposed to risk and, more specifically, the ways that risks are distributed. The distribution of risk refers to the configuration of markets, livelihoods, economies, and other social systems and, in particular, to the ways and by whom risks are experienced or not, by certain actors. For example, when drought occurs, farmers may face a greater amount of risk in terms of being able to feed their families, while those who buy and sell crops, although their income may be affected by drought, often have other options in the market or reserves saved from their previous exchanges. Thus the availability of coping strategies can indicate the distribution of risk. However, beyond understanding the value of coping strategies, the idea of risk distribution is also relational, meaning that each farmer's risk level is related to that of other farmers. One way to think about this situation is to consider how the amount of profit earned by each farmer determines how risk is distributed, in that it provides more or less savings for each. However, risk transfer involves more than just economic or market exchanges; it is linked to how policies such as PSNP are implemented. For example, farmers who work and are thus expecting to receive food aid during a drought are also assuming more risk of the aid arriving late or not at all. Conversely, the aid agency must ensure the food is available in case of drought, assuming some risk that comes with accumulating potentially unnecessary stores of food. In facilitating projects that provide both remuneration and conservation benefits, the PSNP program also helps with some of the community risks due to erosion or fertilizer overuse, thereby reducing some of the potential impact of the hazard. However, the bulk of the risk remains with the farmer, given the other ways, already detailed, that market and other risks are distributed.

Although perhaps not able to offer concrete solutions for resolving these deeper issues of inequalities and distribution of resources and risks, anthropological and/or social science approaches clearly contribute much to our understanding of these dynamic relationships and is key to discern how non-climate-change drivers intersect with climate change and thereby how climate change adaptation projects might aid in improving efforts along these lines.

Conclusions

Like others who have studied Ethiopian agriculture, we argue that transformations may be critical: "Most fundamentally, however, visionary thinking—involving farmers and pastoralists as full and equal participants—is needed about where Ethiopian agriculture is headed in the long run, and how to achieve the structural transformation required to get there" (Devereux, Teshome, and Sabates-Wheeler 2005: 126).

In this chapter we take the perspective that transformation begins with how we think about vulnerability, adaptation, and risk—whether we understand risk as hazards (climate or otherwise) or as a distribution of risk with winners and losers often determined by structural locations of resource control, gender, age, and access (for example, who can take risks, who must take risks). The challenge for us is to rethink climate change adaptation, vulnerability, and risk in the context of agricultural systems that are in transformation. We have illustrated such a dynamic—in places such as Ethiopia, where the status quo of adaptation via coping strategies is not a solution. Given worldwide resource distributions and scarcities, and also looming threats to livelihoods and assets, these adaptive solutions may also become increasingly ineffective worldwide as well.

Notes

1. We would like to thank the many people and organizations who have supported this research and the larger project, including the farmers in Tigray and Oxfam America and its staff. We would also like to thank David Krantz, Eric Holthaus, Michael Norton, Connor Mullally, Gebrehaweria Gebregziabher, Woldeab Teshome, Astor Gebrekirstos, K. Muniappan, and Malgosia Madajewicz for their support of this and related work, as well as Susan Crate and Mark Nuttall for their valuable comments on this chapter. Funding for this research was provided by the Center for Research on Environmental Research (cred.columbia.edu) through NSF cooperative grants SES-0345840 and SES-0951516.
2. The irrigation structure had been funded and built by Oxfam America, who also sponsored the insurance project in Ethiopia.
3. For more information about the project, please refer to http://hdl.handle.net/10.1080/00220388.2014.887685, www.oxfamamerica.org/explore/research-publications/managing-risks-to-agricultural-livelihoods-impact-evaluation-of-the-harita-program-in-tigray-ethiopia-20092012/.
4. See Peterson 2012 for more detail on this study.
5. An evaluation of the HARITA project also suggests that women rent out more rain-fed land (similar to poorer farmers) because they do not have the labor or the oxen to do the ploughing themselves (Madajewicz et al. 2013).
6. All years in the Gregorian calendar; under the Ethiopian calendar the year 2007 began on September 11, 2014 of the Gregorian Calendar.
7. HARITA specifically targets PSNP participants.

References

Carter, M., Little, P., Mogues, T., and Negatu, W. 2004. *Shocks, Sensitivity, and Resilience: Tracking the Economic Impacts of Environmental Disaster on Assets in Ethiopia and Honduras.* University of Manchester: Institute for Development Policy and Management (IDPM).

Collier B., Skees, J., and Barnett, B. 2009. Weather Index Insurance and Climate Change: Opportunities and Challenges in Lower Income Countries. *The Geneva Papers* 34: 401–24.

Davis, M. 2001. *Late Victorian Holocausts: El Niño Famines and the Making of the Third World.* New York: Verso Books.

Devereux, S., Teshome, A., and Sabates-Wheeler, R. 2005. Too much inequality or too little? Inequality and stagnation in Ethiopian agriculture. *IDS Bulletin* 36(2): 121–26.

Gibson-Graham, J. K. 2006. *A Postcapitalist Politics.* Minneapolis: University of Minnesota Press.

Kelman, I., and Gaillard, J. C. 2010. Embedding climate change adaptation within disaster risk reduction. *Community, Environment, and Disaster Risk Management* 4: 23–46.

Madajewicz, M., Tsegay, A. H., and Norton, M. 2013. *Managing Risks to Agricultural Livelihoods: Impact Evaluation of the HARITA Program in Tigray, Ethiopia, 2009–2012.* Report to Oxfam America, http://www.oxfamamerica.org/static/media/files/Oxfam_America_Impact_Evaluation_of_HARITA_2009-2012_English.pdf Accessed September 30, 2014.

O'Brien, K. 2012. Global environmental change II: From adaptation to deliberate transformation. *Progress in Human Geography* 36(5): 667–76.

———. 2013. Global environmental change III: Closing the gap between knowledge and action. *Progress in Human Geography* 37(4): 587–96.

Oxfam America 2013. *R4 Rural Resilience Initiative,* http://www.oxfamamerica.org/static/oa4/r4-rural-resilience-initiative.pdf, accessed September 30, 2014.

Patt, A., Peterson, N., Carter, M., Velez, M., Hess, U., Pfaff, A., and Suarez, P. 2009. Making index insurance attractive to farmers. *Mitigation and Adaptation Strategies for Global Change* 14(8): 737–53.

Peterson, N. 2012. Developing climate adaptation: The intersection of climate research and development programs in index insurance. *Development and Change* 43(2): 557–84.

Teshome, W., Peterson, N., Gebrekirstos, A., and Muniappan, K. 2008. *Microinsurance Demand Assessment in Adi Ha Tabia, Tigray Regional State, Ethiopia.* Final Report to Oxfam America, Boston.

Chapter 24

Pedagogy and Climate Change

Chris Hebdon, Myles Lennon, Francis Ludlow, Amy Zhang, and Michael R. Dove

Pedagogy capable of teaching climate change will need to counter the issue's dominant framings, which are shaped by the politics of science and educational expertise. Climate change does not result from a single cause, nor is it amenable to a one-dimensional fix, despite how common it is to hear simple explanations or dismissals of it. To be successful, pedagogies need to foster abilities to think ecologically and to discern how human agency is implicated in this "natural" disaster (Claus et al. 2015). After all, climate change is but one aspect of global change, which, although it has accelerated over the past 300 years, has a longer history throughout the Holocene (Kirch 2005). Global change involves other alterations such as land-use changes, changes in the nitrogen and phosphorus cycles, ocean acidification, chemical pollution, biodiversity loss, stratospheric ozone depletion, global freshwater depletion, and atmospheric aerosol loading (Rockström et al. 2009). All these anthropogenic global changes and the potential ways to affect their course by definition lead back to human action. Thus, pedagogy is not a secondary activity confined to the classroom but an active verb that describes the way people can deal with the multiple social roots of climate change in order to meet its epochal challenges.

Precisely because global change is not one-dimensional in source or effect, understanding it requires a theoretical lens that engages a holistic human-nature system, not fixating on any one thing (for example, carbon) or the failings of one social group in isolation from their wider context. An adequate pedagogical response should not only trace the roots of anthropogenic climate change to its social infrastructures (Sayer 2012) but also help to make visible and sensible possibilities for counterhegemonic action, or how to shift processes of control into reverse (Nader 1997; Stryker and González 2014).

Hierarchy, a long-time concern for anthropology, is centrally relevant to global change. At the macro level we could note that *Homo sapiens*

Anthropology and Climate Change: From Actions to Transformations (2nd ed.) by Susan A. Crate and Mark Nuttall (Eds.)

collectively appropriate more calories produced on Earth than all other species combined (Barnosky 2009; Speth 2009). In everyday life, hierarchy influences what and how we learn about climate change. Information that is managed and transmitted top-down tends to create not a dialogue or two-way exchange of ideas but a monologue—as in the oft-repeated phrase "we need to educate the public." Hulme (2009: 217) has called this "one-way flow of knowledge and information" the "deficit model" of science education—it presupposes that one group has the answers while others lack them. This one-way model conceives of education as the filling (by the expert) of an empty bucket (the mind of the recipient).

By narrowly focusing on changing the limited knowledge of ordinary citizens, the deficit model draws attention away from what might be lacking in the knowledge and behavior of the powerful and the ways that different groups have different stakes in climate change mitigation. The deficit model may lead one to overlook the extent to which that hierarchical pedagogy, such as between laypeople and socially distant experts, is *internal* to the problem of climate change. Scientists, for example, may not understand why certain laypeople are unreceptive; while some laypeople may come to resent the way that many climate change discussions valorize expert facts over the knowledge and concerns of citizens struggling with everyday issues (Graeber 2007).

Social Climate Change

Dialogical forms of education have been promoted as an alternative to the deficit model since at least Paulo Freire's writing of *The Pedagogy of the Oppressed* (1970). His work has been elaborated in the field of critical pedagogy (Giroux 1983; Shor 1996; Gadotti 2010), which focuses on how personal experience (biography) relates to wider politics (history). In this model students become teachers and teachers become students. Everyone potentially has something to teach and to learn.

Critical pedagogy represents an oppositional strain compared with mainstream pedagogies, preferring being situated over imperial knowledge and rejecting the idea that while the expert speaks the student should be mute. Importantly, it helps one realize that there are many existing traditions of communication in which people with different skills work together: apprenticeship, cooperation, reciprocal work, town-hall-style democracy, community media, and other forms of mutual learning in which difference can be made productive (Berry 2008; Nader 2010, 2013; Hebdon 2013). A scientist may know what climate change means generally across the state of Montana, but Montanan locals may have the practical knowledge of what their specific farms, forests, and cities need (Scott 1998). Such a situation presents an opportunity for co-developing knowledge and actionable solutions—what Ivan Illich called "science by people rather than for them" (Lohmann 2013: 45).

Pedagogies that help us to deal with the complexities of surviving the Anthropocene will need to bridge knowledge and action, forefronting practical responses already underway in the here-and-now and encouraging critical engagement with them. This approach requires not only attention to how power differentials are at the heart of impasses around climate change but also the linking up of what is learned in the classroom with what can be learned from the politics of the wider world. In the following sections we examine pedagogical dynamics between knowledge, action, and inaction. Drawing on examples from international education, climate science, and renewable energy, we argue that failures to mitigate climate change actively have often co-developed with dominant educational approaches. At the same time, in the wake of these failures there is also an increasing awareness of the need for alternative pedagogical priorities.

International Climate Change Education

Publications such as the joint UNESCO and UNEP *Climate Change Starter's Guidebook: An Issue Guide for Education Planners and Practitioners* (2011a) and *Youth Xchange: Climate Change and Lifestyles Guidebook* (2011b) offer insights into how international organizations are framing contemporary global campaigns. These campaigns define environmental education as a strategic resource for the mitigation of and adaptation to future environmental impacts and aim to speak to a global audience to advocate action-oriented pedagogy in both formal and informal settings (UNESCO and UNEP 2011a). Although the ways education is interpreted in these campaigns are appealing as progressive measures, a critical reading of them suggests otherwise. These interpretations fall short in several areas: they ultimately perpetuate an uneven division between developed and developing countries; they obscure the ways that production and consumption are connected globally; and they limit broader critiques of the institutions that perpetuate climate change. Moreover, these frameworks of international education introduce an inaccurate dichotomy, rendering developing countries solely as passive victims who must adapt to deal with the impact of climate change and developed countries as those responsible for producing ethically minded consumers.

Critical theorists, including many anthropologists, have argued that efforts to cultivate public consciousness, environmental or otherwise, often portray dominant ideologies as "common sense" (Gramsci 1971; Agrawal 2005). They remind us that we must remain critical of how education contributes to how we think about difference (Foucault 1980). In this context, the 2011 UNESCO and UNEP education campaign reveals a clear parallel between the UN's two-pronged approach of mitigation (tackling the causes of climate change) and adaptation (dealing with the impacts of climate change) and the assignment of different rights and responsibilities to so-called developed versus developing nations. Whereas developed countries are called on to limit their contribution to ongoing climate change, developing countries are urged to "prepare learners"—in particular, rural and coastal communities along with economically vulnerable groups such as women—"for uncertain futures" (UNESCO and UNEP 2011a: 57). This implied division of labor establishes two different sets of educational frameworks that delineate and limit conversations around the purpose, goals, and potentials for climate change education. Furthermore, this division of responsibility hampers attention to the political and economic interdependence of developed and developing countries, lessening the possibility that education can or should alter these relations.

The discussions of mitigation in these materials focus exclusively on addressing individual lifestyle choices, most directly the reduction of CO_2 (UNESCO and UNEP 2011b). Within this rubric, environmental education aims to emphasize the connection between lifestyle choices and individual consumption, on the one hand, and climate change, on the other, to ultimately produce citizens who are, above all, responsible and conscientious consumers. *Youth Xchange*—a pamphlet aimed at educating younger audiences—focuses almost entirely on ethical consumption. Although its discussions of personal choices such as controlling energy use through smart meters, developing ethical shopping habits, and choosing mass transport (ibid.) are important, it fails to draw out the connections to larger structural systems or question the dominant ideologies of consumer culture and their implication in climate change. It fails to mention any form of collective response—organized civil action, for example.

In contrast, the dramatically different role assigned to developing countries is exemplified by the striking front page image of Chapter 4 of "Education and Climate Change" in the UNESCO's *Climate Change Starter's Guidebook* (2011a). The image is of a group of African school children gathered under a tree, a chalkboard propped up against a truck in this makeshift classroom (ibid.: 54). This picture is a potent reminder that universal access to education remains aspirational in many developing nations. For the vast majority of students like these, the purpose of education is represented as helping them to adapt to inevitable climate change. "New skills," the pamphlet suggests, "may be necessary to live with members of other ethnic groups and/or cope with a changing physical environment" (ibid.: 57). The report goes on to stress the need to prepare for climate change disasters including the "potential of relocation due to environmental conflicts and wars," the challenge of "retention of children in schools despite climate change induced poverty" (ibid.: 58), and, ominously, to "prepare learners for uncertain futures" (ibid.: 57). It is clear that under this rubric the role of the global poor is limited to the passive reduction of harm. This framing reinforces the image of victimized members of developing countries who are powerless to confront the causes of climate change.

The separation of mitigation and adaptation in the UN's current approach to environmental education has, in some senses, taken into consideration the unequal distribution of responsibilities. The program calls on developed countries to be responsible for the consequences of their consumption over many decades and recognizes that poorer communities are more vulnerable to climate change. However, their underlying assumptions about mitigation and adaptation reproduce contemporary relations of dependency between developed and developing countries. They obscure how production and consumption are connected globally and drive climate change while also precluding diverse alliances and solutions. Whereas developed countries have the potential to actively prevent climate change in this framing, developing countries are left to cope with its consequences.

In sum, international efforts to create a "planetary environmental culture" to combat climate change must first refrain from reinforcing existing uneven power relations. An alternative global environmentalism would go beyond discussions of "educated consumption" for developed countries and "skilled survival" for developing countries. Most notably, international environmental pedagogy would benefit from the "praxis" raised by critical pedagogy (Freire 1970; Giroux 1983; Shor 1996). Such a pedagogy could encourage students to reflect on larger structures of power that shape "common sense," produce aspirations for consumption, and stall efforts for both mitigation and adaptation to climate change.

Climate Science and Denial

Climate scientists have led early efforts in the communication and pedagogy of climate change and remain largely dominant. Yet this is a role in which they do not always feel at ease, in part because the norms of natural science often lead scholars to avoid direct political engagement and self-reflection about the politics of science. Of all self-held ideals, dispassionate objectivity is the main cornerstone among scientists. For much of the 20th century, trust accrued to scientists because of their perceived objectivity and political neutrality, and indeed this perception likely contributed to early successes in convincing governments of the need to take seriously the potential for large-scale anthropogenic climate change.

As climate research matured during the 20th century, it shifted from an understanding that humans had the potential to alter climate to a growing certainty that these changes were already underway. Calls for governmental action intensified, and as economists gained a prominent role, it became apparent that many climate scientists felt uncomfortable discussing the specifics of policy within the norms of economics. As Jamieson notes: "Economists often bring views to the table that are foreign or alienating to natural scientists. . . . [Economists] often assume, for example, that any decision involves winners, losers, and trade-offs . . . [and] that cost-benefit considerations are the foundation of public policy-making" (2014: 27).

Many climate scientists took the view that their role was to communicate the objective reality of anthropogenic climate change, at most only suggesting the need for *some* form of action. The premise was that others could formulate and enact appropriate policy (Hulme 2009).

The realization that mitigation policies would affect the short-term profits of powerful economic interests prompted some of the first climate denial campaigns, which sought to undermine the archetypal image of the impartial scientist. These campaigns highlighted (and often misrepresented) scientists' political activities and commitments and exploited climate scientists' self-conceptions to promote division and anxiety regarding the degree to which scientists should engage in advocacy. They promoted images of career-driven climate scientists seeking tenure and preaching (as "climate alarmists") the likelihood of adverse impacts to elevate the importance of their discipline and so benefit from abundant funding. Those climate scientists that have been vocal advocates of active mitigation risk depiction as insular ivory tower elites promulgating economically regressive policies in subservience to green ideologies (Oreskes and Conway 2011; Mann 2012).

To engage in direct advocacy may thus invite attacks on one's professional integrity, and it may also entail uncomfortable judgments about socioeconomic policy and governance that lie beyond the climate scientists' expertise. Within the climate research community, the label "activist scientist" can be one of disapprobrium, and it can be cited as affecting the integrity of the science (for example, Pielke, Jr. 2010). Despite these pressures, because climate scientists carry an acute burden of awareness of the reality of human-induced climatic change, they may still feel obligated to act beyond strictly academic teaching and communication (Nelson and Vucetich 2009).

While it may be tempting to shelter behind the idea of politically detached science, there are lessons that climate scientists can learn from their experience with climate denialism—for example, regarding the dis-utility (and, at a time of possible environmental crisis, even the immorality) of the pedagogic limitations that their own norms and paradigms impose. Elevating science above other forms of social and political activity invites counterattack and suffers by comparison with forms of engagement that are more egalitarian and reflexive. From this perspective the difficulty is not only "educating the public" but also an introspective challenge for climate scientists. Against a background in which current climate science communication strategies have failed to motivate sufficient social and political action (Jamieson 2014), this increased self-knowledge is urgently needed (Barnosky et al. 2014).

A Hermeneutics of Climate Change Studies

Many climate scientists suggest that public skepticism or outright hostility toward the climate change science is due to either the immature state of this science and/or to failure to communicate—to "educate the public" with—what knowledge we have. Many climate change skeptics are also focused on public education, but this often consists of state-based efforts to block climate change education in schools (for example, in Virginia and Oklahoma in the first half of 2014). Both sides of the climate change debate share a belief in the importance of knowledge and education, but there is a difference in how education is viewed—as neutral by climate scientists versus partisan by climate change skeptics.

It is clear that the academic/scientific community in general, and the climate science community in particular, was unprepared for the current debate over climate change, especially in the United States, which has the highest rates of denial in the world (Ipsos MORI 2014). It has been suggested that the modern natural sciences never had a "hermeneutics," a self-reflexive and critical awareness of how they do what they do, which might have better prepared them for this sort of popular miscomprehension. The natural sciences have not had, in Markus's (1987: 9) felicitous phrase, a "neurosis philosophicus." The answer to Heidegger's (2000) question "does science think?" was "no." It has been suggested that for a long time this was not a problem for the modern natural sciences—that they flourished without it. Indeed they did, up until the recent rise on the American political right of skepticism of science in general and of climate (and evolution) science in particular. Now the long-successful norms of the natural sciences are failing them; the failure to "think" has become a crippling handicap. The norms of modern science may thus be imbricated in the modern attacks on it, including the earlier-discussed norms of detachment and avoidance of political engagement. In this sense, climate science and climate denialism may be said to have co-developed.

Not only has climate science/natural science been caught flat-footed, so have we as social scientists. Some of the most prominent and perspicacious scholars seem to be at a loss to explain climate change denialism (although see Fiske's chapter in this volume for a discussion of climate skepticism in the United States). One of the most influential economists in the United States, Paul Krugman (2014), in a recent op-ed, implausibly tries to explain "the venom, the sheer rage, of the denialists," which clearly impresses and disturbs him, in terms of the threat that climate science supposedly poses to the libertarian teachings of Ayn Rand. A more fundamental probing of beliefs on both sides of the debate eludes him.

Scholars who have spent a generation studying the social reproduction of science have mostly baulked when it comes to climate change science. Perhaps most famously, Latour (2004: 227) said he would hesitate to critique the methods by which climate scientists claim authority for their findings, because he deems those findings so important: "The danger would no longer be coming from an excessive confidence in ideological arguments posturing as matters of fact—as we have learned to combat so efficiently in the past—but from an excessive *distrust* of good matters of fact disguised as bad ideological biases!" It follows that Latour would be equally hesitant to rationalize the critiques of climate science methods and findings by the skeptics. All of this is reminiscent of what Ortner (1995) calls "ethnographic refusal"—a reluctance to submit to professional scrutiny the behaviors of those with whom we are in sympathy, such as climate scientists.

In the case of climate change, this reluctance encompasses not only those we support—the climate scientists—but also those we don't support—the climate science deniers. There has been remarkably little ethnographic attention to the skeptics—Ortner's point that we have tended to ignore "the lived worlds inhabited by those who resist" (1995: 187–88) seems apropos here—and similarly little attempt to explain this neglect. In short, there are ample grounds for thinking that we also lack a hermeneutics of the social science of climate change.

Real-world events are challenging the hermeneutics—or lack thereof—of climate change scholarship. These events are demanding a more explicit grappling with the central question of the neutrality—perceived or otherwise—of climate science and education, and indeed of science in general.[1] Gregory Bateson offers a possibly useful perspective on this challenge. Bateson always recommended a broad and systemic view of environmental problems, including the way that we study problems and communicate the results of our studies. As he wrote: "The problem of how to transmit our ecological reasoning to those whom we wish to influence in what seems to us to be an ecologically 'good' direction is itself an ecological problem" (Bateson 1972: 504). The modern history of the development of climate science and the climate policy regime reveals insufficient attention to—in Bateson's words—"those whom we wish to influence" and who, by all available evidence, are very clearly *not* being influenced by us.

Renewable Energy and Professional Denial

One of the big surprises of early anthropological works on the politics of the transition to renewable energy was that some highly trained energy scholars harbored unscientific beliefs and made unsound forecasts that impeded this transition. Economists, for example, often held fast to conventional ways of calculating costs and benefits, which excluded environmental and social "externalities" and so made fossil fuels and nuclear energy appear safer and cheaper than they actually were. Physicists and engineers trained to work within the massive infrastructures of the fossil and nuclear industries were often at a loss to comprehend the social, political, and environmental benefits of downscaling their industries and moving toward more democratic, human-scaled, decentralized systems of renewable energy. Many of these energy researchers were little affected by, if not dismissive of, "political" critiques of what they considered to be their "nonpolitical" analyses.

After working between 1976 and 1979 on a National Academy of Sciences project with 300 energy specialists, Laura Nader (1979: 17) concluded: "the toughest problem will be to get professionals to look inside themselves, to see what their mindset problems are." In subsequent decades, even as fields such as environmental economics became more savvy at calculating externalities into the cost of energy, it became clear that for most energy researchers it was still taboo to consider "the possibility that experts might be part of the problem," since many an expert "thought that he stood outside of the problem" (Nader 2004: 776).

Hermann Scheer, one of Germany's political leaders and a supporter of that nation's *Energiewende* ("energy transition") since its start in 1991, noted that the "widespread [90%] popularity of renewable energy has developed despite decades of extensive denunciation by the traditional power industry and the majority of energy experts" (2012: 2). Energy analysts often played a counterproductive role in the *Energiewende* by producing incorrect forecasts with a veneer of objectivity—for example, saying in 1990 that "'exotic' energies simply don't offer more than a 5% potential" (ibid.: 24). Germany today produces more than 30% of its energy from renewable sources, and 90% of this investment has come from citizens and their municipalities; only 10% has come from the traditional utility sector, despite its greater access to financing (ibid.: 41). Many energy analysts claimed that it would be impossible for distributed energy to play more than a partial and limited role in the national system, claims that persisted even after renewables repeatedly surpassed previous predictions. Mindsets were harder to change than technologies, probably because the two are not separate, as scholars of "technopolitics" have emphasized (Porter 1995; Edwards 2010).

The German case serves as a reminder that just because someone is trained in science does not mean that he or she will always exercise scientific judgment, especially when dealing with issues beyond his or her specialized experience. Denmark in particular has come up with a more "side-by-side" way to deal with this problem of expert bias. Their renewables law mandates that at least 20% of the ownership of any wind power operation must be opened up to everyone living within 4.5 km of the site, thus bringing locals squarely into the planning process *and* providing them with a new source of income. By 2005, 88% of wind projects were owned by citizen-controlled cooperatives (Sovacool and Sawin 2010: 53). This arrangement works because many Danes recognize that citizens and energy professionals have different kinds of expertise, and thus they can work together and learn from each other. In so doing they are helping to achieve Denmark's goal of electrical power generation being 100% renewable by 2030 and transportation—by way of electrical trains and cars—being 100% renewable by 2050. As of 2016, Denmark is ahead of schedule. They offer a clear example of the benefits of shifting decision making—and the educational process—away from the deficit model.

Possibilities

One of the advantages of an anthropological pedagogy of climate change is that in questioning how dominant approaches operate we can also make visible alternative possibilities. A critical stance can help to reveal ways beyond major impasses. In particular we have called attention to the ways that dominant pedagogical approaches have often been imbricated in ongoing failures to actively mitigate climate change. International education campaigns have too often ignored uneven power relations. Critical pedagogy could address this omission and be used in working toward a new form of global environmentalism. The premise of objectivity among climate scientists, which at times has enabled trust, has more recently become crippling as scientists have interacted with questions of state policy and denial movements that question the need for positive climate policy. This not only suggests that the norms of climate science have co-developed with the modern attacks on it; it also identifies a need to think about climate science itself. In Germany and Denmark, expert claims that renewable energy is "not ready for primetime" have often conflicted with citizen priorities for renewable energy now. In Denmark, this situation has led to a linking up of energy professionals and citizens through wind farm cooperatives, enabling mutual learning and benefit.

The co-development of knowledge and inaction, upon further inspection, suggest their opposite—increasing awareness about what educational forms will be required for effective action. In all cases there is a need for critical attention to social power and for reframing educational questions within wider contexts. Rather than simply a problem of carbon in the sky, these critical approaches redirect our attention to how inequalities are involved in the social production of global change from the ground up.

Note

1. A summer 2014 bill introduced into the U.S. Senate proposes to fund a blue ribbon commission to "enhance scientific regard among the American public."

References

Agrawal, A. 2005. *Environmentality*. Durham, NC: Duke University Press.

Bateson, G. 1972. *Steps to an Ecology of Mind*. New York: Ballantine.

Barnosky, A. D., et al. 2014. Introducing the scientific consensus on maintaining humanity's life support systems in the 21st century: Information for policy makers. *The Anthropocene Review* 1(1): 78–109.

———. 2009. Approaching a phase shift in Earth's biosphere. *Nature* 486.

Berry, W. 2008. Faustian Economics. *Harpers* (May).

Claus, A., Osterhoudt, S., Dove, M. R., Baker, L., Cortesi, L., Hebdon, C., and Zhang, A. 2015. Disaster, degradation, dystopia, in R. Bryant and S. Kim (Eds.), *International Handbook of Political Ecology*. Cheltenham: Edward Elgar.

Edwards, P. 2010. *A Vast Machine*. Cambridge, MA: MIT Press.

Foucault, M. 1980. *Power/Knowledge*, C. Gordon (Ed). New York: Pantheon.
Freire, P. 1970. *Pedagogy of the Oppressed*. New York: Continuum.
Gadotti, M. 2010. Reorienting education practices toward sustainability. *Journal of Education for Sustainable Development* 4: 203–10.
Giroux, H. 1983. *Theory and Resistance in Education*. Westport, CT: Bergin & Garvey.
Graeber, D. 2007. An Army of Altruists. *Harpers* (January).
Gramsci, A. 1971. *Selections from the Prison Notebooks*, Q. Hoare and G. N. Smith (Eds. and Trans.). London: Lawrence & Wishart.
Hebdon, C. 2013. Activating pedagogy. *Kroeber Anthropological Society Papers* 102/103.
Heidegger, M. 2000. Gesamtausgabe, Band 16. *Reden und andere Zeugnisse eines Lebensweges (1910–1976)*, H. Heidegger (Ed.). Frankfurt: Vittorio Klosterman.
Hulme, M. 2009. *Why We Disagree about Climate Change*. Cambridge: Cambridge University.
Ipsos MORI. 2014. *Public Attitudes to Science 2014*, www.ipsos-mori.com/Assets/Docs/Polls/pas-2014-main-report.pdf
Jamieson, D. 2014. *Reason in a Dark Time*. Oxford: Oxford University.
Kirch, P. V. 2005. Archaeology and global change. *Annual Review of Environment and Resources* 30.
Krugman, P. 2014. Interests, ideology, and climate, June 8, *The New York Times*.
Latour, B. 2004. Why has critique run out of steam? *Critical Inquiry* 30(2): 225–48.
Lohmann, L. 2013. *Energy Alternatives*. Dorset: Corner House.
Mann, M. 2012. *The Hockey Stick and the Climate Wars*. New York: Columbia University Press.
Markus, G. 1987. Why is there no hermeneutics of natural sciences? *Science in Context* 1(1): 5–51.
Miller, D. 2012. *Consumption and Its Consequences*. Cambridge: Polity.
Nader, L. 1979 [2010]. Barriers to thinking new about energy. *The Energy Reader*. Oxford: Wiley-Blackwell.
———. 1997. Controlling processes. *Current Anthropology* 38(5): 711–37.
———. 2010. Side by side: The other is not mute, in A. Iskandar and H. Rustom (Eds.), *Edward Said*. Berkeley and Los Angeles: University of California Press, pp. 72–85.
———. 2013. *Culture and Dignity*. Malden, MA: Wiley-Blackwell.
Nelson, M. P., and Vucetich, J. A. 2009. On advocacy by environmental scientists. *Conservation Biology* 23(5): 1090–1101.
Oreskes, N., and Conway, E. M. 2011. *Merchants of Doubt*. London: Bloomsbury.
Ortner, S. 1995. Resistance and the problem of ethnographic refusal. *Comparative Studies in Society and History* 37(1): 173–93.
Pielke, R., Jr. 2010. *The Climate Fix*. New York: Basic.
Porter, T. 1995. *Trust in Numbers*. Princeton, NJ: Princeton University Press.
Rockström, J., et al. 2009. A safe operating space for humanity. *Nature* 461.
Sayer, N. F. 2012. The politics of the anthropogenic. *Annual Review of Anthropology* 41.
Scheer, H. 2012. *The Energy Imperative*. London: Routledge.
Scott, J. C. 1998. *Seeing Like a State*. New Haven, CT: Yale University Press.
Shor, I. 1996. *When Students Have Power*. Chicago: University of Chicago Press.
Sovacool, B., and Sawin, J. L. 2010. Creating technological momentum. *Whitehead Journal of Diplomacy and International Relations* 11(2): 43–57.
Speth, J. G. 2009. *The Bridge at the End of the World*. New Haven, CT: Yale University Press.
Stryker, R., and González, R. (Eds.). 2014. *Up, Down, and Sideways*. New York: Berghahn.
UNESCO and UNEP. 2011a. *Climate Change Starter's Guidebook: An Issue Guide for Education Planners and Practitioners*. Paris: UNESCO and UNEP.
———. 2011b. *Youth Xchange: Climate Change and Lifestyles Guidebook*. Paris and Nairobi: UNESCO and UNEP.

Chapter 25

Bridging Knowledge and Action on Climate Change: Institutions, Translation, and Anthropological Engagement

Noor Johnson

To initiate anthropological fieldwork in Nunavut Territory in the eastern Canadian Arctic in 2009, I filled out a research licensing form and emailed it, along with a summary of my proposed project and a sample questionnaire in English and Inuktitut, to the Nunavut Research Institute (NRI).[1] All researchers who want to pursue studies in Nunavut must have a research license before they can begin data collection; the licensing process gives the government some oversight over the focus and location of research projects in the territory (Gearheard and Shirley 2005). In my application I explained that I wanted to study how Inuit knowledge about climate change is mobilized into different decision-making and policy arenas and that I hoped to spend time in Clyde River, a predominantly Inuit community with a population of around 1,000 people in northeast Baffin Island, and in Iqaluit, the capital of Nunavut.

Several weeks later, NRI forwarded a response from the research committee of the Ittaq Heritage and Research Center of Clyde River—or *Kangiqtugaapik* ("a nice little harbor") in Inuktitut—in which they questioned the necessity of my proposed project based on the fact that a considerable amount of work on climate change had already been done in their community. This fact was part of the appeal for my research, since I was interested in what frameworks, understandings, and actions on climate change had developed in the community through these earlier projects. Concerned about repetition, however, Ittaq queried whether or not "the research is of interest and benefit to *Kangiqtugaapingmiut*" (Clyde River residents). Ultimately the committee concluded that they "did not have reason to decline the application," and so they invited me to respond to their questions.

This exchange sensitized me to Clyde River residents' concerns about research practices and also the ways that previous research in the community shaped understandings of climate change. It also demonstrated the role of institutions in mediating processes of knowledge documentation and mobilization, illustrating the institutional networks through which Ittaq interfaced with NRI, which, in turn, is connected to various higher educational bodies around the world through its role in the research permitting process.

Institutions play a central role in organizing, documenting, and disseminating knowledge about climate change (Agarwal, et al. 2009). Framing the licensing process as an ethnographic encounter, rather than a bureaucratic hurdle to be overcome on the way to the field, illustrates the degree to which anthropological fieldwork on climate change is mediated through the same institutional practices that shape social responses to climate change. In the Canadian Arctic context, federal and territorial institutions provide funding for climate change research and adaptation programs through grant-making processes, while local organizations serve as translators between goals and priorities at the community level and those imposed through research and governance practices.

As a region, the Arctic has received significant attention in global scientific and social science engagements with climate change (Orlove et al. 2014), and the eastern Canadian Arctic has been a hub for research examining the human dimensions of these changes (Gearheard 2002; Henshaw 2006). Scholarly work on climate change in Inuit contexts, however, has had a fairly localized "place-based" (Crate 2011) focus, documenting the observations of Inuit elders and hunters of environmental change (Huntington and Fox 2005; Laidler, Dialla, and Joamie 2008), or assessing climate change impacts and adaptation needs using a standardized vulnerability framework (Ford and Smit 2004; Pearce et al. 2012). Shifting the frame to focus on the bureaucratic practices of Inuit institutions and institutional networks provides a counterpoint to representations of climate change that frame indigenous actors as either knowledgeable about or vulnerable to climate change impacts owing to their close association with and dependence on land-based activities (Cameron 2012).

As Peter Rudiak-Gould notes, translating climate change between different domains "necessarily entails transformation" (2012: 47). In this chapter I consider the climate change translations and transformations enacted in two institutional contexts, focusing on how different perspectives on change were negotiated or reconciled, and on the agency of individual spokespersons in these processes. First I discuss the role of community institutions in Clyde River in facilitating climate change research and adaptation projects, focusing in particular on grant and report writing as institutional technologies. I then consider translation in the representational work of the Inuit Circumpolar Council, an organization that advocates on behalf of Inuit in the Arctic Council, the United Nations, and other international policy bodies. I focus on the ways that local observations of change were reframed to fit into a human rights petition, considering both the creativity involved in translating climate change into the framework of human rights, as well as some of the drawbacks of these transformations in terms of Inuit agency.

I conclude with some reflections on the role of anthropologists as actors who are, like our interlocutors, embedded in institutional networks and practices as we study and represent climate change. I examine some of the ways that my research became entangled in the institutional practices I set out to study and reflect on the need to bring these entanglements—and the acts of translation that anthropologists enact—more squarely into focus in anthropological analyses of climate change as a socio-natural process.

Community Institutions and Climate Change in the Proposal Economy

Reclaiming ownership of knowledge production has been an important part of decolonizing relations between Inuit and *qallunaat*(non-Inuit) in Nunavut. In Clyde River, residents founded the Ittaq Heritage and Research Centre to mediate between the interests of visiting researchers and those of community residents and to promote community involvement in research, monitoring, and cultural heritage projects. Ittaq has installed local sea-ice monitoring stations and portable weather stations (Weatherhead, Gearheard, and Barry 2010; Gearheard et al. 2013) and developed and tested GPS units to document hunters' observations of environmental change while also capturing the routes they traveled and how they changed over time (Gearheard et al. 2011).

Ittaq's parent organization, the Ilisaqsivik Society, runs a range of social programs focusing on health and wellness, many of which incorporate Inuit knowledge and land-based activities. While climate change is not a significant focus of Ilisaqsivik's programming, their land programs offer opportunities for elders to share observations and knowledge with youth, and they have developed sea-ice safety workshops focused on how to travel safely in uncertain and changing conditions. From the perspective of Clyde River residents, what sets Ittaq and Ilisaqsivik's initiatives apart is that they are initiated and facilitated by community members, and they provide information that informs individual and household decisions in support of safe travel on the land.

While community institutions implement these programs, federal and territorial government agencies provide the needed funds through grant-making processes. In North America, divestment from directed assistance to local governments and organizations has been a hallmark of neoliberal reform in the social sector. Core or block funding for nonprofits has been replaced by a project funding model in which nonprofit staff must juggle grant applications and reporting requirements from multiple sources, making it much harder for these organizations to meet their overhead expenses (Baines 2010; Stern and Hall 2010; Harney 2011). In this environment, grant writing has become the primary mechanism that local institutions use to secure the funds needed keep their doors open, lights and heat on, and staff paid, a phenomenon that Pamela Stern and Peter Hall (2010) refer to as "the proposal economy."

The proposal economy is evident in the climate change adaptation initiatives of the Canadian government, particularly those aimed at engaging First Nations and Inuit communities. Funding for climate adaptation, channeled through three federal agencies, Natural Resources Canada, Health Canada, and Aboriginal Affairs and Northern Development Canada, is awarded to specific projects through a series of internal and external grant making processes.[2] These initiatives translate and translocate the idea of "adaptation," established in international political negotiations (Orlove 2009), into community settings through programs and research projects. To illustrate the local implications of these practices in action, I turn to a women's retreat on climate change facilitated by Ilisaqsivik.

Report Writing and Other Technologies of Translation

In 2009 Ilisaqsivik received funding from Health Canada's Climate Change and Health Adaptation program to conduct an Inuit women's berry-picking retreat. The goals of the retreat were to support women in enlarging and strengthening their support networks and to document their knowledge and observations of environmental change in an informal way while spending time on the land. Two *qallunaat* researchers helped to prepare the proposal and worked with community members on the format of the retreat, while Inuit staff members of Ilisaqsivik did most of the retreat planning and facilitation. After spending five months in Clyde River earlier that year, I went along as an observer and visiting researcher.

During the retreat, 21 women from Clyde River and the community of Qikiqtarjuaq, located about 372 km southeast of Clyde River, spent five days together harvesting crowberries and the small blueberries that cover the Arctic landscape in late summer. They participated in conversations in the morning and evening, discussing questions that the Ilisaqsivik staff and researchers had developed ahead of time. During the rest of the day, they walked the landscape alone or in groups of two or three, plastic buckets in hand, picking berries and conversing informally. Retreat participants focused on how the land offered a source of support and well-being when other parts of their lives felt difficult to control. They shared stories about past trips on the land, locations where berries were most plentiful, and the different hardships associated with life on the land and in the settlement. They talked about their struggles with spouses, their worries and hopes for their children, the pain they felt about addiction and suicide in the community, and the difficulties they had finding money to feed their families. The women did not volunteer information about climate change or even the broader category of environmental change, and the facilitators and researchers did not force the topic. Instead, the process was iterative, using the opportunity to be on the tundra together where environmental knowledge and social processes were interwoven (Johnson 2012).

After the retreat, in accordance with grant requirements, Ilisaqsivik submitted a report to Health Canada that summarized the retreat activities and emphasized the holistic vision of change that had emerged from the conversations of the women during the retreat. One of the group conversations on the retreat, for example, had been centered on the question: "What do you think it means to be a good person, or a good woman?"—a question intended to prompt people to reflect on Inuit women's lives in their totality. The report stated: "One might not think that 'being a good person' would fit within a climate change centered project, but it is precisely by being a good, healthy person, in a healthy community, that we will be able to deal with any change that comes our way. It was this core belief that created the foundation for the women's design of the retreat."

A Health Canada staff member responded to the final report with a request for more specific information linking climate change and health adaptation, asking: "Could you share the data that was collected? What was the role of researchers while out on the land? What are the changes being observed by the women?" and concluding: "There are a lot of details missing, and I cannot approve the final activity report at this time."

Health Canada and the community members involved in the retreat had two very different conceptions of how climate change should be understood and prioritized. While the women focused on change as it emerged from the total environment of family, animals, tundra, sea ice, and settlement, Health Canada staff wanted the project to yield clear connections between "health" on the one hand and the "environment" on the other. Their interest in identifying clear linkages between health and environment reflected the fact that this program was, itself, embedded in the federal government hierarchy and was vulnerable to funding cuts. The staff members were hoping to assemble evidence that could convince bureaucrats across scales of government to keep funding flowing to communities through their program.

How were these different understandings negotiated? Ilisaqsivik prepared a second report, drawing on a survey administered after the retreat was over. The survey's clear emphasis on country food, health, and environmental change met Health Canada's need for "data" that could be reported up the federal chain. While accommodating the funder's request, the revised report also critiqued the assumption, embedded in the framing provided by Health Canada, that "what women wanted to talk about regarding climate change was observations of change." It rearticulated the women's relational conception that engaging with and responding to change "means being a strong woman, having strong families, and joining together as a community."

Local institutions such as Ilisaqsivik and Ittaq play a critical role in translating between the bureaucratic norms of federal and territorial agencies and the local visions and goals of Clyde River residents. When it comes to climate change, the differences between these scales of experience are significant—they carry very different ideas about the nature of change and the kinds of activity that can strengthen community-level responses. In this example, the report document served as a technology through which Ilisaqsivik was able to mediate between these different visions and understandings of change. It enabled Health Canada to stabilize their understanding of health as centrally related to climate change, but it also served as a vehicle of resistance to this framing on the part of Ilisaqsivik staff.

In the language of networks, institutions are "nodes" that connect dispersed actors and consolidate particular understandings of change (Mathews 2009). Through practices of grant and report writing, individuals translate between nodes, facilitating agreement or finding ways for diverse perspectives to coexist within climate change research and adaptation initiatives. These practices can be seen as a kind of brokerage, which David Mosse and David Lewis suggest involves mediating between "different rationalities, interests, and meanings, so as to produce order, legitimacy, and 'success' and to maintain funding flows" (Mosse and Lewis 2006: 16).

Proposals are documentary artifacts that circulate through institutional networks in defined ways. As Matthew Hull notes, documents have received little attention in anthropological work in part because it is easy to "see them as giving immediate access to what they document" (Hull 2012: 253) rather than as technologies that mediate relations and mobilize particular ideas through networks of circulation. Considering documents such as grants and reports as mediators that shape relations and enable resource flows also facilitates a more nuanced understanding of the work of institutions such as Ilisaqsivik. For Ilisaqsivik and many other community institutions, documents such as grants and reports are critical technologies that support the legitimacy of the institution's work in the community by providing resources to fund programs like the women's retreat. In the era of the proposal economy, these resources underwrite nearly all institutional practices, providing funds for facilities and salaries in addition to programmatic support.

Bridging Local and Global: Translating Inuit Experiences for Global Audiences

While the example of Ilisaqsivik and the retreat report illustrate the role of local institutions as brokers in climate change projects, institutions also play a role in mobilizing local observations and experiences of change into regional, national, and global governance arenas. Here, too, individual actors, including employees and researchers, serve as translators, "the people in the middle," as Sally Engle Merry (2006) has described them, who try to interpret local experience in the context of the legal and scientific frameworks that operate at global scales. The Inuit Circumpolar Council (ICC), an institution that represents Inuit in the Arctic Council and in a number of United Nations bodies including the UN Framework Convention on Climate Change (UNFCCC), has used a variety of techniques to mobilize Inuit knowledge and represent Inuit experience in these venues.

In 2005, for example, ICC developed a legal petition in collaboration with two environmental law organizations, which was submitted to the Inter-American Commission for Human Rights (IACHR). Drawing on both scientific knowledge and Inuit knowledge, the petition was among the first legal documents to make climate change a human rights issue. To prepare, testimonies were collected from 63 people in 15 communities across the Alaskan and Canadian Arctic, including 6 residents of Clyde River. The legal team asked residents a series of questions focusing on traditional uses of the land and skills related to hunting and survival. They then carefully selected particular narratives from interview transcripts to support the evidentiary chain they were trying to construct to demonstrate the direct impacts of climate change on the ability of Inuit to pursue traditional livelihood activities.

Take, for instance, the following exchange between the petition interviewer and Apak, a man in his early 70s:

> INTERVIEWER: Did you used to use igloos earlier in your life more?
>
> APAK: Yes, very much so, it was our main form of shelter when you are traveling. While you were out, as soon as it would start to get dark outside you would start searching for snow that was suitable for igloo.
>
> INTERVIEWER: Why don't you rely on igloos anymore?
>
> APAK: Because we have great accommodations from qallunaat (non-Inuit) such as tents and other items that are easier to use.
>
> INTERVIEWER: Is it more difficult to find the right ice and snow conditions to build the igloo now?
>
> APAK: Yes, it's harder to find snow suitable for an igloo, and I want to ask a question. Are people aware of the wind from the North? It is now shifted, our prevailing winds from the North. It's almost . . . turned. (IsumaTV 2009)

In his interview, Apak shared observations about environmental change that supported the petition's main point that climate change is significantly affecting Inuit engagement with the Arctic environment. At the same time, his perspective on the relationship between environmental and social change was more nuanced than the petition's legal framework would allow. He explained that while climate change had affected the quality and availability of snow needed to build igloos, he stopped building igloos by choice because of the availability of tents. When the interviewer attempted again to construct a narrative about the importance of igloos, Apak tried to redirect the conversation to focus on the changes that he had observed. By contrast, the petition referenced the impact of climate change on igloo building in a section entitled "Deteriorating ice and snow conditions have diminished the Inuit's ability to travel in safety, damaging their health, safety, subsistence harvest, and culture" (Watt Cloutier 2005: 39–48). Here, igloo building knowledge was presented as "an important component of Inuit culture" and a crucial technology for safe travel on the land.

While the IACHR ultimately rejected the petition, it garnered widespread media coverage for ICC and helped to introduce into the framework of campaigns for climate justice the idea that climate change is a human rights issue (Osofsky 2009). The petition represented a creative engagement with a bureaucratic process that, in ways similar to the report that Ilisaqsivik submitted to Health Canada, required local understandings be reframed to meet a standardized logic and format. Unlike the two-way translation of ideas and resources in the women's retreat, the translational work of the petition was designed to render local experiences and observations in a format of interest to a very particular audience: activists, law makers, and others seeking ways to ground the diffuse, global idea of "climate change" into particular narratives of suffering and loss.

In addition to the petition, ICC spokespersons have used a number of strategies to translate the highly diverse ideas and experiences of Inuit about climate change into clear narratives to deploy in global settings. They have written position papers that weave together data from satellite models with in situ observations of sea-ice loss, hosted side events and press briefings at United Nations climate change meetings to promote Inuit knowledge as a resource for climate policy, and argued the need to expand the current framework for adaptation assistance under the UN Framework Convention on Climate Change. This framework, the Adaptation Fund, currently facilitates transfers of funding from developed to developing country states; ICC argued that the funding should include mechanisms to support Inuit and other groups at a subnational level who live in developed countries yet lack adequate adaptation support.

ICC is an institution in the middle; its purpose is to make connections between the concerns of dispersed Inuit communities and the policy priorities of global institutions like the UNFCCC. ICC's role, then, is defined in many ways by the necessity to translate, which in global climate policy requires identifying and broadcasting compelling, local narratives that capture negotiators' diffuse attention and direct it northward. Their hope is that in doing so, policy makers and climate activists might come to understand the interplay between melting ice, Inuit livelihoods, rising sea levels, and vanishing island nations. Perhaps, then, decision makers within the UNFCCC would develop a framework for action that would preserve what former ICC Chair Sheila Watt-Cloutier dubbed the "Inuit right to be cold" (Watt-Cloutier 2015).

Bridging Knowledge and Action through an Anthropology of Climate Change Institutions

I first learned of ICC's work in 2005 by reading media reports of the human rights petition. Given my interest in knowledge mobilization across governance scales, I approached ICC with the hope of supporting their efforts while simultaneously conducting institutional ethnographic research. As with Ittaq, my first exchange with ICC staff was mediated by email and required a formal introduction of my research goals and questions, along with an offer to assist in any way that I might be useful. My opportunity to do just that came when an ICC staff member left on maternity leave in September, 2009, and I joined ICC-Canada in their Ottawa offices to help with preparations for the 15th Conference of Parties (COP) of the UNFCCC in Copenhagen.

One of the challenges of engaging simultaneously in practices of research and advocacy is that, as Annelise Riles has pointed out, the networks and networked practices that we participate in routinely become largely invisible to us because they "share with our interpretive tools a singular aesthetic and a set of practices of representation" (2000: 1). These tools include many of the knowledge practices discussed in this chapter: writing grant proposals, documenting data, and drafting meeting documents. To conduct ethnographic work in the context of translational networks requires two primary tasks: first, to reconceive of the space of fieldwork by identifying ways to study institutional practices that go beyond document analysis and that situate documents within a field of interaction and practice, and, second, to learn to see these practices with new eyes, to turn the network "inside out" so that what is familiar and routine becomes more distant.

In my collaboration with ICC I participated in translation and representation by adopting institutional technologies and narrative strategies that were familiar to the institution. While in my research I tried to resist succumbing to the bureaucratic logic of institutions by making their practices a subject of study, in working with ICC I helped to reinforce these practices by matching content to preexisting forms and by engaging with a repertoire of representations that included adopting standardized narratives emphasizing the vulnerability of Inuit communities to melting ice and snow. I drafted statements, press releases, and internal briefing papers on the state of the climate negotiations, modeling my work on earlier ICC documents and statements. For COP 15, for example, I helped prepare a document entitled "Inuit Call to Global Leaders: Act Now on Climate Change in the Arctic," which was circulated through ICC's networks, released to the media, and disseminated at several side events hosted by ICC in Copenhagen.

The statement outlined a number of issues that the organization hoped would be taken up in the climate negotiations. Many of these were fairly high level and noncontroversial, such as regulation of greenhouse gas emissions, or designating avoidance of climate impacts on the Arctic as a benchmark for effectiveness of global policy. Others were more specific to the ongoing work of ICC, such as the recommendation to engage traditional knowledge in future climate assessments through the Intergovernmental Panel on Climate Change (IPCC).

One recommendation, in particular, however, went against the grain of the established institutional norm of the UNFCCC: the recommendation to change the global framework for adaptation funding to include Inuit and "other vulnerable populations and communities living in developed nations" (ICC 2009). This recommendation echoed statements made by Mary Simon, then President of Inuit Tapiriit Kanatami, Canada's national Inuit organization (representing communities in Nunavut, the Inuvialuit Settlement Region in the western Canadian Arctic, Nunavik in northern Quebec, and Nunatsiavut in northern Labrador), who had called for the creation of an adaptation fund for Inuit. Strong statements about the need for adaptation assistance had also been made by the Iñupiat communities of Kivalina and Shishmaref in Alaska, where storm surges combined with diminishing sea ice are causing severe erosion, and state and federal resources have been inadequate to assist with relocation and reconstruction needs (Marino 2012; see also Marino and Schweitzer's chapter in this volume).

As an international Inuit organization, ICC's constituents live in four countries and have varied, sometimes divergent, ideas about what the changing climate means for Inuit identity and the future flourishing of Inuit communities. At COP 15, some of this diversity inadvertently came to the fore when Jimmy Stotts, the acting Chair of ICC, gave an interview with news media in which he suggested that Inuit should be exempt from emission reduction standards established through the global negotiation process (CBC 2009). As the week progressed, other Inuit leaders exchanged conflicting views on whether oil and gas development, particularly in the offshore marine environment, should play a role in the devolution agendas of Inuit-led governments and territories (Nuttall 2012). This controversy about oil and gas development reflects the double standards of global indigenism, which requires indigenous peoples to display a commitment to land-based identities in their representational strategies (Niezen 2003). These contradictions reflect what Stewart Kirsch (2007) refers to as the "risks of counterglobalization"—his term for indigenous politics that make use of global networks—which create pressure for indigenous communities to position themselves as antidevelopment.

In the fraught political environment of the UNFCCC, the vast gap between what is known about the changing climate and the paltry steps that have been taken to secure emission reductions grows larger every year. In this venue, the political stakes are high—those with the most compelling scientific evidence and personal narratives, the most dramatic images, or the most creative use of activist theater stand the best chance at capturing media coverage or the fleeting attention of negotiators. Categories of "victim" and "perpetrator" are presented as oppositions, avoiding the more complex ways in which individuals and collectives can contribute to warming the climate while also experiencing its effects (Hughes 2013). There is little room here for nuance or for competing notions of a desirable future.

In this context representations of Inuit as vulnerable to climate change because of their participation in traditional activities such as hunting and traveling on the land invite public interest and generate opportunities for Inuit to share their perspectives and experiences with global audiences. Yet in these venues, they find that they must draw on a very limited conception of identity, one that emphasizes subsistence but ignores the many ways that Inuit communities are tied to global economic production and governance processes (Cameron 2012; Haalboom and Natcher 2012). This representation not only ignores the vested interests that some Inuit have in oil and gas development but also diminishes the significant work that Inuit do as grant writers, nonprofit workers and managers, government employees, and political representatives. Meanwhile, vulnerability narratives have thus far failed to move global adaptation policy to support Inuit adaptation needs.

As a collaborator in ICC's work to prepare for COP 15 in Copenhagen, I observed and participated in their representational strategies first-hand. My participation in the practices of inscription and documentation that I write about in this chapter inevitably honed my interest in the role these artifacts play in global networks and also heightened my awareness of the limited repertoire of narrative options available in these venues. In the "Inuit Call to Global Leaders" document, one of the action points that I helped to draft was to "recognize the vulnerability of Inuit and other indigenous peoples by adopting a mechanism for adaptation assistance to vulnerable groups, communities, and countries" (ICC 2009: 1) that would prioritize assistance based for those most vulnerable and devolve funding to the community level as much as possible. The document emphasized the responsibility of developed nations not only to provide assistance to developing countries but also "to ensure that vulnerable communities *within their own borders* have the resources, knowledge, and technology needed to adapt" (ICC 2009: 4, emphasis in the original). Although the language of vulnerability facilitates a very limited depiction of Inuit identity, it also offers one of the few semantic windows into the UNFCCC process for those, like the Inuit, who have largely found themselves unable to substantively participate in or access resources for assistance emerging from this venue.

Conclusion

Attention to the role of bureaucratic networks, documents, and translation remains underdeveloped in social science engagements with climate change. Arctic communities must increasingly adapt not only to changes in the tundra and sea ice but also, and perhaps more critically, to the bureaucratic and social infrastructures that shape and limit possibilities for action. As the examples of the women's retreat and the Inuit petition demonstrate, climate change is imagined and enacted through networks that link community-level institutions and actors with government bureaucracies and global norms such as human rights. These networks contain multiple visions and understandings of change, which are navigated through acts of translation and brokerage.

Translation is never neutral; indeed, it inevitably involves prioritization of certain perspectives or modes of explanation over others (West 2005), and is enacted within uneven institutional networks (Engle Merry 2006). Translators, I would argue, are largely aware of these political stakes and of the limitations of representation in policy venues. They make creative use of various tools including personal narratives and documents to bridge scales and build links across domains of experience. These translations are what ultimately bring local experience into global constructions of environmental change, and vice versa.

Anthropologists working on climate change inevitably become part of broader networks and communities of practice as we conduct our research; our inquiries, collaborations, and publications become part of the repertoire of representations that various actors use to advance diverse agendas. Yet these networked practices of representation often remain outside the scope of anthropological inquiry. Making institutions and bureaucratic documents such as grant proposals, reports, and petitions a larger focus of ethnographic analysis of climate change significantly broadens opportunities for fieldwork and facilitates reflexivity about the networks that we are already a part of. Only by doing so will anthropologists come to understand more fully the diverse, often mundane, and sometimes creative ways that climate change is understood and acted on through networked practices of translation.

Notes

1. The name Aboriginal Affairs and Northern Development Canada was changed to Indigenous and Northern Affairs Canada in early November 2015, following the election of a new government led by the Liberal Party.
2. I am grateful for funding support from the Vanier Canada Graduate Fellowship Program, the National Science Foundation, and Foreign Affairs and International Trade Canada. I am particularly grateful for the collaboration of the Ittaq Heritage and Research Centre and the Ilisaqsivik Society of Clyde River and the Inuit Circumpolar Council, which gave me access to the networked practices I have discussed in this chapter. Shari Gearheard and Martha Dowsley provided significant intellectual support and invited me to join the women's retreat. Finally, thanks to the community members, scientists, and government employees who granted interviews and asked good questions that helped to refine my project.

References

Agarwal, A., Perrin, N., Chhatre, A., Benson, C., and Kononen, M. 2009. *Climate Policy Processes, Local Institutions, and Adaptation Actions: Mechanisms of Translation and Influence*. Social Development Papers, No. 199: Social Dimensions of Climate Change. Washington, D.C.: The World Bank.

Baines, D. 2010. Neoliberal restructuring, activism/participation, and social unionism in the nonprofit social services. *Nonprofit and Voluntary Sector Quarterly* 39, 1: 10–28.

Cameron, E. S. 2012. Securing Indigenous politics: A critique of the vulnerability and adaptation approach to the human dimensions of climate change in the Canadian Arctic. *Global Environmental Change* 22(1): 103–14.

CBC. 2009. *Exempt Northerners from Emission Cuts: Inuit Leader*. CBC Online, December 10, www.cbc.ca/news/canada/north/exempt-northerners-from-emission-cuts-inuit-leader-1.848239.

Crate, S. A. 2011. Climate and culture: Anthropology in the era of contemporary climate change. *Annual Review of Anthropology* 40(1): 175–94.

Ford, J. D., and Smit, B. 2004. A framework for assessing the vulnerability of communities in the Canadian Arctic to risks associated with climate change. *Arctic* 57(4): 389–400.

Gearheard, S. 2002. These are things that are really happening: Inuit perspectives on the evidence and impacts of climate change in Nunavut, in I. Krupnik and D. Jolly (Eds.), *The Earth Is Faster Now: Indigenous Observations of Arctic Environmental Change*. Fairbanks: Arctic Research Consortium of the United States, pp. 12–53.

Gearheard, S., Aporta, C., Aipellee, G., and O'Keefe, K. 2011. The Igliniit Project: Inuit hunters document life on the trail to map and monitor Arctic change. *Canadian Geographer/Le Géographe Canadien* 55(1): 42–55.

Gearheard, S., Holm, L. K., Huntington, H., Leavitt, J. M., and Mahoney, A. R. 2013. *The Meaning of Ice: People and Sea Ice in Three Arctic Communities*. Dartmouth, NH: International Polar Institute.

Gearheard, S., and Shirley, J. 2007. Challenges in community-research relationships: Learning from natural science in Nunavut. *Arctic* 60(1): 62–74.

Haalboom, B., and Natcher, D. C. 2012. The power and peril of "vulnerability": Approaching community labels with caution in climate change research. *Arctic* 65(3): 319–27.

Harney, N. D. 2011. Neoliberal restructuring and multicultural legacies: The experiences of a mid-level actor in recognizing difference. *Ethnic and Racial Studies* 34(11): 1913–32.

Henshaw, A. 2006. Pausing along the Journey: Learning landscapes, toponymy, and environmental change. *Arctic Anthropology* 43(1). 52–66.

Hughes, D. M. 2013. Climate change and the victim slot: From oil to innocence. *American Anthropologist* 115(4): 570–81.

Hull, M. S. 2012. Documents and bureaucracy. *Annual Review of Anthropology* 41(1): 251–67.

Huntington, H. P., and Fox, S. 2005. The changing Arctic: Indigenous perspectives, in *Arctic Climate Impact Assessment: Scientific Report*. Cambridge: Cambridge University Press, pp. 61–98.

Inuit Circumpolar Council (ICC). 2009. *Inuit Call to Global Leaders: Act Now on Climate Change in the Arctic*. Inuit Circumpolar Council, www.iccalaska.org/servlet/content/other.html, accessed January 20, 2015.

Isuma, T. V. 2009. Nunavut, Clyde River, Qaqqasiq Apak, August 27, 2005. Interview for legal petition on climate change, www.isuma.tv/international-legal-action-climate-change/nunavut-clyde-river-qaqqasiq-apak-august-27-2005-tape-1, accessed January 20, 2015.

Johnson, N. 2012. "Healing the land" in the Canadian Arctic: Evangelism, knowledge, and environmental change. *Journal for the Study of Religion, Nature, and Culture* 6(3): 300–18.

Kirsch, S. 2007. Indigenous movements and the risks of counterglobalization: Tracking the campaign against Papua New Guinea's Ok Tedi Mine. *American Ethnologist* 34(2): 303–21.

Laidler, G. J., Dialla, A., and Joamie, E. 2008. Human geographies of sea ice: Freeze/thaw processes around Pangnirtung, Nunavut, Canada. *Polar Record* 44(4): 335–61.
Marino, E. 2012. The long history of environmental migration: Assessing vulnerability construction and obstacles to successful relocation in Shishmaref, Alaska. *Global Environmental Change* 22(2): 374–81.
Mathews, A. S. 2009. Unlikely alliances: Encounters between state science, nature spirits, and Indigenous industrial forestry in Mexico, 1926–2008. *Current Anthropology* 50(1): 75–101.
Merry, S. E. 2006. Transnational human rights and local activism: Mapping the middle. *American Anthropologist* 108(1): 38–51.
Mosse, D., and Lewis, D. 2006. Theoretical approaches to brokerage and translation in development, in D. Lewis and D. Mosse (Eds.), *Development Brokers and Translators: The Ethnography of Aid and Agencies*. Sterling, VA: Kumarian Press, pp. 1–26.
Niezen, R. 2003. *The Origins of Indigenism: Human Rights and the Politics of Identity*. Berkeley and Los Angeles: University of California Press.
Nuttall, M. 2012. Imagining and governing the Greenlandic resource frontier. *The Polar Journal* 2(1): 113–24.
Orlove, B. 2009. The past, the present, and some possible futures of adaptation, in W. Neil Adger, I. Lorenzoni, and K. O'Brien (Eds.), *Adapting to Climate Change: Thresholds, Values, Governance*. New York: Cambridge University Press, pp. 131–63.
Orlove, B., Lazrus, H., Hovelsrud, G. K., and Giannini, A. 2014. Recognitions and responsibilities. *Current Anthropology* 55(3): 249–75.
Osofsky, H. M. 2009. The Inuit Petition as bridge? Beyond dialectics of climate change and Indigenous people's rights, in W. C. G. Burns and H. M. Osofski (Eds.), *Adjudicating Climate Change: State, National, and International Approaches*. Cambridge: Cambridge University Press, pp. 272–91.
Pearce, T., Ford, J. D., Caron, A., and B. P. Kudlak. 2012. Climate change adaptation planning in remote, resource-dependent communities: An Arctic example. *Regional Environmental Change* 12(4): 825–37.
Riles, A. 2000. *The Network Inside Out*. Ann Arbor: University of Michigan Press.
Rudiak-Gould, P. 2012. Promiscuous corroboration and climate change translation: A case study from the Marshall Islands. *Global Environmental Change* 22(1): 46–54.
Stern, P., and Hall, P. V. 2010. The proposal economy. *Critique of Anthropology* 30(3): 243–64.
Watt-Cloutier, S. 2005. *Petition to the Inter-American Commission on Human Rights Seeking Relief from Violations Resulting from Global Warming Caused by Acts and Omissions of the United States*. Center for International Environmental Law, www.ciel.org/Publications/ICC_Petition_7Dec05.pdf, accessed January 20, 2015.
———. 2015. *The Right to Be Cold: One Woman's Story of Protecting Her Culture, the Arctic, and the Whole Planet*. Toronto: Allen Lane.
Weatherhead, E., Gearheard, S., and Barry, R. G. 2010. Changes in weather persistence: Insight from Inuit Knowledge. *Global Environmental Change* 20(3): 523–28.
West, P. 2005. Translation, value, and space: Theorizing an ethnographic and engaged environmental anthropology. *American Anthropologist* 107(4): 632–42.

Chapter 26

Escaping the Double Bind: From the Management of Uncertainty toward Integrated Climate Research

Werner Krauss

Climate change has made for a spectacular career, culminating in the Nobel Prize for Peace in 2007 awarded to the Intergovernmental Panel on Climate Change (IPCC) and Al Gore. But there has been a price to pay for establishing an ongoing master narrative based on scientific consensus, tipping points, and thresholds; so-called climate skeptics have hijacked the basic scientific concepts of uncertainty and skepticism and turned them into an argument against the implementation of climate politics. As a consequence, climate science has increasingly become politicized; the close vicinity of climate research and politics has raised suspicion concerning the objectivity and neutrality of science. Controversies surrounding the iconic hockey stick curve, the hacked e-mails from climate scientists (Climategate), and errors in the IPCC report (Himalayagate) have not helped. Under pressure, climate science has deployed diverse strategies to regain public trust, such as the inclusion of uncertainty into the working reports of the IPCC and in public communication. But there is more to politicization and uncertainty than a management problem; the debate about their role in the climate debate has left its traces in climate science and raised new questions: how do climate scientists deal with this permanent double bind of maintaining fidelity to scientific standards, while the object of their research is politically charged? If the linear model of science speaking truth to power is a failed conception of science communication, then what are the alternatives? Using the example of climate science in Hamburg, this chapter presents an ethnographic account of the concepts and strategies deployed by a group of scientists who actively face the challenges imposed by this double bind.

From early on, climate science in Hamburg established a loose network with social scientists; they embraced concepts informed by policy advice and social science studies—namely, "the honest broker" and post-normal science. The implementation of an interdisciplinary climate blog served (and still serves) as a testing ground and as a form of extending the conversation about climate change. Based on my ethnographic account, in this chapter I argue that climate science as a social practice is changing and situates itself differently in society. Adjusting climate science in a highly politicized environment is an often ambiguous and open process, and, more often than not, new double binds emerge and replace old ones. But there is no way back to an innocent state of science, and in the future social sciences will play a greater role in the production of knowledge.

In what follows, I take the double bind inherent in climate communication as a starting point to situate my own ethnographic approach in the current anthropological debate about climate research. After presenting my research site and the main actors, I follow the process of networking that resulted in the adoption of the concepts of the honest broker and post-normal science. On this basis, I critically discuss the potentials and restrictions of these concepts. In the conclusion, I argue that despite the idiosyncrasies of this specific example, climate science may well increasingly accept politicization and uncertainty as part and parcel of its mode of existence.

Escaping the Double Bind

Throughout his career, the late climate scientist Stephen Schneider discussed the problem of advocacy and climate science; how can one stay loyal to scientific standards and add value in science communication? Scientific knowledge serves as the legitimization of climate politics and its goals; as a consequence, climate science comes increasingly under pressure and ends up in a permanent double bind. Once the object of research is politically charged, scientific statements immediately turn into political arguments. In anthropology, the issue of the "double bind" has a long history that goes back to Gregory Bateson. He was interested in the question of how sciences deal with paradoxes and uncertainty, and he moved on to research the effects of double binds in family interaction. Throughout the history of their discipline, anthropologists were faced, for example, with the problem of changing the very cultures they were at the same time representing. Only recently, Sarah Green (2014) reflected on these "anthropological knots" and recounted a classical Zen *koan* that goes back to Bateson; it exemplifies a typical double bind and how to get out of it:

> [A Zen Buddhist master] holds a stick over the head of the student and says, 'If you say this stick is real, I will strike you with it; if you say it is not real, I will strike you with it; if you say nothing, I will strike you with it.' A student can escape the double bind by reaching up and taking the master's stick away." (Green 2014: 8)

In more recent environmental research, Kim Fortun used the concept of double bind as her starting point in *Advocacy after Bhopal*; she finds the victims of the industrial disaster, the "enunciatory groups," permanently challenged with impossible alternatives by corporations and nation-states as potential addresses for compensation; as an anthropologist, she is familiar with facing permanent double binds herself:

> Double binds proliferated. I learned languages of law and bureaucracy, while learning how badly these languages represent everyday life. I learned to speak in terms of environmentalism, while learning how badly environmentalism represents the Third World poor. I learned the many truths of theoretical critiques of representation, on the ground—while producing one representation after another. (Fortun 2001: 53)

Fortun actively deals with this double bind in turning advocacy consciously into her way of fieldwork while permanently exposing the inherent contradictions. In her book *How Climate Change Comes to Matter: The Social Life of Facts*, Candis Callison builds on the work of Fortun and their mutual advisors:

> This book uses multisited ethnographic methods suggested and pioneered by Marcus, Fischer, and Fortun to get inside how climate change becomes meaningful in diverse and specific groups and how this underlying double bind *of maintaining fidelity to science and expanding beyond it* is negotiated by groups that are both central and peripheral to evolving discussions about how to communicate climate change. (Callison 2014: 5)

Callison's study reflects this shift of perspective in recent anthropological research from improving science communication toward the question of how "the meanings of climate change are established through attention to multiple discourses, assemblages (institutions, actors, networks), and vernaculars where situated knowledge, advocacy, activism, ethics, and morality become apparent" (2014: 12). In this context climate change is understood as an emergent form of life, and the research focus shifts toward the inquiry of how climate change comes to matter in diverse social formations; in her book she discusses science journalism, religious groups, scientists, and carbon managers.

In this chapter I present the emergence of a specific social formation, a group, in climate science that deviates from mainstream science. This group of scientists is sometimes labeled "climate realists" or "climate pragmatics," and they set themselves apart from what they call "alarmism" or "climate catastrophism" as well as from climate skepticism; instead, they situate science differently by applying the concepts of "the honest broker" (Pielke, Jr. 2007) and of "post-normal science" (Funtowicz and Ravetz 1993) as their trademark. In various functions as "their" anthropologist, as a collaborator, coblogger and coauthor, I accompanied this process over a long period of time, starting in 2001. Needless to say, working in this interdisciplinary field creates double binds, too—anthropology and its qualitative and interpretative methods serve as "the other" of scientific research; once you gain credibility on the one side, you easily lose it on the other. Nonetheless, I interpreted my task as being an interlocutor to my informants or, more optimistically, as a diplomat between different modes of existence as suggested by Latour (2013).

The Research Site and Fieldwork

Hamburg has a long tradition in meteorology and is today one of the hot spots in international climate research. The major contributors of the University of Hamburg's Cluster of Excellence "Integrated Climate System Analysis and Prediction" (CliSAP) are the Max-Planck-Institute for Meteorology, the German Climate Computing Center, and the Helmholtz Research Center with its Institute for Coastal Research. The director of the Institute, Hans von Storch, is also one of the speakers of this Cluster of Excellence, and he advocated early on for the participation of social sciences in climate research. At the beginning of the new millennium, he willingly opened the doors of his institute for my research on "the tribe of climate scientists"; from here we developed, over several years, various forms of collaboration and engaged in a continuous conversation about the politics of climate research. The concepts of the honest broker and of post-normal science served as a common field of interest, with the establishment of a climate blog, *Die Klimazwiebel* (*The Climate Onion*), as a test ground for the extension of the debate and the inclusion of new publics. The German North Sea coastline as field site for coastal, climate, and ethnographic research is our common point of reference. In workshops and common publications, as lead and contributing authors for the IPCC (Jones et al. 2014) and as authors of a book on "the dangerous vicinity of climate research and politics" entitled *Die Klimafalle* (*The Climate Trap*) (von Storch and Krauss 2013), we discussed the prospects and contours of a new understanding of integrated climate research. As an anthropologist, I became an active member of a mostly informal and loose network bringing together climate science with social and political scientists, political analysts and philosophers of science, policy advisors and journalists.

Hans von Storch started his career at the Max-Planck-Institute for Meteorology, and he soon developed a critical stance toward his colleagues in respect to the undifferentiated interpretation of climate change as "catastrophe." From early on, he distanced himself from premature interpretations of individual weather events as signs of climate change, and he was critical about an increasing tendency in climate science to link each and every event to climate change. Obviously, there were social drivers informing scientific studies and changing the interpretation of the results accordingly.

His collaboration with the sociologist Nico Stehr was a major turning point and intensified his latent interest in climate science as a social practice; in the 1990s they started to work on the cultural history of climate change and of climate research. The main contribution of this continuing collaboration was putting climate research into a cultural, historical, and political perspective and consistently calling for the participation of social sciences. They wrote about the long tradition of climate determinism in geography and the political abuse of climate alarmism in history; in doing so, they tried to put current climate science and its practices into context. As one of the main protagonists of the hockey stick controversy and the discussion about Climategate, von Storch showed himself to be concerned about the loss of public trust in science and its diminishing integrity in respect to Mertonian norms.

The contact with social science helped von Storch to situate himself and his institute differently in the field of climate science. He created his own niche in the polarized climate debate, with the stereotypical camps of "alarmists" on the one and "skeptics" or "deniers" on the other side. Hans von Storch never hesitates to speak to the media, and he maintains close relationships with journalists. Like many leading climate scientists he is a good interviewee; he polarizes consistently while developing a position of his own as a "climate realist." Establishing a loose network with social sciences helps to define and to maintain this niche; as it turned out, the concepts of "the honest broker" and of post-normal science served well to outline the contours of his position and finally turned into a trademark of the integrated climate science in Hamburg.

The Honest Broker

The polarized climate debate was easily diagnosed in this framework: catastrophism and raising the alarm on the one side and scientific ignorance on the other—a view that daily newspapers often reinforce in their reporting on climate change. But how to ban such hysterical voices, and how can science be protected from being used as a support for these respective agendas? For the appropriation of a niche between skeptics and alarmists, theories and concepts from the social sciences play an important role. The honest broker is a traveling concept that is precise enough to be applied in specific situations and broad enough to cover diverse interests and fields. Furthermore, it has almost magical qualities in providing forms of classification to order the cacophonic choir of voices in the climate debate.

The concept of the honest broker comes from the political scientist Roger Pielke, Jr., one of the leading and most disputed researchers in the emerging field of interdisciplinary climate research. He is well known for opposing, using statistical means, the often premature link between extreme weather events and rising costs of disaster to climate change; especially in the United States, this is a highly politicized issue, with Pielke as one of the most prominent protagonists. This position made him an almost natural candidate for the Eduard Brückner prize, named after a famous German climate researcher from the turn of the 19th to the 20th century and awarded from the German association of climate scientists. Hans von Storch was the main initiator of this prize and head of the committee in 2006, when the prize was awarded to Pielke for outstanding achievements in interdisciplinary climate research, thus setting another milestone in the integration of social sciences into climate research. Both profited in equal terms from this long-term relationship, with Pielke providing the conceptual foundations for the emerging field of integrated climate research as initiated in Hamburg.

His main contribution is his typology of five modes of engagements of science in society (Pielke 2015). This typology is highly idealized and abstract, but the individual categories perfectly illustrate the different positions in the politicized field of climate research. As in a theatre, the figures of the *pure scientist*, the *issue advocate*, the *science arbiter*, and the *honest broker* are ideally suited to stage the drama of climate science.

The fifth category, the *stealth advocate*, is perfect to begin with. This is a scientist who presents scientific data in a way that fits his hidden political agenda. But who is this bad guy in reality? Is it the skeptic paid by the oil industry who fakes data, or is it the climate scientist who denies uncertainties and preaches evidence to the public? In any case, this schematic typology bears a considerable tension. This is also true for the *pure scientist*, who is interested only in science and research—a state of science that is highly idealized among scientists but hardly achievable. Pielke wisely adds that once the objects of research are highly politicized a neutral stance is no longer possible; without including the political context, even pure equations automatically gain political traction and turn into open or stealth advocacy.

The *science arbiter* is perhaps the most common figure in expert advisory committees; science arbitration provides answers that can be addressed empirically with the tools of science. This role is familiar in many situations and highly accepted in society. The *issue advocate* is perhaps the most prominent role and can easily be imagined as a Greenpeace activist or a scientist working for an oil company—both often dismissed as not trustworthy in terms of proper science. Pielke is more indulgent when it comes to the *issue advocate* who seeks to reduce the scope of available choice, often to a single preferred outcome among many possible outcomes; that is, the scientist argues openly in favor of a specific solution or agenda. Callison (2014) introduces the related term of the "near-advocate," probably one of the most prevalent roles assumed by climate scientists. Finally, there is the *honest broker of policy alternatives*, who seeks "to clarify the scope of possible action so as to empower the decision maker." This mode of engagement changes the role of the climate scientist in society, and it does so in a fundamental way: science no longer takes the lead in defining how to deal with the challenges imposed by climate change but serves to project the possible outcomes, the feasibility and the range of possibilities in the framework set by political and societal decisions. The main idea is to integrate science into a democratic framework.

For von Storch and the Hamburg school of climate research, the honest broker is one of the central nodes in the emerging network of climate science, policy advice, and social sciences. The schematic way of this characterization of five modes of engagement leaves enough room for natural sciences to adopt the concept and to interpret it in their own way without risking losing the authority attributed to the hard sciences and big data; climate change is still science-based, but the categorization of the multitude of voices serves as a magical ban: conversation about climate change can both be extended and controlled. Science does no longer determine the political debate, but it still is in power by defining the scope and thus framing the debate.

The Honest Broker as Patron Saint: Extending the Combat Zone into the Blogosphere

One of the main features of the climate debate is the extension of the combat zone into the blogosphere; blogs have played a central role in the climate discussion ever since the hockey stick debate (Krauss 2012), providing a public for dissenting voices or those from outside climate science. In 2009, the year of Climategate and the failed climate summit COP 15 in Copenhagen, von Storch implemented a climate blog on the Internet, *Die Klimazwiebel*; "Following the paradigm of the 'honest broker' we write about climate research and its interaction with politics" (http://klimazwiebel.blogspot.com/). The honest broker as the patron saint provides safe ground both for steering a conversation and motivating the extension of the conversation about climate change beyond the narrow confines of science as well as the polarized climate debate. For me, "the risk of diplomacy" as mentioned by Latour (2013: 52) became a full reality when I was asked to participate as one of the editors of the blog, together with the sociologists Dennis Bray and Rainer Grundmann and the climate scientist Eduardo Zorita. Starting a climate blog marked an important step from the conceptual approach to its performance in the semi-official world of the blogosphere. It is unknown terrain in science communication and as such a real test field for future developments. *Die Klimazwiebel* intends to bridge not only the gap between science and politics but also the one between social and natural sciences as well as between science and the public. In terms of the polarized climate debate, the main feature of *Die Klimazwiebel* is to provide a space for skeptical voices as part of the strategy to overcome the polarized debate and to extend the conversation.

Unavoidably, the modes of engagement provided by the honest broker concept are easily transformed into moral categories in these discussions; Pielke himself is fully aware that the word "honest" easily catches attention. In the heat of the discussion, dissenting opinions are automatically disqualified as "dishonest" or as "stealth activism." In these cases, *Die Klimazwiebel* develops specific group dynamics that gain their own traction, creating a new double bind. Hans von Storch, as globally renowned German climate scientist, professor, and founder of the blog, is almost automatically identified by non-academic commentators as the impersonation of the honest broker; dissenting views are easily dismissed as dishonest or stealth advocacy. These semantic dynamics are deeply rooted in German national culture, where academic education has great symbolic value and is identified with high social status. Linking climate and questions of cultural identity heats up the discussions and is one of the causes of elevated blood pressure as a typical health risk for bloggers.

In any case, the application of the concept of the honest broker has opened up the climate debate and situated climate science differently in society; furthermore, it has helped uncertainty and skepticism to find their way back into the climate debate. When applied as in the context of *Die Klimazwiebel*, climate science rubs with society, causes frictions, and finally becomes a case for post-normal science (Hulme 2007).

Post-Normal Science: Managing Uncertainty

How to manage uncertainty effectively? How to deal with the politicization of climate science, and how to get rid of the permanent paradoxes and double binds? The concept of post-normal science was suggested by Sylvio Funtowicz and Jerry Ravetz to provide answers to these questions, and like all traveling concepts it is based on a catchy credo: post-normal science applies when knowledge is uncertain, stakes are high, values are in play, and decisions are urgent. When Jerry Ravetz promoted post-normal science as a means of managing uncertainty and politicization, the concept already had a long history.

In her article about the origins of the concept of post-normal science, Silvia Tognetti (1999) goes back to the anthropologist Gregory Bateson, who found normal science in the Kuhnian sense in a state of permanent double bind. Problem and policy-driven science like climate science has difficulties in dealing with new sources of uncertainty that "typically fall outside the paradigm" and thus limit their relevance for "real world problem solving" (Tognetti: 700). She argues that science cannot be defined "in isolation of social context"; like a person, science exists in relation to a social context and thus is permanently confronted with new sources of uncertainty. The management of uncertainty is at the heart of post-normal science; according to Jerry Ravetz (2010), this was the case with climate science. After Climategate, rumors abounded: allegedly, there was a witch-hunt on skeptics; scientific journals practiced gate-keeping, trying to keep skeptical views out, and peer-review in climate science turned out to be pal-review. These were strong accusations, and there was hardly a chance to provide evidence for one or the other side. The evangelical tone and the attitude of "the science is settled" of many scientists, thus excluding uncertainty from the debate, indeed prevailed especially before climate summits such as COP 15; Jerry Ravetz sarcastically spoke of a "war on climate" comparable to the war on terror or drugs.

In 1999, Dennis Bray and Hans von Storch argued that climate science is "an empirical example of post-normal science" (Bray and von Storch 1999); they stated that, based on their surveys among climate scientists, consensus on anthropogenic climate change is as much a result of normative assumptions as of science. One decade later, Climategate was the incentive for the founders of the concept of post-normal science, Jerry Ravetz and Silvio Funtowicz, to turn their attention to climate science. Ravetz chose a skeptical blog to publish his highly controversial statement *Climategate: Plausibility and the Blogosphere in the Post-Normal Age* (2010). For him, the hacked e-mails offered a shocking insight into the scientific production of policy-relevant knowledge. With a reference to Pielke he stated:

> There are deep problems of the management of uncertainty in science in the policy domain that will not be resolved by more elaborate quantification. In the gap between science and policy, the languages, their conventions and their implications, are effectively incommensurable. It takes determination and skill for a scientist who is committed to social responsibility, to avoid becoming a "stealth advocate" (in the terms of Roger Pielke, Jr.). When the policy domain seems unwilling or unable to recognise plain and urgent truths about a problem, the contradictions between scientific probity and campaigning zeal become acute. It is a perennial problem for all policy-relevant science, and it seems to have happened on a significant scale in the case of climate science. The management of uncertainty and quality in such increasingly common situations is now an urgent task for the governance of science. (2010)

In 2012 Ravetz organized a workshop in Lisbon about "Reconciliation in the Climate Debate." His intention was to invite skeptics and advocates alike to find out if there is a common ground both sides can agree on. While prominent skeptical bloggers such as Steve McIntyre and Judith Curry joined the workshop, there were only a few participants who represented the mainstream, with von Storch and I representing the middle ground, dubbed by skeptical colleagues as "lukewarmers." The result of the workshop was that skeptics do not consider themselves necessarily as a group, so they refused to make a common statement. They agreed only on "the monster of uncertainty" (van der Sluijs 2005) as being an integral part of climate science that cannot be excluded or exorcised. Following Lisbon, the networking continued, and we organized a follow-up workshop in Hamburg entitled "Climate Science in a Democratic Society." The workshop served well as an example for integrated climate research with social scientists, policy advisers, anthropologists, and journalists outnumbering climate scientists. The focus was on discussing uncertainty in respect to practical applications such as regional climate services, adaptation, and the

IPCC. Instead of politicizing climate science and depoliticizing adaptation (Beck 2011), climate science was brought back into democracy—at least, in theory. Post-normal science is simultaneously theory, method, and ethnographic description; while its real nature remains diffuse, it serves perfectly to outline the contours of an integrated climate science.

Conclusion

Climate change is not a monolithic fact; instead, it is permanently negotiated between science and other systems of knowledge. This is an insight that anthropology can contribute to climate science, with the anthropologist as an informed interlocutor and a diplomat between different systems of knowledge and modes of existence (Latour 2013).

It is a long way from Stephen Schneider's discussion of the double-ethical-bind to the polarized and heated debate following the scandals, controversies, and failures of the linear model of science communication. When climate science and the IPCC finally started to manage uncertainty, they did so without challenging the overall framework of science leading the climate debate and framing climate politics. But as I have shown with the example of climate research in Hamburg, there are also attempts to escape the double bind and to define climate research in the framework of an integrated science. In Hamburg, the tentative adoption of traveling concepts such as the honest broker and post-normal science is an attempt to situate science differently in society and to understand climate as both physically and culturally constituted. Of course, this is a highly idealized version of my ethnographic account. In everyday reality, my interlocutors from climate science still want both to purify science and to integrate social sciences and the public, thus permanently repeating the double bind: the more they want to keep science pure and free of context, the more climate science becomes politicized. Furthermore, governance strategies and international programs still favor natural over social sciences and humanities. Thus climate science still takes the lead and defines climate change primarily as a physical phenomenon, while social sciences are mandated to communicate this knowledge and to incite transformation on this basis. But the honest broker as a patron saint and post-normal science as illegitimate child will constantly provide encouragement to finally escape the double bind and to move from managing uncertainty toward integrated science In any case, the future of both climate and its science is still uncertain.

References

Beck, S. 2011. Moving beyond the linear model of expertise? IPCC and the test of adaptation. *Regional Environmental Change* 11: 297–306.

Bray, D., and von Storch, H. 1999. Climate science: An empirical example of postnormal science. *Bulletin of the American Meteorological Society* 80(3): 439.

Callison, C. 2014. *How Climate Change Comes to Matter: The Communal Life of Facts*. Durham, NC: Duke University Press.

Fortun, K. 2001. *Advocacy after Bhopal: Environmentalism, Disaster, New Global Orders*. Chicago: Chicago University Press.

Funtowicz, S. O., and Ravetz, J. R. 1993. Science for the post-normal Age. *Futures* 25(7): 739–55.

Green, S. 2014. Anthropological knots: Conditions of possibilities and interventions. *Hau: Journal of Ethnographic Theory* 4(3): 1–21.

Hulme, M. 2007. The appliance of science. *The Guardian*, 14 March, www.theguardian.com/society/2007/mar/14/scienceofclimatechange.climatechange, accessed April 25, 2015.

Jones, R. N., A. Patwardhan, S. J. Cohen, S. Dessai, A. Lammel, R. J. Lempert, M. M. Q. Mirza, and von Storch, H. 2014. Foundations for decision making, in C. B. Field, V. R. Barros, D. J. Dokken, K. J. Mach, M. D. Mastrandrea, T. E. Bilir, M. Chatterjee, K. L. Ebi, Y. O. Estrada, R. C. Genova, B. Girma, E. S. Kissel, A. N. Levy, S. MacCracken, P. R. Mastrandrea, and L. L. White (Eds.), *Climate Change 2014: Impacts, Adaptation, and Vulnerability. Part A: Global and Sectoral Aspects—Contribution of Working Group II to the Fifth Assessment Report of the Intergovernmental Panel on Climate Change*. Cambridge: Cambridge University Press, pp. 195–228.

Krauss, W. 2012. Ausweitung der Kampfzone: Die Klimablogosphäre. *Forschungsjournal Soziale Bewegungen* 25(2): 83–89.

Latour, B. 2013. *An Inquiry into Modes of Existence: An Anthropology of the Moderns*. Cambridge, MA: Harvard University Press.

Pielke, Jr., R. 2007. *The Honest Broker: Making Sense of Science in Policy and Politics*. Cambridge: Cambridge University Press.

———. 2015. *Five Modes of Science Engagement*, http://rogerpielkejr.blogspot.com.tr/2015/01/five-modes-of-science-engagement.html, accessed April 26, 2015.

Ravetz, J. 2010. *Plausibility and the Blogosphere in the Post-Normal Age*, http://wattsupwiththat.com/2010/02/09/climategate-plausibility-and-the-blogosphere-in-the-post-normal-age/, accessed April 28, 2015.

Schneider, S. H. 2005. *Mediarology: The Roles of Citizens, Journalists, and Scientists in Debunking Climate Change Myths*, http://stephenschneider.stanford.edu/Mediarology/Mediarology.html, accessed April 29, 2015.

Tognetti, S. 1999. Science in a double-bind: Gregory Bateson and the origins of post normal science. *Futures* 31: 689–703.

van der Sluijs, J. 2005. Uncertainty as a monster in the science-policy interface: Four coping strategies. *Water Science Technology* 52(6): 87–92.

von Storch, H., and Krauss, W. 2013. *Die Klimafalle: Die gefährliche Nähe von Politik und Klimaforschung*. München: Hanser.

Epilogue: Encounters, Actions, Transformations

Susan A. Crate and Mark Nuttall

Our intent with the first edition of *Anthropology and Climate Change* was to produce a comprehensive volume assessing the roles anthropologists play in research, policy, and practice concerned with climate change, using specific case studies to illustrate anthropology's scope and depth in its engagement. We also wanted to map out where the discipline was headed as it was initiating new research and policy-oriented approaches. And finally we wanted the volume to examine the dynamics of various epistemologies, ontologies, and practices, whether related to understanding local knowledge and perceptions of climate change, human-environment relations, or climate science and decision making, with the objective of locating theoretical frames, research approaches, and applied practices. In short, we wanted to open and expand dialogue among anthropologists to questions concerning the extent of our role in climate change not only as researchers but also as advocates, pedagogues, communicators, and activists.

The first edition was premised on anthropological encounters with climate change, arising largely from research in particular places, but also the forms of action that anthropologists were taking and the public voice they were finding. Since the first volume's publication, scientific research on climate variability and change and the regional and global assessments that draw on and synthesize the main findings of that research have continued to confirm that Earth's climate is changing—to the extent that we may be transgressing a "safe operating space for humanity" (Rockström et al. 2009; Steffen et al. 2015)—and anthropologists have continued to encounter and report on the sociocultural manifestations and effects of such change in their field sites, from rural to urban contexts. Whether this work is with people in drought-affected regions or areas of increasing precipitation leading to flooding, in places where glacial retreat threatens water supplies or where extreme weather conditions lead to crop failure and pasture scarcity, in coastal areas experiencing greater storm surges and erosion, on islands affected by sea-level rise, or in urban environments

---

challenged by the increasing reality of food deserts, anthropologists have established research projects of depth and complexity in a diversity of topics and issues, ranging from current and projected levels of exposure to climate-related sensitivities, to understanding the limits and restrictions to adaptive capacity. Indeed, the contributors to this second edition illustrate the vastness and global dimensions of anthropological concern with climate change and the continued relevance of anthropological encounters.

Clearly anthropology is more than ethnography, and localized fieldwork is not what defines the discipline alone, even if it ever did (Ingold 2008; Comaroff and Comaroff 2012; Fardon et al. 2012). Nonetheless, both the first and this second edition underscore the importance of long-term ethnographic research in specific places, to track human interactions with the environment and human responses and adaptation to change over time, among other things, and not just in the sense of how ethnography has been defined as a description of "the lives of people other than ourselves" (Ingold 2008: 69). The chapters in this book illustrate how anthropologists encounter and work with issues of pressing contemporary concern and urgency in diverse contexts. Climate change is already transforming environments, having profound social and cultural consequences and challenging adaptation strategies, even meaning displacement and relocation, in some cases of people and communities who have long-established residences in specific places. For example, in parts of Africa livelihoods are under threat, people are facing economic losses and are dependent on emergency relief food and water supplies (although, of course, such situations have many causes other than climate change). In the Himalayan region, the high Andes, and the European Alps glaciers are melting because of an increase in temperatures, which affects agriculture, water supply, and the natural resources on which people depend in both the immediate and distant contexts. In northern Alaska, the Canadian Arctic, and Greenland, hunters are traveling on increasingly thinning ice, while at the same time the key marine species they depend on—seals, walrus, narwhals, and polar bears—are moving away from the areas in which they are traditionally hunted, as they in turn respond to changes in local ecosystems. In the Amazon region, climate change combined with deforestation, forest fragmentation, and the transformation of tropical rainforest into dry grassland savannah leads to critical loss of biodiversity and severe droughts and has put indigenous and local livelihood strategies under increased stress. In the northern reaches of Siberia permafrost degradation is destabilizing entire tracts of land important to the herding and farming practices of communities there, challenging their already unique adaptation of horse and cattle breeding and reindeer herding in the far north. Although these cases are highly local, in all of them the rest of the globe is implicated—in the loss of biodiversity, in the world's freshwater supply, and in global human security. Our book is not a compendium of the local and regional effects of global climate change, but it does reinforce how critical it is to understand these impacts on people's everyday lives and the encounters and interplay between the local, the regional, and the global.

The uncertainty that characterizes much scientific knowledge about climate change and its effects is partly explained because of the large-scales at which scientists gather their data. People feel the effects of climate change in the way the weather is experienced, and their observations and knowledge of climate effects in specific geographical localities are of critical importance for scientific analysis and decision making. Local knowledge, combined with local people's experiences of the environment, may also prove to be better placed to deal with uncertainty. Anthropological attentiveness to such knowledge contributes not only to the discipline's concern with a "critical understanding of human being and knowing in the one world we all inhabit" (Ingold 2008: 69) but also to the discussion about the applied practices that can move us from impartial witnesses of our field collaborators' experiences of unprecedented environmental change into the realm of action-oriented researchers. There may be an ethnographic component to much of the work we do on climate change, but our concern is above all with an enquiry that is fundamentally anthropological in its concern with social worlds and the nature of human being.

Climate change is a complex interplay of physical processes and environmental, historical, social, cultural, and economic factors. Its effects are highly variable and regionally specific and are significant for people and for local and regional economies in many different ways. Anthropologists draw attention to the regional texture of climate change and the diversity of different experiences, demonstrating that some environments and peoples are more exposed to climate change and are significantly more vulnerable to, and affected by, its impacts and long-term consequences. Indigenous peoples, for instance, often depend on natural resources for their livelihoods, and they often inhabit diverse but fragile ecosystems. At the same time indigenous peoples are among the world's most marginalized, impoverished, and vulnerable peoples, with many living on the edges of the mainstream societies of nation states. Hence, while indigenous peoples in some places may often bear the brunt of climate change, they have minimal access to resources to cope and respond. For indigenous peoples around the world, climate change threatens cultural survival, brings different kinds of risks and opportunities, and undermines indigenous human rights (Nuttall and Nilsson 2008). The consequences of unprecedented changes in ecosystems have significant implications for the use, protection, and management of wildlife, fisheries, and forests, affecting the customary uses of culturally and economically important species and resources. Despite the impact of climate change on indigenous peoples and their livelihoods, international experts most often overlook indigenous rights and also the potentially invaluable contributions from traditional knowledge, innovations, and practices in the global search for climate change solutions. In national, regional, and international processes, such as the UN Framework Convention on Climate Change (UNFCCC), where climate change mitigation policies are discussed, negotiated, and designed, indigenous peoples have found it extremely difficult to get their voices heard and their concerns taken into consideration. This situation stands in stark contrast to their experiences with the Convention on Biological Diversity (CBD), where the International Indigenous Forum on Biodiversity (IIFB) is an advisory body to the Convention. The UNFCCC, however, does not provide a similar institutional and discursive space for indigenous peoples. To date, the concerns and views of indigenous peoples—especially gendered and generational perspectives—have not been seriously addressed in climate negotiation processes in all climate decision making and actions. The issues of human rights and environmental justice—and not just for indigenous peoples—continue to dominate much of the discourse surrounding international decision-making processes.

Anthropologists can fill a valuable role as cultural interpreters of both Western science and local knowledge, focusing on not so much what is known but how it is known (Murphy 2011; Bartels et al. 2013). This approach focuses understanding local perceptions of change more on the process of interpreting how communities frame their experiences based in their immediate experience and opens ways to bring a framing of issues of global change into local perspectives. At the same time, indigenous and other place-based communities have extensive and intimate knowledge of their surroundings, which itself is invaluable knowledge for Western science about exactly how global change is having a diversity of on-the-ground effects. Anthropologists can play a key role as the interface and facilitators of knowledge exchanges, bringing local inhabitants into conversations with regional scientists both to benefit more holistic understandings for all stakeholders and to bolster affected communities' resilience by framing more complex global processes within local understandings (Crate and Fedorov 2013).

But anthropologists are also careful to disentangle the effects of climate change and social, environmental and economic change and seek to understand climate change in the context of multiple stressors. While narwhals in northern Greenland may be affected by shifting pack ice and changing weather conditions, for instance, the hunters who pursue them do not necessarily see the changing climate as something that makes their lives harder. Yes, it may contribute to difficulties in accessing traditional hunting areas, especially during winter, when the sea ice does not form the way it used to. But the extensive and relentless seismic surveys that have taken place in late summer and early autumn over the last few years as part of oil exploration, and also the quota system for narwhal hunting that is based on scientific knowledge and ignores local understandings of animals and the environment, have a much more immediate presence in the everyday lives of communities. They effectively erode the basis of subsistence hunting lifestyles more than changing ice conditions or increasingly stormy weather patterns do (see Nuttall this volume). As our previous volume and this edition (and the many other collections, monographs, articles, and reports published over the past decade or so) demonstrate, anthropologists can no longer be accused of "fiddling while the globe warms" (Rayner 1989); rather, they have made significant contributions to action and have produced finely tuned assessments of what a climate change anthropology looks like and where it is headed (Fiske et al. 2014).

Along these lines, this volume reemphasizes encounters and actions but also sheds new light on transformations, thereby emphasizing not just the shifts in geographies, geologies, and social worlds but also those in the discipline of anthropology and its interfaces with other social sciences, the humanities, and the natural and physical sciences. Attempts to integrate human/qualitative approaches into natural science/quantitative ones fall short, as echoed in the recent literature: "the potential fruits of interdisciplinary exchange are far greater than, and altogether different in character [from], those implied by most recent clarion calls for the reformatting of GEC (Global Environmental Change) science" (Castree et al. 2014: 763). Most GEC research fails to engage human aspects because it does not take a bottom-up approach, understanding and potentially addressing the more immediate needs of on-the-ground communities; "for millions of people around the world, there are issues more immediate, more urgent, than addressing political economic equity or designing some new social or technological utopia" (Rajan and Duncan 2013: 70). Increasingly researchers across the disciplines are calling for transformative approaches that move away from the tendency of the last decade of global change research to promote the kind of adaptation that accommodates change and that moves toward rigorous research and actions that contest change and create alternatives (O'Brien 2012).

The interdisciplinary process is highly iterative, and it expands, interacts, and itself transforms. For example, in recent years, the use of the term *tipping point* has refigured perceptions of climate change and influenced the media response to scientific reports of climate change and the public imagination concerning it. Although the idea of abrupt change in climate systems and ecosystems is not new, a growing body of scientific evidence points to ecosystems, whether they are large marine ecosystems, deserts, or small ponds, having tipping points—meaning that the system undergoes sudden changes resulting from destabilization arising from the influence of external factors. The long-standing assumption that ecosystems are simple, linear, and reversible seems now to be flawed, and scientists increasingly talk about the nonlinear dynamics and tipping points in nature. Timothy Lenton and colleagues use *tipping point* to refer to a critical threshold at which a tiny perturbation can qualitatively alter the state or development of a system. A tipping point, they argue, occurs when a change in an ecosystem triggers a strongly nonlinear response in the internal dynamics of that system, qualitatively changing its future state. They introduced the term *tipping element* to describe large-scale components of the Earth system that may pass a tipping point. Large-scale tipping elements, they claim, have been identified in Earth's climate system that may, in this century, pass a tipping point under human-induced global change. At the smaller scale of ecosystems, some tipping points have already been observed, and more are anticipated in future (Lenton et al. 2008). However, once we have reached these tipping points and crossed thresholds, it is suggested, we will have reached a point of no return, and there will be tremendous social and economic consequences arising from these ecosystem transformations.

Although the term *points of no return* was used in the 1990s, and scientists talked about breaking points in the early 2000s, the term *tipping point* is virtually absent from climate science publications before 2007 or so—indeed, we can probably date the current use of the term to 2005, when NASA scientist James Hanson used it. Borrowed from physics, it has a longer history of usage in the political and social sciences, particularly in sociological theorizing about social change, having been applied in the late 1950s and early 1960s to explain and account for demographic shifts in suburban America neighborhoods. Now, the science literature is dripping with references to tipping points, suggesting a different approach to visualizations of time and space, confidence and certainty about the acceleration of the pace and nature of climate change, and different horizons of expectation. According to the tipping-points narrative, the future is approaching us with increasing speed, and although the tipping point will mean we will cross a point of no return, the future remains nonetheless characterized by its unknown quality, and to gaze into a world beyond the tipping point is to be confronted with something outside our collective experience (Nuttall 2012). Lenton (2011) has suggested it is possible that some approaching tipping points can carry early warning signals. The methods and prospects for gaining such early warning of approaching tipping points in the Earth system are not yet precise, but anthropologists could and should play a significant role in their development and refinement, particularly as it could be argued that the science of tipping points sometimes veers close to climatic determinism, reducing the explanation of complex social behavior to physical or biological models and accounting for social change in a positivistic fashion (Nuttall 2012).

Lenton and colleagues have suggested that there are indications that certain dramatic social and cultural change in the past can be attributed exclusively to abrupt climate change, a view that anthropologists challenge. In his chapter in the first edition of this book, Fekri Hassan argues that we are far from certain of the probable impact of climate change on the trajectory of human history—our methodologies and interpretive strategies are limited, and this fact raises questions over the claim that tipping points in ecosystems precipitate social change by themselves. Hassan's work elucidates the intractable and complex interrelationships between climate and human societies and makes a plea to overcome simplistic notions of determinism and indeterminacy. Case studies from North Africa and southwest Asia show that climate change did play a major role in the origins of agriculture and the emergence of state societies; however, the impact of any climate event depends on local ecological settings and the scale and complexity of local social and cultural settings, pointing to multistage, long-term transformations rather than abrupt switches resulting from climate forcing (Hassan 2009). And as Arlene Miller Rosen argues, societies do not interact directly with their environment but with their perceptions of that environment. The environment, one of many actors in determining social change, plays a less important role than do perceptions of nature. Societies overcome environmental shifts in a diversity of ways, and failure to do so may signal a breakdown in one or more social and political subsystems. Success or failure is most often related to internal factors: social organization, technology, and the perception of environmental change (Rosen 2007). Furthermore, can we even talk about a social system in the same way we can talk about ecological systems? Is it feasible and valid to apply the term *tipping points* to both the physical world and the social world in the same way? Do glaciologists who talk of tipping points to describe changes in the Greenland ice sheet and sociologists who talk of tipping points to describe demographic change in urban neighborhoods mean the same thing? Research on tipping points needs to be grounded firmly and securely in better knowledge of the complexity of social worlds and social relations, as well as the complexity of human-environment relations, and how these worlds and relations intersect with regional and global processes.

As Kirsten Hastrup points out in her chapter in Part 1, there is a critical need for a greater understanding of the human responses to climate change, which unfold at "the interface between natural and social histories that always outstretch a particular moment or place." This need, she argues requires that anthropologists turn their attention to the processes by which climate change is configured and to how a motley combination of knowledge forms enters into its making. It remains that climate change assessments are framed by the natural and physical sciences and that anthropologists are seen by scientists to be useful in contributing to scientific understanding of climate change—often in larger multidisciplinary projects we and our local research partners provide information that endorses the science—rather than influencing its direction and scope and playing a greater role in our understanding of ecological and social futures. As Marieke Heemskerk pointed out several years ago, anthropological methods can contribute to speculation about the future, because they incorporate what most extrapolations and forecasts based on science alone lack: (1) uncertainty and surprise, (2) people's own mental models of the future, and (3) a detailed understanding of specific cultures and the diversity within these cultures (2003). We should not ignore the sociocultural setting, and we should try to make sense of it, and we should not ignore the fact that people will, we can perhaps safely assume, still be born, live, eat, work, and die; form complicated social relationships and fight and argue with one another; have beliefs and attitudes; create, nurture, and maintain worlds of meaning and symbols; and make choices about their lives.

One particular distinguishing feature of climate change science, specifically when it feeds into regional and global assessments with the intention of providing scientific information to decision makers, has been its careful and often guarded use of a terminology of likelihood and probability that goes hand in hand with the caveat that, although climate change is happening and is going to be more pronounced and potentially transformative, there are nonetheless elements of uncertainty about its extent and its effects (Nuttall 2012). Discussion of future events and conditions must take into account the likelihood that these events and conditions will occur. This accounting requires a precise vocabulary of possibility, and therefore a lexicon of terms is used in describing this likelihood. Climate science assessments have to be confident and must reflect consensus, but they also have to highlight areas of disagreement and uncertainty. They tread a fine line between stressing likelihood and uncertainty and stating with a degree of confidence what the future could look like. So, the terminology of likelihood emphasizes that "in 20 years, temperatures could rise," or "it is highly probable that Arctic sea ice may disappear during summer in the coming decades."

At the heart of climate science is the climate scenario—not a prediction but a simulation or a storyline intended as a plausible representation of the future climate that is consistent with scientific assumptions about future human activities and their impact on the atmosphere and climate. There is no one accepted climate scenario—there are many of them providing a range of storylines about a number of possible futures. And, along with the lexicon of likelihood, underlying many climate assessments has been the uncertainty of what causes climate change—how to disentangle human-induced change from natural climate variability is a scientific challenge. These assumptions, together with guarded feelings about likelihood and uncertainty, have been characteristic of climate change science until relatively recently and are increasingly challenged by the Anthropocene narrative.

Climate models become even less effective when they attempt to assess and represent the complexity of everyday social, cultural, political, and economic life for purposes of probabilistic analysis. There may be widespread consensus that climate change has anthropogenic causes, yet how the world and its climate are transformed by human action remains a critical area of research. Policy and decision-making processes depend on a "sound scientific basis" for their success in terms of the consensual understanding needed for how to take action on climate change. The unreliability of models provides a way to question the legitimacy of science. The uncertainty of climate change science, together with the unreliability of models for representing social and ecological interactions, remains an obstacle to moving forward and addressing environmental dilemmas. There is perhaps at least one thing that is certain about scientific uncertainty concerning the effects of climate change—it affects the policy and decision-making process, and arguments have been put forward for a reassessment of the role of climate models for use in the development of policy and societal decision making (Stainforth et al. 2007).

Environmental discourses have precipitated shifts within the social sciences that are linked to the science agenda, and a number of social sciences have reached a point of convergence from different points of departure. It is here that we begin to see how interdisciplinarity might work. Of course, we need to be grounded in our own respective scientific/academic disciplines, and from different points of departure we explore and meet on common ground. It is on this common ground, for example, that a growing body of knowledge and explanation of ecosystem processes, or sustainability, or climate change can be found. Scientific knowledge, as the case of climate change science shows us, is contested and subject to major indeterminacies. Science is far from being able to offer clear-cut accounts of the world and physical processes, and scientists continue to struggle with uncertainties generated by climate models (and, anyway, the kinds of problems encountered by science, such as climate change, do not fit within neat disciplinary boundaries). As we pointed out in the Epilogue to our first volume, understanding the processes driving social, cultural, economic, and environmental changes and assessing their effects demands the development of innovative multidisciplinary collaboration and research methodologies, for example, through the integration of different kinds of knowledge—scientific, social scientific, and local. This collaboration enhances understanding of the complex interaction between human societies and the environment, and the value of this cooperation is also underscored by its relevance for policy making. Collaboration between the natural and social sciences and the humanities is an essential step for improved communication and further collaboration between researchers, communities, stakeholders, and policy makers. It is only through this transdisciplinary interdependence that we will come to holistic understandings on which we can move into the transformative changes we need to preserve and perpetuate the biocultural diversity of our Earth's system. As this book makes clear, anthropologists have been taking their essential first steps in this larger process.

References

Bartels, W.-L., Furman, C. A., Diehl, D. C., Royce, F. S., Dourte, D. R., Ortiz, B. V., Zierden, D. F., Irani, T. A., Fraisse, C. W., and Jones, J. W. 2013. Warming up to climate change: A participatory approach to engaging with agricultural stakeholders in the southeast U.S. *Regional Environmental Change* 13(1): 45–55.

Castree, N., Adams, W., Barry, J., Brockington, D., Buscher, B., Corbera, E., Demeritt, D., Duffy, R., Fely, U., Neves, K., Newell, P., Pellizzoni, L., Rigby, K., Robbins, P., Robin, L., Rose, D., Ross, A., Schlosberg, D., Sorlin, S., Whitehead, M., and Wynne, B. 2014. Changing the intellectual climate. *Nature Climate Change* 4: 763–68. DOI: 10.1038/nclimate2339.

Comaroff, J. L., and Comaroff, J. 2012. Foreword: Thinking anthropologically, about British social anthropology, in R. Fardon, O. Harris, T. H. J. Marchand, M. Nuttall, C. Shore, V. Strang, and R. A. Wilson (Eds.), *The SAGE Handbook of Social Anthropology*. London: Sage, with the Association of Social Anthropologists of the Commonwealth, pp. xxviii–xxxiv.

Crate, S., and Fedorov, A. 2013. A methodological model for exchanging local and scientific climate change knowledge in northeastern Siberia. *Arctic* 66(3): 338–50.

Fardon, R., Harris, O., Marchand, T. H. J., Nuttall, M., Shore, C., Strang, V., and Wilson, R. A. (Eds.). 2012. *The SAGE Handbook of Social Anthropology*. London: Sage, with the Association of Social Anthropologists of the Commonwealth, pp. xxviii–xxxiv.

Fiske, S. J., Crate, S. A., Crumley, C. L., Galvin, K., Lazrus, H., Lucero, L. Oliver-Smith, A., Orlove, B., Strauss, S., and Wilk, R. 2014. *Changing the Atmosphere. Anthropology, and Climate Change. Final report of the AAA Global Climate Change Task Force,* December 2014. Arlington, VA: American Anthropological Association, 137 pp.

Hassan, F. A. 2009. Human agency, climate change, and culture: An archaeological perspective, in S. A. Crate and M. Nuttall (Eds.), *Anthropology and Climate Change: From Encounters to Actions*. Walnut Creek, CA: Left Coast Press, Inc., pp. 39–69.

Heemskirk, M. 2003. Scenarios in anthropology: Reflections on possible futures of the Suriname maroons. *Futures* 35(9): 931–49.

Ingold, T. 2008. Anthropology is *not* ethnography. *Proceedings of the British Academy* 154: 69–92.

Lenton, T. M. 2011. Early warning of climate tipping points. *Nature Climate Change* 1: 201–09.

Lenton, T. M., Held, H., Kriegler, E., Hall, J. W., Lucht, W., Rahmstorf, S., and Schellnhuber, H. J. 2008. Tipping elements in the Earth's climate system. *Proceedings of the National Academy of Sciences* 105(6): 1786–93.

Murphy, B. L. 2011. From interdisciplinary to inter-epistemological approaches: Confronting the challenges of integrated climate change research. *The Canadian Geographer / Le Géographe canadien* 55(4): 490–509.

Nuttall, M. 2012. Tipping points and the human world: Living with change and thinking about the future. *Ambio* 41(1): 96–105.

Nuttall, M., and Nilsson, C. (Eds.). 2008. Climate change and Indigenous peoples. Thematic issue, *Indigenous Affairs* 1-2/08.

O'Brien, K. 2012. Global environmental change II: From adaptation to deliberate transformation. *Progress in Human Geography* 36(5): 667–76.

Rajan, S. R., and Duncan, C. A. M. (Eds.). 2013. Introduction: Ecologies of hope. *Journal of Political Ecology* 20: 70–79.

Rockström, J., Steffen, W., Noone, K., Persson, A., Chapin, F. S., Lambin, E. F., Lenton, T. M., Scheffer, M., Folke, C., Schellnhuber, H. J., et al. 2009. A safe operating space for humanity. *Nature* 461(7263): 472–75.

Rosen, A. M. 2007. *Civilising Climate: Social Responses to Climate Change in the Ancient Near East*. Walnut Creek, CA: AltaMira Press.

Stainforth, D. A., Allen, M. R., Tredger, E. R., and Smith, L. A. 2007. Confidence, uncertainty, and decision-support relevance in climate predictions. *Philosophical Transactions of the Royal Society* 365: 2145–61.

Steffen, W., Richardson, K., Rockström, J., Cornell, S. E., Fetzer, I., Bennett, E. M., Biggs, R., Carpenter, S. R., de Vries, W., de Wit, C. A., Folke, C., Gerten, D., Heinke, J., Mace, G. M., Persson, L. M., Ramanathan, V., Reyers, B., and Sörlin, S. 2015. Planetary boundaries: Guiding human development on a changing planet. *Science*. DOI: 10.1126/science.1259855, 15th January.

INDEX

Contributors

About the Editors

Susan A. Crate is Professor of Anthropology in the Department of Environmental Science and Policy at George Mason University. An environmental and cognitive anthropologist, she has worked with indigenous communities in Siberia since 1988. Her recent research has focused on understanding local perceptions and adaptations of Viliui Sakha communities in the face of unprecedented climate change—a research agenda that has expanded to Canada, Peru, Wales, Kiribati, and the Chesapeake Bay. Crate is the author of numerous peer-reviewed articles and one monograph, *Cows, Kin, and Globalization: An Ethnography of Sustainability* (AltaMira Press 2006), and she is coeditor of *Anthropology and Climate Change: From Encounters to Actions* (Left Coast Press, Inc. 2009). Crate also served on the American Anthropology Association's Global Climate Change Task Force.

Mark Nuttall is Professor and Henry Marshall Tory Chair of Anthropology at the University of Alberta and Fellow of the Royal Society of Canada. He also holds a visiting position as Professor of Climate and Society at Ilisimatusarfik/University of Greenland and the Greenland Climate Research Centre at the Greenland Institute of Natural Resources. He has carried out extensive research in Greenland, Alaska, Canada, Finland, and Scotland and is co-PI of the EU-funded project ICE-ARC (Ice, Climate, and Economics—the Arctic Region in Change). He is editor of the landmark three-volume *Encyclopedia of the Arctic* (Routledge 2005) and author or editor of many other books, including *The Scramble for the Poles: The Geopolitics of the Arctic and Antarctic*, coauthored with Klaus Dodds and published by Polity in 2016.

About the Contributors

Eduardo S. Brondizio is Professor of Anthropology, Indiana University Bloomington, codirector of the Anthropological Center for Training and Research on Global Environmental Change (ACT), and member of the Ostrom Workshop in Political Theory and Policy Analysis. Brondizio's research combines longitudinal studies of Amazonian small farmers and regional transformation, and research on global environmental change.

Prof. Brondizio is also Member of the Science Committee of *Future Earth* and the International Geosphere-Biosphere Program (IGBP), and co-Editor-in-Chief of *Current Opinion in Environmental Sustainability*.

Michael R. Dove is the Margaret K. Musser Professor of Social Ecology, School of Forestry and Environmental Studies; Curator, Peabody Museum of Natural History; Professor, Department of Anthropology; Fellow, Whitney Humanities Center; and co-coordinator of the joint doctoral program in Anthropology and Environmental Studies, Yale University. His most recent books are *The Anthropology of Climate Change* (Wiley/Blackwell 2014); *Climate Culture* (coeditor J. Barnes, Yale University Press 2015), and *Science, Society, and Environment* (coauthor D. M. Kammen, Routledge 2015).

Timothy J. Finan is Professor of Anthropology in the School of Anthropology, University of Arizona. A development anthropologist, he has been on the faculty since 1981 and served as the Director of the Bureau of Applied Research in Anthropology from 1994–2009. His research in climate anthropology spans three continents and four decades, and he has worked on climate change and disaster management in Bangladesh since 2001. His academic experience includes the study of vulnerability, resilience, and adaptive governance.

Shirley J. Fiske is an environmental anthropologist specializing in coastal, marine, and natural resource issues. She worked as a program and research manager in the National Oceanic and Atmospheric Administration and in the U.S. Senate as a senior legislative aide, and is currently Research Professor in the Anthropology Department at University of Maryland College Park. She is the coeditor of numerous publications and commentaries. Most recently she chaired the American Anthropological Association's Global Climate Change Task Force (2011–2014).

Donna Green is a senior research scientist in the Climate Change Research Centre, University of New South Wales, and an associate investigator in the Centre of Excellence for Climate Systems Science. In this position she leads a program that explores how indigenous and non-indigenous knowledge can be used to understand how climate change affects remote communities in northern Australia. Her research focuses on human-environment interactions, specifically on social and economic vulnerability, adaptation, and risk.

Kirsten Hastrup is Professor of Anthropology at the University of Copenhagen. In recent years she has worked with hunters and their families in Northwest Greenland and has studied the interlinked changes in climate and community. She has written extensively on the history and anthropology of Iceland and has published critical explorations of the philosophical and epistemological foundations of anthropology. She led the European Research Council-funded project Waterworlds, and among her recent books is the edited volume *Anthropology and Nature* (Routledge 2014).

Chris Hebdon is a Ph.D. student in the joint doctoral program in anthropology and environmental studies, Yale University. His research interests include environmental anthropology, ethnography of expertise, and the anthropology of energy. His dissertation examines the politics of energy transition in Latin America with a focus on the Andean region and specifically Ecuador.

Anne Henshaw has worked with the Oak Foundation since September 2007 as a marine conservation program officer for projects in the North Pacific and the Arctic. She has carried out anthropological fieldwork in the Canadian Arctic and before joining the Oak Foundation was a visiting professor in the Sociology and Anthropology Department at Bowdoin College from 1996–2007 and director of Bowdoin's Coastal Studies Center from 2000–2007.

Clare Heyward is a Leverhulme Early Career Fellow at the University of Warwick, working on the project *Global Justice and Geoengineering*. Before joining Warwick, she was James Martin Research Fellow on the Oxford Geoengineering Programme, University of Oxford. She is interested in issues of global distributive justice and intergenerational justice, especially those connected to climate change.

Jerry K. Jacka is an assistant professor of anthropology at the University of Colorado at Boulder. He has carried out anthropological fieldwork in Papua New Guinea and in East and West Africa. His research focuses on the intersection of development, natural resource management, and the political ecology of land use and land cover change. He is author of *Alchemy in the Rain Forest: Politics, Ecology, and Resilience in a New Guinea Mining Area*, (Duke University Press 2015).

Tori L. Jennings is adjunct professor and anthropology coordinator in the Department of Philosophy, Religious Studies, and Anthropology at University of Wisconsin-Stevens Point. The primary focus of her research since 2003 has been the sociocultural and political dimensions of climate and climate change in the Cornish Peninsula. She has publications in interdisciplinary journals and edited volumes including *Weather,*

Climate, & Society and *Adapting to Climate Change: Thresholds, Values, and Governance* (Cambridge University Press 2009).

Noor Johnson is Senior Advisor in the Office of International Relations at the Smithsonian Institution, where she works on issues related to climate change, wildlife trafficking, biodiversity conservation, and cultural heritage. She holds an appointment as Research Assistant Professor of Environment and Society at Brown University. Her research focuses on knowledge politics and practices in relation to climate change, offshore oil and gas development, and community-based observing in the Arctic.

Werner Krauss is a cultural anthropologist and fellow at the Cluster of Excellence, Integrated Climate System Analysis and Prediction (CliSAP), Understanding Science in Interaction (USI), University of Hamburg. He was a contributing author to the IPCC report WGII, AR5. He conducted multisited fieldwork in Northern Germany on climate science, climate services, coastal landscapes, and renewable energy. He has published widely on environmental conflicts and is an editor of the interdisciplinary climate blog *Die Klimazwiebel*.

Heather Lazrus is an environmental anthropologist and a project scientist at the National Center for Atmospheric Research in Boulder, Colorado. Using the theories and methods in the anthropological toolkit, she investigates the cultural mechanisms through which all weather and climate risks are perceived, experienced, and addressed. Her research contributes to improving the utility of weather forecasts and warnings, reducing social vulnerability to atmospheric and related hazards, and understanding community and cultural adaptations to climate change.

Myles Lennon is a Ph.D. student in the joint doctoral program in anthropology and environmental studies at Yale University. His research focuses on the convergence of techno-scientific knowledge and participatory organizing practices in the U.S. climate movement. He previously worked on energy efficiency policy with urban planners, state government, community development organizations, and labor unions in New York.

Francis Ludlow is a Postdoctoral Fellow with the Yale Climate and Energy Institute and Department of History, Yale University. He has previously held fellowships with the Rachel Carson Center for Environment and Society of the Ludwig-Maximilians-Universität, Munich; the Harvard University Center for the Environment; and the Trinity Long Room Hub, University of Dublin. Francis is a climate historian employing written and natural sources to reconstruct past climate and examine the influence of climatic changes, extreme weather, and natural hazards on past societies.

Julie Koppel Maldonado obtained her doctorate in anthropology, conducting research with tribal communities in coastal Louisiana on experiences of environmental change. She has consulted for the UN Development Programme and World Bank, was a lead author on the U.S. National Climate Assessment's Indigenous Peoples, Land, and Resources Chapter, co-organizes Rising Voices gatherings, and was lead editor and organizer for the *Climatic Change* Special Issue and book, *Climate Change and Indigenous Peoples in the United States: Impacts, Experiences, and Actions*.

Elizabeth Marino is a research associate and lead of the social science program at Oregon State University – Cascades. She is interested in the relationships among climate change, vulnerability, slow and rapid onset disasters, human migration, and sense of place. Her research focuses on how historically and socially constructed vulnerabilities interact with risk. She is author of *Fierce Climate, Sacred Ground: An Ethnography of Climate Change* (2015), based on her dissertation work.

Anthony Oliver-Smith is Professor Emeritus of Anthropology at the University of Florida. He held the Munich Re Foundation Chair on Social Vulnerability at the United Nations University in Bonn, Germany from 2005–2009. He has done research and consultation on disasters and involuntary resettlement in Peru, Honduras, India, Brazil, Jamaica, Mexico, Japan, and the United States since the 1970s. He won the Bronislaw Malinowski award for lifetime achievement from the Society for Applied Anthropology in 2013.

Daniel Osgood leads the Financial Instruments Sector Team at IRI in Columbia University. Tens of thousands of farmers have purchased index insurance contracts he has helped to design through farmer-driven processes, with significant development effects. He studies uncertainty in decision making, environmental valuation, remote sensing, index insurance, and economic development. He has been involved in global policy processes such as the UNFCCC and has had press coverage in *The Guardian, Nature*, *The New York Times*, and Reuters.

Karsten Paerregaard is Professor of Anthropology at the School of Global Studies, University of Gothenburg. His research is focused on Peruvian rural-urban and global migration and climate change, water scarcity, and environmental tensions in Peru. His publications include *Linking Separate Worlds: Urban Migrants and Rural Lives in Peru*, (Berg 1997), *Peruvians Dispersed: A Global Ethnography of Migration,* (Lexington 2008) and *Return to Sender: The Moral Economy of Peru's Migrant Remittances*, (University California Press 2015).

Kristina Peterson is an applied social scientist doing boundary-crossings that support and prepare scientists and community members to work together. She is affiliated with the Lowlander Center, the Gender and Disaster Network, the Thriving Earth Exchange, the First Peoples' Conservation Council, Rising Voices, and the Indigenous Climate-Change Network. She is a Society of Applied Anthropology Fellow. Her awards include the Distinguished Service to Rural Communities honor from the Rural Sociology Association and the William Gibson Environmental Award. She currently teaches at the University of New Orleans, Urban/Regional Studies, formerly with Hazards Center, UNO-CHART.

Nicole D. Peterson is Assistant Professor of Anthropology at the University of North Carolina–Charlotte and is affiliated with the Center for Research on Environmental Decisions. She has studied sustainability, economic development, and food systems in Mexico's Baja Peninsula, Ethiopia, and Charlotte and is leading an NSF-funded research coordination network on social sustainability.

Md. Ashiqur Rahman is a provost's postdoctoral scholar in the Department of Anthropology at the University of South Florida. He recently completed his Ph.D. from the University of Arizona, where he focuses on the role of governance in shaping vulnerability to climate change in coastal Bangladesh. His previous works focused on conservation and society, governance and livelihood, and power and natural resource management. He published numerous articles in peer-reviewed journals including *Nature Climate Change*.

Steve Rayner is James Martin Professor of Science and Civilization and Director of the Institute for Science, Innovation, and Society (InSIS) at Oxford University and is a Professorial Fellow of Keble College. He has served on various U.S., U.K., and international bodies addressing science, technology, and the environment, including Britain's Royal Commission on Environmental Pollution, the Intergovernmental Panel on Climate Change and the Royal Society's Working Group on Climate Geoengineering.

David Rojas obtained his Ph.D. degree in anthropology from Cornell University. Since 2009 he has carried out research in Brazilian Amazonia studying the influence of Amazonian experts and non-experts in global climate policy discussions. He is currently Assistant Professor of Latin American Studies at Bucknell University.

Peter Rudiak-Gould is Assistant Professor Status-Only in the Department of Anthropology, University of Toronto. An environmental anthropologist, he has conducted extensive fieldwork on the human dimensions of climate change in the Marshall Islands. He is the author of the ethnography *Climate Change and Tradition in a Small Island State* (Routledge 2013) and coeditor of a forthcoming volume on the reception of climate science in Pacific Island societies.

Peter Schweitzer is Professor at the Department of Social and Cultural Anthropology at the University of Vienna and a Professor Emeritus at the University of Alaska–Fairbanks. His theoretical interests range from human-environmental interactions to kinship and identity politics, and his regional focus areas include the circumpolar North and the former Soviet Union. Schweitzer is past president of the International Arctic Social Sciences Association and is currently involved in a variety of Arctic sustainability projects.

Michael Sheridan teaches anthropology and environmental studies at Middlebury College. He first got into African Studies as a Peace Corps volunteer in Kenya and has done ethnographic fieldwork in Tanzania and Cameroon. His major interests include environmental management and power, with particular focus on irrigation, forestry, and sacred sites. Recently he has been expanding a mostly Africa-based comparative ethnobotany project to include the Caribbean and Oceania.

Sarah Strauss is professor of anthropology at the University of Wyoming. Her research focuses on environmental and health issues, values, and practices. She has conducted ethnographic fieldwork in India, Switzerland, and the United States. Books by Strauss include *Positioning Yoga* (2004); *Weather, Climate, Culture* (2003, edited with Ben Orlove); and *Cultures of Energy* (2013, edited with Stephanie Rupp and Thomas Love).

John Urry is Distinguished Professor of Sociology at Lancaster University. He is a Fellow of the Academy of Social Sciences, has an Honorary Doctorate from Roskilde, and is founding coeditor of *Mobilities*. He has published 40 books, including *Mobilities* (2007), *After the Car* (2009), *Mobile Methods* (2011), *The Tourist Gaze 3.0* (2011), *Climate Change and Society* (2011), *Societies Beyond Oil* (2013), *Offshoring* (2014), and *Cargomobilities* (2015).

Richard Wilk is Distinguished Professor of anthropology at Indiana University, where he runs the new Food Research Center and a Ph.D. program in Food and Culture. Trained as an economic and ecological anthropologist, he has researched many different aspects of global consumer culture. His recent work has turned toward the global history of food and sustainable consumption. His most recent books include *Rice and Beans* (coedited with Livia Barbosa) and *Teaching Food and Culture* (coedited with Candice Lowe Swift).

Amy Zhang is a Ph.D. candidate in the joint anthropology and environmental studies program at Yale University. Her research interests include environmental anthropology, urban ecology, and environment and development in the global south. Her dissertation examines the waste management crisis in Guangzhou, China, and traces how the material practices of informal collectors and the growing contentions by citizens and environmentalists challenge the state's vision of development, modernization, and planned cities.

88233142R00250

Made in the USA
Middletown, DE
08 September 2018